The Physician's Guide to Delusional Infestation

Gale E. Ridge

Editor

The Physician's Guide to Delusional Infestation

 Springer

Editor
Gale E. Ridge
Department of Entomology
The Connecticut Agricultural Experiment Station
New Haven, CT, USA

ISBN 978-3-031-47034-9   ISBN 978-3-031-47032-5   (eBook)
https://doi.org/10.1007/978-3-031-47032-5

This Springer imprint is published by the registered company Springer Nature Switzerland AG
The registered company address is: Gewerbestrasse 11, 6330 Cham, Switzerland

Paper in this product is recyclable.

Preface

The World Health Organization [1] defines health as "a state of complete physical, mental and social well-being and not merely the absence of disease and infirmity." When patients suffer from delusional infestation (DI), they may be free of pathogens and parasites, but they still have a medical condition that can often incapacitate them with infirmity.

Peace is when reality and expectations align. For those who live with and care for DI patients, this can be elusive. DI is often a spectrum of multiple complaints with common themes. The most common presentation involves unwelcome cutaneous sensations, often described as feelings of insects or other unwanted creatures biting and crawling on or in the skin. Other patients with DI believe that they have enteric parasitic worms and infestations of companion animals, objects, or family members, and many share their delusions with others. Many DI patients provide medical histories or display physical findings that do not conform to any known disorder. Their accounts are often perseverative and tangential. Moreover, they often present specimens which they believe prove the presence of unwanted infesting organisms, but thorough examinations of such materials reveal them to be nonspecific household debris, environmental detritus, or medically harmless insects or other organisms.

Medicine is a demanding profession with more than 13,000 identifiable diseases, injuries, conditions, and syndromes with often a unique response to each [1]. In addition, the United States Food and Drug Administration (FDA) has approved more than 6,000 drugs, and the American Medical Association (AMA) recognizes >4,000 medical and surgical procedures, each having its own risks, needs, and considerations. Extreme complexity is the norm for almost everyone in medicine. When a DI patient walks through the door, the usual situation is that nothing in the doctor's training has prepared them for the constellation of complaints they are about to hear. DI occupies a blind spot in medicine—and most specialists find the circumstances so fraught and challenging that no medical discipline seeks the opportunity to claim DI as one of "its own." The purpose of this guide is to explore this malady with the hope of shining a light on this dark unnoticed corner of medicine.

This book is written by doctors for doctors. DI patients are often high-functioning individuals with complex issues, and their presentations can confound many physicians. Self-diagnosis of parasitism with a potent overlayer of psychological investment creates an ocean of uncharted waters, where a physician's ship might run aground on a patient's reef of delusion. Identifying patients with possible DI and charting a course toward appropriate care are the missions of the book. The intent of this guide is not to vilify DI sufferers, but to bring an understanding of DI to the field of medicine and to open the door. This guide is the sum of our knowledge to help doctors better understand DI and to know what to do.

Reference

1. World Health Organization (WHO). International Classification of Diseases 11th revision. (ICD-11) (159). 2019.

New Haven, CT   Gale E. Ridge

Acknowledgments

We wish to thank the following people: Dr. Kirby C. Stafford II for his support as well as Drs. Peter Lepping, Nancy Hinkle, Bobbi Pritt, and Ms. Katherine Dugas, for their contributions and advice. In addition, we would like to thank Johanna Searles, Adam Rogers, Jinjer Atkins, and Lanette Barber whose stories they shared with us for this book. Thank you!

Contents

Editor and Contributors

About the Editor

Gale E. Ridge, Ph.D., M.S. first career was as a concert pianist and teacher. In the early 1990s, she became a single parent of three children and switched to biological sciences. She obtained an M.Sc. from Southern Connecticut State University in 1998 and Ph.D. from the University of Connecticut in insect evolution and taxonomy in 2008. Dr. Ridge is an international expert and researcher in the human-feeding bed bug, *Cimex lectularius* L., and delusional infestation. Dr. Ridge is Chair of the Connecticut Coalition Against Bed Bugs and was instrumental in passing a law in 2016 directed at landlords and tenants, the "Public Act Concerning the Rights and Responsibilities of Landlords and Tenants Regarding the Treatment of Bed Bug Infestations." She is an insect taxonomist and director of the insect identification laboratory and manager of its insect collection at The Connecticut Agricultural Experiment Station, New Haven, Connecticut (USA). Her taxonomic work has revealed a complete series of internal pterothoracic characters used to identify Heteropteran (true bugs) species. Dr. Ridge is a sitting member of the EPA FIFRA Scientific Advisory Panel based in Washington, DC.

About the Authors

Nicholas Brownstone, M.D. received a BA degree in psychology from Cornell University and attended medical school at Rutgers Robert Wood Johnson. He trained for over 3 years as a resident in plastic and reconstructive surgery before switching to a career in dermatology. Dr. Brownstone completed a fellowship at the University of California, San Francisco Department of Dermatology in Psoriasis, Phototherapy and Clinical Research, and a fellowship in melanoma at the National Society for Cutaneous Medicine/Mount Sinai Icahn School of Medicine. He is currently a dermatology resident at Temple University Hospital in Philadelphia, Pennsylvania (USA).

Lyle Buss, M.S., B.S. is a Senior Biological Scientist at the University of Florida, Florida (USA), Department of Entomology and Nematology. He manages the Insect Identification and Photography labs and provides insect diagnostic services for county extension agents and residents of Florida. Lyle maintains and enhances the insect image collection and takes pictures of insects at the request of faculty and students in the department. He assists with student training in diagnostics via his insect diagnostics class.

Elliott Harrison Campbell, M.D. is a dermatology resident at Mayo Clinic, Rochester, Minnesota (USA). He has an interest in dermatologic surgery and psychocutaneous medicine, including publishing on evidence related to N-acetylcysteine and skin picking disorders as well as management recommendations for delusional infestation. Dr. Campbell will continue his training at Mayo Clinic as a fellow in micrographic surgery and dermatologic oncology (MSDO).

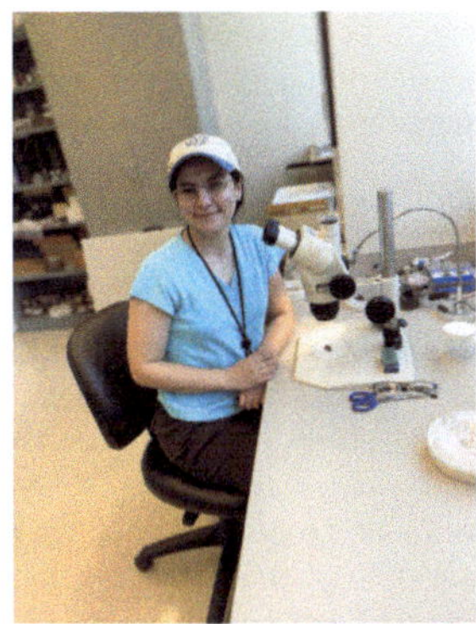

Katherine Dugas, M.Sc. received a bachelor's degree in biology with a minor in art at Connecticut College, New London, Connecticut, USA, in 2005, and received a master's degree in plant sciences and entomology from the University of Rhode Island, Kingston, Rhode Island, USA, in 2008. Katherine worked on several studies as a student, including the effect of *Phragmites* control on the lower Connecticut River and the relationship between microclimate relative humidity and black-legged tick, *Ixodes scapularis*, abundance in Rhode Island (USA). She started at The Connecticut Agricultural Experiment Station (CAES), New Haven, Connecticut (USA), in 2008 as part of the tick program, but her role later expanded to serving as the Connecticut State Survey Coordinator for the Cooperative Agricultural Pest Survey (CAPS) program, conducting pest and disease surveys in important agricultural commodities such as nursery ornamentals, Christmas trees, and forest products. She currently is a research technician assisting in both the CAES Insect Information and Plant Disease Information Offices in New Haven, where she provides insect and plant disease identifications for Connecticut residents, municipalities, and businesses.

Dirk Elston, M.D. is Professor and Chairman of the Department of Dermatology and Dermatologic Surgery at the Medical University of South Carolina, Charleston. He is Editor of the Journal of the American Academy of Dermatology and has served as President of both the American Academy of Dermatology and the American Society of Dermatopathology. He has served as Honorary Professor at China Medical University in Shenyang, China; Guest Professor at the Xiangya School of Medicine, Central South University in Changsha, China; Visiting Professor, Peking University School of Medicine, Beijing, China; and Adjunct Professor, Tongji University School of Medicine, Shanghai, China. He is an honorary member of the German and French Dermatological Societies.

Dr. Elston is a graduate of Jefferson Medical College and did his dermatology residency at Walter Reed Medical Center and a dermatopathology fellowship at the Cleveland Clinic. He is the author of over 600 peer-reviewed publications and 75 textbook chapters. He is the Associate Editor in Chief of eMedicine dermatology, one of the lead authors of Andrews' Diseases of the Skin, and Editor in Chief of the Requisites in Dermatology series of textbooks. He received the 2020 American Academy of Dermatology Gold Medal, 2008 Walter Nickel Award for Excellence in Dermatopathology Education, and 2013 Founders' Award of the American Society of Dermatopathology. Other teaching awards include the UIC, Rush and Cook County Dermatology Residents Teaching Award, St. Luke's-Roosevelt Volunteer Teacher of the Year Award, Brooke Army Medical Center Department of Medicine Outstanding Teacher Award, Brooke Army Medical Center Department of Pathology Outstanding Teacher Award, and the Darl Vanderplueg Excellence in Teaching Award.

Erika Engelhaupt, M.S., B.S. is a writer and editor with considerable journalism experience on top of a deep science background, who is passionate about reporting and explaining all areas of science. She is known for developing new content ideas and bringing an engaging voice to stories. Erika Engelhaupt is a freelance science journalist whose work has appeared in National Geographic, Science News, Scientific America, NPR, and many other newspapers and magazines. She has written on delusional infestations for *National Geographic's* website and in her book, *Gory Details: Adventures from the Dark Side of Science.* Erika has published peer-reviewed scientific articles in journals including *Proceedings of the National Academy of Sciences.* She has edited award-winning coverage for *National Geographic* website and award-winning issues of *Science News*, including 2012 Eddie for Best Digital Magazine and 2020 Eddie for Best News Coverage (consumer) for coronavirus coverage.

Kathleen Flanagan is a medical student in the class of 2024 at the Uniformed Services University of the Health Sciences in Bethesda, Maryland (USA). Before medical school, Kat was a language teacher at schools in the United States and Japan. As a medical student, she has already seen two patients with "our condition," namely delusional infestation, and, like all of us, finds it a curious, fascinating, and perplexing disorder.

Nell Frackowiak is a premedical student in the International Baccalaureate program class of 2024. She is interested in all things science-related and making the world a better place. In her free time, Nell enjoys spending time with her friends and family, rowing with her team, and baking.

Nancy C. Hinkle, Ph.D. is a veterinary entomologist working on arthropod pests and ectoparasites of cattle, poultry, wildlife, and pets. Because she has studied chicken mites for 30 years, her name frequently pops up when individuals suffering from delusional infestation (DI) do a web search for "bird mites." Since receiving a grant from the Florida Entomological Society as a graduate student in the 1990s, Dr. Hinkle has dealt with people suffering from DI, both in her former position at the University of California, Riverside, and for the past 20 years at the University of Georgia, Athens, Georgia (USA). As a veterinary entomologist, she is particularly intrigued by cases in which animals are presumed to share the human's infestation and to serve as a source of reinfestation.

John Y. M. Koo, M.D. is Professor of Dermatology at the University of California at San Francisco (UCSF) School of Medicine, California, USA. Dr. Koo received his medical degree from Harvard Medical School in Boston, Massachusetts (USA), where he was the Harvard National Scholar, an honor given to only one per graduating class. Dr. Koo is board certified in both psychiatry and dermatology. Dr. Koo has authored or co-authored more than 300 peer-reviewed articles and books. Dr. Koo has been listed in *Best Doctors in America* continuously for several decades. He is the recipient of 2014 American Skin Association (ASA) Annual Research Award, Best Speaker Award at Fall Clinical Symposium, and Best Mentor Award at his home institution of UCSF Medical Center. He was also the third recipient of Lifetime Achievement Award by the National Psoriasis Foundation. Lastly, in the 2022 Newsweek survey of thousands of dermatologists across the United States, Dr. Koo was voted as the #1 most prominent dermatologist in California and the #7 most prominent dermatologist across the entire United States.

Peter Lepping, Ph.D. went to the University of Münster, Germany. He moved to Britain in 1995 and did his postgraduate education in and around Liverpool and North Wales. He worked as a community consultant psychiatrist in Wrexham from 2004 to 2017. Since then, he has been working as a Liaison Consultant in the Wrexham Maelor Hospital. He was the Associate Medical Director for ethics, capacity, and consent for the Betsi Cadwaladr University Local Health Board. He is part of the new North Wales Centre for Mental Health and Society. He has a master's in medical ethics and has won several research prizes. His research interests include violence and aggression, delusional infestation, capacity, medical ethics, translational research, systematic reviews, and various other projects. He has been appointed Honorary Professor by Bangor University, Wales, and by Mysore Medical College and Research Institute, Mysore, India. He has published over 160 papers and book chapters. He serves on the editorial boards of various psychiatric journals.

Michelle Magid, M.D., M.B.A. is President of Austin PsychCare. She is an Associate Professor at the University of Texas, Dell Medical School, and Texas A&M Health Science Center, where she teaches residents and medical students on psychodermatology, mood and anxiety disorders, personality disorders, psychiatric diagnoses and treatments, and business basics for the starting physician.

Dr. Magid has been named colleague of the year and teacher of the year and has been honored as a Texas Monthly "Super Doctor" for the past 9 years. She has over 100 invited lectures and publications and was inducted into the American College of Psychiatrists in 2014, for her significant contributions to the field of psychiatry. She is a distinguished fellow for the American Psychiatric Association. She is one of very few physicians to receive a Brain and Behavior Institute research grant to investigate novel interventions for the treatment of depression and anxiety.

In private practice, she is a regional expert in anxiety and mood disorders. Dr. Magid is also an executive coach for high-profile doctors, lawyers, film executives, and politicians who want to "up their game." In addition to executive coaching, she is well published in areas of emotional intelligence and speaks nationally on "Treating the Difficult Patient," "Interpersonal Dynamics in the Work Setting," and "Effective Communication Styles."

Dr. Magid is an international expert on psychodermatology, the interface between the mind and the skin. She is co-author of the international book, "Practical Psychodermatology."

Dr. Magid attended medical school at Boston University, Boston, Massachusetts (USA), where she was captain of the Division I tennis team and named an All-American. She completed her residency in psychiatry at Mayo Clinic and her MBA at the University of Texas in Dallas.

Blaine A. Mathison, B.S., M (ASCP). is currently a Research Scientist with the Institute of Clinical and Experimental Pathology at ARUP Laboratories in Salt Lake City, Utah, USA, and Adjunct Instructor for the Department of Pathology at the University of Utah. Mr. Mathison has been doing diagnostic parasitology for over 20 years, including assignments at the Arizona State Public Health Laboratory (Phoenix, Arizona), the Phoenix Veterans Administration (Phoenix, Arizona, USA), and the Division of Parasitic Diseases and Malaria at the Centers for Disease Control and Prevention, Atlanta, Georgia (USA). Mr. Mathison publishes and lectures regularly on a variety of topics related to both parasitology and entomology, including taxonomy, biology, diagnostics, pathology, and case reports. His specialties include unusual and zoonotic helminth infections, histopathology of parasitic infections, and arthropods of medical importance. In 2018, Mr. Mathison was the recipient of the ASM Scherago-Rubin Award, which recognizes contributions by non-doctoral level microbiologists to the field of clinical microbiology.

Scott A. Norton, M.D., M.P.H., M.Sc. is a Clinical Professor of Dermatology and Pediatrics at the George Washington University, Washington, D.C. (USA). He recently retired as Chief of Dermatology at Children's National Hospital, also in Washington. In 2008, he completed a distinguished 25-year career as a physician in the United States Army. He served as Chief of Dermatology at Walter Reed Army Medical Center for many years and as Professor of Dermatology and Preventive Medicine at the Uniformed Services University of Health Sciences in Bethesda, MD.

In addition to pediatric dermatology, Dr. Norton's interests are in tropical medicine, infectious diseases, and global health. While with the federal government, his overseas medical experiences included duties in Zambia, Egypt, Ghana, Mauritania, Senegal, Haiti, Jamaica, Guyana, Bolivia, Peru, Marshall Islands, Fiji, Palau, the Federated States of Micronesia, and the Philippines, where he worked on clinical, educational, humanitarian assistance, and public health projects. He has additional certification in Tropical Medicine and Traveler's Health from the American Society of Tropical Medicine and Hygiene. He continues to serve as a consultant in dermatology and in tropical medicine to the State Department, the Peace Corps, and the CDC.

Dr. Norton earned his undergraduate degree at Tulane University in New Orleans, where he graduated Phi Beta Kappa. After college, he received a Fulbright Scholarship to New Zealand to conduct field research in plant-animal coevolution, earning a Master of Science with Distinctive Honours from Victoria University in Wellington, New Zealand. He then returned to Tulane University for degrees in medicine and in public health, where he was selected for Alpha Omega Alpha and Delta Omega, the medical and public health equivalents of Phi Beta Kappa.

After the events of 9/11 and the anthrax scare in late 2011, Dr. Norton worked with the CDC to prepare for threats of bioterrorism. He helped develop the CDC's algorithm for the evaluation of suspected index cases of smallpox and monkeypox, treatment algorithm for cutaneous anthrax, national guidelines for smallpox vaccination, and protocols for managing adverse events after smallpox vaccination. Other academic interests include ethnobotany (traditional uses of plants), medical entomology, and medical anthropology.

Dr. Norton has more than 100 indexed publications, including papers in many journals outside the dermatology literature, such as *New England Journal of Medicine, JAMA, Lancet, British Medical Journal, Clinical Infectious Diseases, Emerging Infectious Diseases, Vaccine*, and *MMWR*. He is a former Assistant Editor of the *Journal of the American Academy of Dermatology* and remains a reviewer for that journal, the *New England Journal of Medicine*, and many other journals.

Anne Louise Oaklander, M.D., Ph.D., F.A.A.N., F.A.N.A. is a Harvard Medical School Associate Professor of Neurology and Assistant in Neurology and Neuropathology at Massachusetts General Hospital, Boston, Massachusetts (USA), where she directs their nerve unit. After undergraduate studies at Columbia and Cornell, she received M.D. and Ph.D. (neuroscience) degrees from the Albert Einstein College of Medicine. After Rutgers Neurology residency and Peripheral Nerve fellowships at the Johns Hopkins University School of Medicine, she joined Hopkins' faculty. At MGH, she directs a federally funded research team that investigates peripheral neuropathy. Her 130+ publications document evidence of small-fiber pathology in 40% of fibromyalgia patients, and her lab characterized early-onset small-fiber neuropathy in children and young adults. She proposed B-cell-mediated inflammation/autoimmunity as the most common cause and published the first large series showing benefit from immunotherapy. She co-authored the first clinical trial for HSAN-1 and founded and directs the Neuropathy Commons website. A Fellow of the American Academy of Neurology and the American Neurological Association, her research has been profiled in Science and Scientific American Mind and by PBS and YouTube. She has served on NIH's Research Advisory Council; on review and advisory panels for the NIH, the FDA, the AAN, and the Institute of Medicine; and on multiple journal editorial boards, currently Neurology® Neuroimmunology and Neuroinflammation.

Richard J. Pollack, Ph.D. is a public health entomologist/zoologist/parasitologist serving Harvard University in Boston, Massachusetts (USA), as a Senior Environmental Public Health Officer. He also operates the consulting venture, IdentifyUS LLC, that provides services to other institutions, businesses, government entities, clinicians, and the general public. He has reviewed tens of thousands of samples of real and imagined pests and parasites, whether they bite, sting, infest, irritate, or otherwise cause actual or perceived harm to persons, other animals, dwellings, library/museum accessions, and commodities.

Bobbi S. Pritt, M.D., M.Sc. is a Professor of Laboratory Medicine and Pathology at Mayo Clinic, where she serves as the Chair of the Division, and the Medical Director of the Clinical Parasitology Laboratory. Dr. Pritt obtained her medical degree at the University of Vermont College of Medicine, followed by a residency in anatomic and clinical pathology. She then completed a fellowship in Medical Microbiology at Mayo Clinic and an M.Sc. in Medical Parasitology at the London School of Hygiene and Tropical Medicine in London, England. She also received a Diploma in Tropical Medicine and Hygiene from the Royal College of London, England. Dr. Pritt's primary areas of practice are the laboratory diagnosis of parasitic and vector-borne diseases, including the identification of medically important arthropods, and the pathology of infectious diseases. She has published over 200 manuscripts, 4 textbooks, and 25 book chapters on these topics.

Jason S. Reichenberg, M.D., Ph.D. is a Professor in Dermatology at Dell Medical School at the University of Texas at Austin, Texas (USA). His clinical interests include psychodermatology, the interaction between the mind and the skin. He has lectured internationally on this subject and is a co-editor of the textbook Practical Psychodermatolgy. He was the clinical lead on the design of a noninvasive device to detect skin cancer research, which won the "SciFi No Longer Award" at South by Southwest in 2015. He continues to work on NIH-sponsored research on using this light-based technology.

He is currently President of Ascension Medical Group, a multispecialty practice of over 1000 providers in Texas, where he has focused on improving the quality of clinical care. He received his dermatology training at Mayo Clinic and his M.B.A. from the University of Texas at Dallas.

Johnathan M. Sheele, M.D., M.P.H., M.H.S. is an Associate Professor of Emergency Medicine at Mayo Clinic in Jacksonville, Florida (USA). He is fellowship-trained in international emergency medicine, has a master's in public health, and conducts research in infectious diseases and bedbugs.

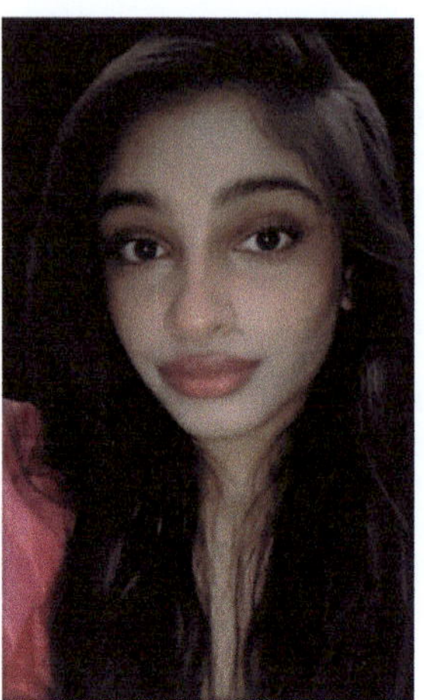

Sahithi Talasila, B.Sc. graduated from the University of Pennsylvania (USA) with a B.S. in Pre-Medicine *Summa Cum Laude* in 2021. She is currently training to become a physician at the Sidney Kimmel Medical College, Thomas Jefferson University, Philadelphia, Pennsylvania (USA), and is expecting to graduate in 2025. She is a research assistant in the Department of Dermatology and Cutaneous Biology and is looking at growth rates and immunosuppressive indices for MBL2 tumors in TLR4-/- hosts and identifying the expression of immune-suppressive cytokines in tumor-localized macrophages in TLR4-/- hosts as well as reviewing the dermatological history of 120 keloid patients to determine the investigated treatment outcomes of ILK compared to ILK and cryotherapy, including changes in pruritus, pain, and keloid size.

Part I
Introduction

1.1 Dedication

We dedicate this work to all who've suffered from infestations, infections, and irritations, whether they resulted from causes that could be objectively documented or from those that have been elusive or imagined. May they—and their families, associates, clinicians, and other caregivers—keep an open mind and be endowed with an abundance of patience, sensitivity, fortitude, and caring as they pursue the best course of action for all involved. We hope the information and guidance we provide in this book may help resolve or better manage such difficult medical mysteries and challenges.

Richard J. Pollack, PhD.

Introduction

Gale E. Ridge

As physicians, we need to approach Delusional Infestation (DI) patients with empathy and compassion. Before starting to manage the disorder, physicians and non-medical professionals should ask themselves, "Do I have the time, resources, patience, and compassion to help these patients?" In many ways, physicians and patients are navigating a difficult condition in opposite directions, and both need support. This book brings the two together. The more the disorder is understood, the better the outcome for both physicians and patients.

DI patients who complain of being bitten often self-excoriate and present samples as proof of a true infestation. When the samples are examined and no pathogens or parasites are found, it leaves physicians unable to provide a diagnosis many patients find acceptable. This effect can be applied to other forms of DI, such as those who believe that they have body-invading parasites or pathogens and those who share or project DI onto acquaintances, relatives, animals [1, 2], or objects and possessions. All too often, these patients react by feeling not believed, frustrated, and angry. Some feel humiliated and helpless. Many retreat into exile, distrust, and isolation, including sundering familial bonds. Their trust in the medical profession becomes broken.

All possess two common themes, unwelcome parasitism, and themed magical thinking to explain the unexplainable. If not psychiatric in origin, the delusion often is a misperception of an unexplained discomfort that ultimately evolves into DI. Most somatic DI cases will have two elements, the psychiatric emotional investment masking an unidentified medical condition.

The genesis of the book evolved from a symposium held in November 2020 as part of the annual meeting of the Entomological Society of America (ESA). The

G. E. Ridge (✉)
Department of Entomology, The Connecticut Agricultural Experiment Station,
New Haven, Connecticut, USA
e-mail: gale.ridge@ct.gov

G. E. Ridge (ed.), *The Physician's Guide to Delusional Infestation*,
https://doi.org/10.1007/978-3-031-47032-5_1

meeting was originally planned for Toronto, but because of the SARS-CoV-2 (COVID-19) pandemic, it was held virtually. Experts from several disciplines met to discuss DI and further our understanding of the condition. It became apparent that there was a need to know more about DI in medicine, and it was decided to write a book to help physicians and nonmedical professionals manage the disorder. Although the various medical disciplines share the same goals of helping patients with DI, each possesses a particular perspective and offers its own insights and treatment strategies. These differing medical disciplines are brought together in a multidisciplinary approach because DI demands it.

DI has wide ranging impacts and numerous costs for everyone involved. Many patients find themselves wandering through diverse disciplines in medicine and philosophy, seeking help. Defining the disorder has garnered much debate in many circles, because though patients have much the same narratives, what causes DI can be varied and hidden. This is detective work, and the purpose of this book is to provide clues and tools to help explain and manage the disorder. The authors' mission is to help professional communities who encounter DI and expedite care. The book is a road map with signposts to guide everyone to a safe place where expertise and suffering might meet on common ground for the purpose of healing.

How to Use the Book.
The book is divided into three parts. Part I introduces DI with its definition, various manifestations, history, and useful literature, Part II provides material for a differential diagnosis and treatment, and Part III is a collection of chapters written by the 13 principal authors and their collaborators. Included is a chapter on common medically significant arthropods. It provides detailed information with images about some common arthropods of medical concern. This includes behavior, biology, distribution, and urban myths.

The book is designed to be read at three levels. There is the main text which goes in-depth about the disorder and text boxes that provide supplemental information or condensed summary material taken from the main text. Given the time constraints many physicians face, this information is intended for in-office use when working in real time with DI patients. The third level is information physicians might use *with* their patients, for example, explaining the life cycles and distributions of known parasites, or features of scabs or injuries which might be misunderstood by patients. Providing sound information for patients who have endured a diet of misinformation has shown to be supportive and helpful in the healing process.

Part III provides in-depth material from the professional perspectives and experiences of each author. Observations and experiences include what works and does not work with DI patients. Additionally, some information provided by the authors in Part III has been selected and used in Part II to assist differential diagnosis, treatment, and care. The authors hope this book will make a difference in the lives of patients who are struggling with the disorder, as well as physicians and other health professionals who work with them, affected families, friends, and colleagues.

Thank you.

Erika Engelhaupt (journalist and editor), Anne Louise Oaklander (neurology and neuropathology), John Sheele (ED physician), Richard Pollack (entomology and parasitology), Bobbi Pritt (clinical parasitology), Blaine A. Mathison (pathology), Scott Norton (dermatology), Kathleen Flanagan (medical student pursuing residency in psychiatry), Dirk Elston (dermatology and dermatopathology), John Koo (psychiatry and dermatology), Nicolas Brownstone (dermatologist), Sahithi Talasila (dermatology), Michelle Magid (psychodermatology) and Jason Reichenberg (dermatology and psychodermatology), Nell Frankowiak (medical student), Peter Lepping (psychiatry), Nancy Hinkle (veterinary entomology), Gale E. Ridge (entomology), Lyle Buss (entomology), Katherine Dugas (artist and graphic designer), and Elliott Campbell (dermatologist).

1   US Emergency Hotlines, Support Groups, and Other Important Information

1	National suicide prevention hotline	(800) 273-8255
	National suicide prevention hotline for hearing impaired (Deaf hotline)	(800) 799-4889
2	Suicide and mental health crisis hotline	988
3.	Substance abuse and mental health services admin	(800) 662-HELP (4357)
		Online: samhsa.gov/find-help/ national-helpline
4.	The samaritans crisis hotline:	116 123
5.	National domestic violence hotline	English (800) 799-7233
	National domestic violence hotline	Spanish (800) 942-6908
6.	Veterans crisis hotline	(800) 273-8255
7.	National grad crisis line	(877) 472-3457
8.	Childhelp national child abuse hotline	(800) 422-4453
9.	National Alliance on Mental Illness (NAMI)	(800) 950-NAMI (6264)
10.	Alcoholics anonymous	Online: aa.org
11.	Narcotics anonymous	Online: na.org
12.	Crisis text:	Online: Text HOME to 741741
13.	United states elder abuse hotline	(866) 363-4276
14.	Families anonymous (addiction)	(800) 736-9805
15.	Well spouse foundation	(800) 838-0879
17.	Griefshare (coping with grief)	(800) 395-5755 USA and Canada). International (919) 562-2112
18.	National eating disorders association: phone and text	(800) 931-2237 info@ nationaleatingdisorders.org **Time open**. Mon-Thurs. 11:00-21:00, Fri. 11:00-17:00

2 Worldwide Emergency Hotlines and Websites for Crisis Care

Country	Hotline organization	Website	Phone number
Argentina	Centro de Asistencia al Suicida	www.asistenciaalsuicida.org.ar	(011) 5275-1135 or 0800 345-1435
Australia	Lifeline Australia	www.lifeline.org.au	13 11 14
Austria	TelefonSeelsorge Österreich	www.telefonseelsorge.at	142
Belgium	Centre de Prévention du suicide	www.preventionsuicide.be	0800 32 123
Belgium	CHS helpline	www.chsbelgium.org	02 648 40 14
Belgium	Zelfmoord 1813	www.zelfmoord1813.be	1813
Brazil	Centro de Valorização da Vida	www.cvv.org.br	188
Canada	Crisis services Canada	988.ca	988
Chile	Ministry of health of Chile	www.hospitaldigital.gob.cl	6003607777 or *4141
China	Beijing suicide research and prevention center	www.crisis.org.cn	800-810-1117 or 010 82951332
Costa Rica	Colegio de Profesionales en Psicología de Costa Rica	psicologiacr.com/aqui-estoy	2272-3774
France	SOS Amitié	www.sos-amitie.com	09 72 39 40 50
Germany	TelefonSeelsorge Deutschland	www.telefonseelsorge.de. siterate.org	0800 1110111 or 0800 9110222
Hong Kong	Suicide prevention services	www.sps.org.hk	2382 0000
India	iCall helpline	icallhelpline.org	9152987821
Ireland	Samaritans Ireland	www.samaritans.org/how-we-can-help	116 123
Israel	[Eran ער״ן]	www.eran.org.il	1201
Italy	Samaritans Onlus	www.samaritansonlus.org	800 86 00 22
Japan	Ministry of education, culture, sports, science and technology	www.mext.go.jp	81-0120-0-78310
Japan	Ministry of health, labour and welfare of Japan	www.mhlw.go	0570-064-556
Malaysia	Befrienders KL	www.befrienders.org	03-76272929
The Netherlands	113Online	www.113.nl	0800-0113
New Zealand	Lifeline Aotearoa incorporated	www.lifeline.org.nz	0800 543 354
Norway	Mental Helse	mentalhelse.no	116 123
Pakistan	Umang Pakistan	www.umang.com.pk/	(92) 0311-7786264
The Philippines	Department of health—Republic of the Philippines	doh.gov.ph/NCMH-Crisis-Hotline	0966-351-4518
Portugal	SOS Voz Amiga	www.sosvozamiga.org	213 544 545 **Time open.** 15:30-00:30 963 524 660 912 802 669

Country	Hotline organization	Website	Phone number
Russia	Фонд поддержки детей, находящихся в трудной жизненной ситуации [fund to support children in difficult life situations]	www.ya-roditel.ru	8-800-2000-122
Singapore	Samaritans of Singapore	www.sos.org.sg	1-767
South Africa	South African depression and anxiety group	www.sadag.org.za	0800 567 567
South Korea	중앙자살예방센터 [Korea suicide prevention center]	www.spckorea.or.kr	1588-9191
Spain	Teléfono de la Esperanza	www.telefonodelaesperanza.org	717 003 717
Switzerland	Die Dargebotene hand	www.143.ch	143 **Time open.** 24 hours/day. or 0800 143000 **Time open.** 18:00-21:00
Taiwan	国际生命线台湾总会 [International Lifeline Taiwan Association]	www.life1995.org.tw	1995
Ukraine	Lifeline Ukraine	lifelineukraine.com	7333
United Kingdom	Samaritans	www.samaritans.org/how-we-can-help	116 123
United States	988 Suicide & Crisis Lifeline	988lifeline.org	988

3 Part I. Delusional Infestation Definition and Forms, History, and Useful Literature

The definition we provide here is a compilation of published material supported by the American Psychiatric Association and the British Association of Dermatologists [3].

Delusional Infestation Definition

Delusional infestation (DI) can arise as a primary (de novo) or secondary delusion.

Primary (de novo) DI is a psychiatric illness characterized by patients holding a monothematic fixed belief of an infestation of their skin, body, or immediate environment, which is not supported by objective medical evidence. The intensity of the delusional beliefs is dynamic and variable over time. They are not merely "overvalued ideas" (see glossary), and the delusional beliefs may be shared between individuals (shared delusion, folie à deux, etc.). Perceived pathogens include living organisms such as parasites, bacteria, viruses, worms, and insects. Organic or non-organic infiltration by fibers, threads, or other forms of inanimate particles or objects (including the DI type Morgellons) known or unknown to medical science, are also reported.

Secondary DI make up 60% of DI cases [38]. For an ICD-11 [4] diagnosis of delusional disorder, symptoms should be present for more than one month and without meeting criteria for schizophrenia or another major mental illness. The delusion may be secondary to organic pathology, other medical, neurological [5, 6], or psychiatric disorders; environmental sensitivities, and/or substance abuse, or as side effects of various prescribed and unprescribed medications. The psychopathology aspect "is limited to the delusion and abnormal tactile sensations related to the delusional theme", while "patients are otherwise entirely mentally healthy and argue rationally if they discuss issues other than their infestation" [1]. When dealing with secondary delusions there is a spectrum of progressive illness from acute to chronic. The patient may also show progressive thinking from overvalued ideas to somatic preoccupation and true delusions [39]. A comprehensive list of medical disorders that can trigger DI is found in Table 1.

The symptoms of DI can impact all aspects of everyday life including work, relationships, and quality of life [7]. It constitutes a high disease burden and is a debilitating disorder that causes significant suffering to the individual and those associated with them [8, 9]. Many people with DI lose insight and challenge the lack of objective findings of infestation. Therefore treatment compliance is variable, and suffering can be significantly protracted. Longer duration of untreated illness is associated with poorer outcome [10]. Antipsychotic medication is highly effective [11].

More specific research indicates that DI patients may have a different pathoetiology compared to similar disorders. It suggests that DI has some unique characteristics. Following is a description written by Bewley, Lepping, and Taylor [12]:

> "A number of MRI studies have looked at the etiology of DI. DI patients show lower grey matter volume in thalamic, striatal, insular, and medial prefrontal brain regions in contrast to patients with non-somatic delusional disorder (such as paranoid delusions) and healthy controls. These differences were consistently detected at regional and network levels. This indicates that patients with delusional infestation may have a different etiology than other patients with other non-delusional disorders, making DI a unique disorder. The available data also support the notion that dysfunctional somatosensory and peri personal networks could mediate somatic delusions in patients with DI. This would support the idea that errors of probabilistic reasoning are involved in symptom formation in DI, with patients preferring unlikely explanations over likely ones because of changes to neuronal pathways."

Additionally, a review of 15 studies between 1980 and 2014 supports the Bewley, Lepping, and Taylor description, which showed most delusional disorder patients possessed brain atrophy. They found white matter lesions, particularly in the temporal-parietal or frontal lobes of the brain [13]. The research also found that dysfunction in the somatosensory and neural peripheral networks could promote somatic delusions. Damage or dysfunction of any of the regions in the brain, which serve the functions of self-awareness, judgment, visuospatial control, physical sensation, learning, and problem solving, may lead to delusional beliefs, visual and tactile hallucinations, and disordered reasoning [1, 5, 7–9, 14–19].

Table 1 Medical disorders, diseases, and conditions which could trigger patients to think they have parasitic activity on or in the body. Sensations of formication, generalized itching, and/or burning accompanied by red spots, and/or rashes, or visual hallucinations can be undiagnosed underlying medical, drug-induced, or environmental conditions that lead people to believe that they are being attacked by mites, insects, or other parasites. The compiled list is based on case reports of at least one confirmed case [21, 25, 30, 31, 33, 34, 37, 40–58, 59]

Agent Orange long-term effects; peripheral neuropathy
Aging
AIDS/HIV
Alcohol withdrawal
Alcoholism
Algorithms (lacks critical thinking with potent power of suggestion)
Allergies: Food, tree, grass, weed, dust mites, mold
Alzheimer's dementia
Anemia including pernicious anemia
Antibiotics
Antiepileptics
Anxiety
Atopic dermatitis
Autoimmune disease incl. Multiple sclerosis, lupus erythematosus, Wegener's granulomatosis
Body dysmorphic disorder (BDD)
Brachioradial pruritus
Brain cysticercosis
Brain grey matter deficiency
Cancer
Cannabis use
Carbon monoxide poisoning
Carcinoid syndrome
Caterpillar dermatitis (brief exposure to toxic hairs)
Cerebrovascular accident
Cerebrovascular disease
Chicken pox (history of infection)
Cholestasis (reduced bile flow from liver)
Chronic folliculitis
Chronic renal failure
Chronic urticaria
Cirrhosis (late-stage liver disease)
Cocaine, esp. acute intoxication
Congestive heart failure
Darier's disease
Dementia
Dementia due to vitamin B12 deficit
Dengue fever
Dental infection
Depression
Dermatitis artefacta
Dermatitis herpetiformis (autoimmune disease)
Dermatophagia
Diabetes mellitus
Dopamine agonists

(continued)

Table 1 (continued)

Dry skin
Eating disorders
Encephalitis (brain inflammation)
Endocrine abnormalities (blood biochemical regulation)
Endogenous eczema
Fiberglass dermatitis
Fibromyalgia
Filariasis (sources: Food, insect, particular habitats)
Fluoride poisoning
Folate deficiency
Folliculitis
Grover's disease
Hallucinations, esp. visual
Heavy metal toxicity
Hemochromatosis (liver disease)
Hepatic disease (alcoholic fatty liver disease)
Hepatitis B and C
Herpes zoster infection (shingles)
Hodgkin's lymphoma
Human immunodeficiency virus
Huntington's disease
Hyper-awareness of normal nerve end firing
Hypertension
Hyperthyroidism
Hypoglycemia (low blood sugar)
Hypoperfusion in temporal and parietal lobes of the brain
Hypothyroidism
Hypovitaminosis, including B12 deficiency
Illegal drug use especially stimulants (cocaine, cannabis, amphetamines)
Illegal drug use withdrawal
Insect phobia
Interferon therapy
Internet searches with confirmation bias
Intestinal flora and fauna imbalance
Iron deficiency
Irritable bowel syndrome linked to urticaria
Isolation
Itch (neurological abnormalities)
Jaundice
Kidney damage
Lack of sleep and exercise as contributing factors
Lactogenic disease
Lichen simplex chronicus
Liver damage

Table 1 (continued)

Lymphocytic leukemia, lung cancer, polycythemia rubra vera, multiple myeloma, or neoplasia)
Lymphoma
Malaria, trypanosomiasis, dengue
Mast cell activation syndrome
Meningitis (central nervous system membrane inflammation)
Menopause (perimenopausal pruritus)
Mental retardation
Munchausen by proxy (mental disorder of fake illness to garner attention)
Munchausen disease
Nasal MRSA carrier
Neoplasia (abnormal tissue growth)
Neurocysticercosis
Neuropsychiatric delusion caused by some tropical diseases
Neurotic excoriation disorder
Niacin overdose
Nocturnal pruritus
Notalgia paresthetica (intense burning/itching in the inner shoulder blade and spine)
Obsessive compulsive disorders (OCDs)
Obesity, poor circulation, and other comorbidities as contributing factors
Onchocerciasis
Oncostatin M (OSM), high levels cause chronic itch
Onychophagia
Opioids
Paper sensitivities
Paresthesia (pins and needles)
Parkinson's disease
Parkinsonian medication
Pathological skin picking (PSP)
Pemphigoid (dermal blistering and rashes)
Pesticide exposure
Poison ivy exposure causing long-term formication
Polycythemia vera (elevates red blood cells)
Polypharmacy
Poor nutrition
Power of suggestion in DI by proxy relationships
Prescription drugs
Prurigo nodularis
Pruritic conditions aggravated by psychological conditions
Psoriasis

(continued)

G. E. Ridge

Table 1 (continued)

Psychodermatosis such as onychotillomania
Psychosomatic pruritus
Pulmonary diseases (heart and lung disease)
Pyrethroid exposure
Reduced grey matter in the brain
Renal diseases
Rheumatoid arthritis
Schistosomiasis
Schizoaffective disorder
Schizophrenia
Scleroderma
Shock and/or trauma with possible PTSD
Sjögren's syndrome
Small fiber polyneuropathy (peripheral neuropathy through somatosensory pathways)
Somatoform dissociation with pruritus (functional itch disorder)
Spinal cord injury
Staphylococcal furunculosis infection
Statins
Steroids
Stinging nettles
Stress
Stroke
Structural lesions in the striatum (putamen) of the brain
Subdural hematoma
Syphilis
Thiamine deficiency
Trauma to the head
Traumatic brain injury
Travel to tropics within a year of symptom onset
Trichotillomania disorder
Trigeminal trophic system
Tropical diseases (neurocysticercosis)
Trypanosomiasis
Tuberculosis
Uremia (kidney and/or bladder disease; hyperuricemia)
Urticaria
Vitamin B12 deficiency
Vitamine D deficiency
Wallenberg syndrome
White matter lesions of the brain
Zinc deficiency

The Central Nervous System: Primary Delusions (Psychiatric Conditions)

True psychiatric illness as defined by the British Association of Dermatologists [3] is:

> "A monosymptomatic hypochondriacal psychosis characterized by delusions, somatosensory abnormalities, behavioral alteration, cognitive distortions."

Etiology may include genetic susceptibility, deleterious life events, activation of dopamine and aberrant itch pathways, dysfunction of the frontal, temporal, and/or parietal cortices, and dysfunction of the dorsal stratum, thalamus, and/or cerebellum [20–25].

Psychiatric and dementia differentials: 80% of patients are in a monosymptomatic delusional state. Very few experience real schizophrenia unless drug-induced or from stroke. These can be resolved. Depression, anxiety, OCD, or personality disorders, especially borderline patients, are very difficult to diagnose. Patients with dementia will often dig at their skin.

Schizophrenia may be suspected if behavior is odd or bizarre. This may include hallucinations with confused thoughts that progress into more and more severe illness. Persecution is a common ideation. Schizophrenia often manifests at a younger age than most DI cases, usually in the late 20's. In some cases Schizophrenia can engender DI. The DSM-IV-TR diagnostic criteria should be met.

The Body: Secondary Delusions (Somatic Conditions)

Secondary DI is triggered by many undiagnosed underlying medical disorders or conditions that cause symptoms of pruritus, erythema, urticaria, paresthesia, or formication of the skin [26, 27]. Legal or illegal drugs, iatrogenic, and polypharmacy can also induce dermal symptoms [28–37]. Additionally, secondary DI can take the form of perceived deep tissue "parasitism" with believed myiasis and body infestation by worms.

Approximately 60% of DI patients experience secondary DI [38]. Some patient experiences are transitory (a few days), while others linger for years. DI can but not always start with real misunderstood undiagnosed underlying medical disorders that over time gain psychiatric investment. When dealing with secondary DI, there is a spectrum of progressive illness from mild to chronic. This includes progressive thinking from overvalued ideas, somatic preoccupation, and delusional state to a terminal delusional state [39]. A comprehensive list of medical disorders, diseases, and conditions that can trigger DI is found in Table 1. Additionally, Table 3 has a comprehensive list of papers that address DI issues by subject.

Other Predominantly Delusional Infestation Psychiatric Disorders

These are chronic psychotic disorders in which comparatively stable delusions or hallucinations are the main clinical features, but do not justify a diagnosis of schizophrenia. If these delusions persist, the diagnosis should be changed to persistent delusional disorders, paranoid reactions, or psychogenic paranoid psychosis. Here, five forms are described, folie à deux, folie à trois, folie à familiar [60], double DI, and DI by proxy [20, 60, 61].

Induced delusional infestation (IDI) (*folie à deux, folie à trois, folie à familiar*): Folie à deux, folie à trois, and folie à familiar are shared psychotic disorders involving two or more people with close emotional, professional, or social ties. One person, "the inducer" with the DI, abets a manifestation of the disorder in others who are associated with them. Once those who have been influenced by the inducer are separated or taught they have been influenced, they usually experience an abatement of delusions. Approximately 10% of patients have shared forms of delusion [12]. A further description of shared DI is found in the DSM-IV (1994) (Box 1). Tey et al. [21] and Papoiu et al. [62] showed that the power of suggestion can promote pruritus. By viewing videos of other people scratching, patients with atopic dermatitis as well as healthy subjects "experienced more intense itch" [21].

Folie à deux example case: Names have been changed in the following email to protect patient privacy. Text is original:

> "Well, Doc. Me and George handled the situation. After three months of excruciating pain. We decided if it hurts get rid of it. He poured peroxide on it, and I shaved of the bumps with a knife. Found a tiny hole on a slight bias. Poured some shit on it and threw the little white maggots in the sink. Looks like a fruit fly larva only smaller. Thank you. Kenneth."

Double-delusional infestation (DDI): Someone who has been inferred by the inducer as infested and who cannot speak for themselves. These would be children and animals. They are perceived to be infesting the inducer as well as infesting themselves. Lepping et al. [63] suggest that DDI is defined as "patients who believe that they *and* someone or something else is infested, but the other person or pet/ animal does not or cannot share that belief." In the case of animals, veterinary clinicians are poorly equipped to deal with these kinds of cases [64].

Additionally, worrisome is that many animals are treated by their owners with pesticides and other chemicals such as bleach. Their owners often treat themselves with the same chemicals. Further discussion on how veterinarians might manage suspected DDI, or Delusional Infestation by Proxy (DIP) clients and their animals can be found in "Veterinarians and Delusional Infestation" and Chap. 6 "Animal Suffering and Euthanization Due to Human Psychosis" written by Nancy Hinkle.

Delusional Infestation by Proxy (DIP). The inducer thinks that someone else including animals [65], children, vulnerable adults [61], or objects are infested, and they are not. This could lead to serious child safety issues [61, 66]. Fisher [67] described a 6-year-old boy whose mother insisted that he had a skin and scalp infestation. He was eventually removed from her custody and placed into the care of grandparents, and the mother underwent psychiatric treatment. With animals including pets, DIP could lead to baseless demands to euthanize them [64, 68]. Additionally, the environment, buildings, cars, particular rooms, and wild or other domesticated animals could also be included in the list. Some patients experience both DDI and DIP at the same time [62].

Box 1. Folie à deux [69]

Shared Psychotic Disorder. The essential feature of shared psychotic disorder (folie à deux) is a delusion that develops in an individual who is involved in a close relationship with another person (sometimes termed the "inducer" or "the primary case") who already has a psychotic disorder with prominent delusions (Criteria A). Usually, the primary case in "shared psychotic disorder" is dominant in the relationship and gradually imposes the delusional system on the more passive and initially healthy second person. Individuals who come to share delusional beliefs are often related by blood or marriage and have lived together for a long time, sometimes in relative isolation. If the relationship with the primary case is interrupted, the delusional beliefs of the other individual usually diminish or disappear. Although most seen in relationships of two people, "shared psychotic disorder" can occur in a higher number of individuals, especially in family situations in which the parent is the inducer and the children adopt the parent's delusional beliefs to varying degrees.

Associated Features and Disorders.

Aside from the delusional beliefs, behavior is usually not otherwise atypical in "shared psychotic disorder." Impairment is often less severe in individuals with "shared psychotic disorder" than in the primary case (the inducer).

Prevalence.

Little systematic information about the prevalence of "shared psychotic disorder" is available. This disorder is rare in clinical settings, although it has been argued that some cases go unrecognized.

Course.

Without intervention, the course is usually chronic because this disorder most commonly occurs in relationships that are long-standing and resistant to change. With separation from the inducer/primary case, the healthy person's delusional beliefs disappear, sometimes quickly, sometimes slowly.

Understanding the Morgellons Phenomenon, a Subtype of Delusional Infestation

Delusional infestation is a disorder with a history of era-dependent variable nomenclature. Since the mid-1890s, there have been over 40 different names given to DI (Tables 2 and 3). This is because DI does not arise from one single cause or closely related causes but from a broad spectrum of psychological, somatic, and environmental causes. Additionally, given the psychological aspect of the disorder and patient dislike of being labeled with a mental illness, euphemisms were also adopted.

Table 2 Historic names for delusional infestation. Amended from RW. Freudenmann and P. Lepping (2009)

Year of publication	Author's name	Name used for delusional infestation	New concepts
1896	Perrin (F)	Des névrodermies parasitophobiques	Matchbox sign
1920	Gamper (D)	Der Psychosen des Rückbildungsalter	First hypothesized etiology (thalamic dysfunction)
1921	Pierce (USA)	Entomophobia	
	Myerson (USA)	Acarophobia	
1923	Giacardy (F)	Un cas d'acarophobie familiale	Shared psychotic disorder described
1925	Grøn (FIN)	Les dermatophobies	Psychogenic etiology
1928	MacNamara (USA)	"Cutaneous … hallucinations"	Hallucinations plus secondary delusions; affected pets
1929	Schwarz (D)	Circumscripte Hypochondrie	Occurs as a depressive symptom
1930	Mallet and male (F)	Délire cénesthesique	
1934	Smith (USA)	Hallucinations of insect infestation	
1935	Wilhelmi (D)	Ungezieferwahn	
1938	Ekbom (S)	Der präsenile Dermatozoenwahn	Primary delusional disorder
	Hase (D)	Pseudoparasitismus	Illusion plus secondary delusions
1944	Davis	Insect hallucination	
1946	Wilson and Miller (USA)	Delusion of parasitosis	Four different etiologies
1949	Harbauer (D)	Dermatozoenwahn (Ekbom)	Presenile or in depression, response to ECT (first), illusions or hallucinations occur

Table 2 (continued)

Year of publication	Author's name	Name used for delusional infestation	New concepts
1954	Bers and Conrad (D)	Die chronische taktile Halluzinose	Multiple etiologies, mostly organic, with emphasis on hallucinations
	Böttcher (D)	Das Syndrom des wahnhaften Ungezieferbefalls	First to gather cases from pest control officers
1960	Döhring (D)	Ungezieferwahn	Sample of 77 cases reported by a microbiologist
1962	Liebaldt and Klages (D)	Isolierte taktile Dermatozoenhalluzinose	Only postmortem histology (thalamocortical disconnection)
1965	Tullett (UK)	Delusions of parasitosis	Separate entity, "monosymptomatic hypochondriasis"
1966	Schrut and Waldron (USA)	Delusory parasitosis (entomo-, acaro-, or dermatophobia)	Psychoanalytical model (unconscious sexual guilt)
1970	Hopkinson (CAN)	Delusions of infestation	First response to antidepressant in major depression
1975	Ganner and Lorenzi (D)	Der Dermatozoenwahn	Primary and secondary delusion
	Riding and Munro (UK)	Monosymptomatic hypochondriacal psychosis	First response to pimozide (case 3)
1978	Annika Skott (S)	Delusions of infestation; Dermatozoenwahn (Ekbom syndrome)	First real study, "primary delusion," organic (>50% of cases)
1979	Frithz (S)	Delusions of infestation	First study with depot antipsychotics
1982–1986	Ungvari and Vladar (H), Hamann and Avnstorp (S)	Dermatozoenwahn, delusion(s) of infestation	Open and placebo-controlled studies with pimozide, very small sample size of 11
1983	Lyell (UK)	Delusions of parasitosis	First and largest survey; DI starts from senile pruritus
1985	Berrios (UK)	Delusional parasitosis	Variety of etiologies
1986	Bourgeois (F)	Syndrome d'Ekbom et délires d'infestation cutanée	Large survey of French dermatologists
1987	Renvoize et al. (UK)	The syndrome of delusional infestation	Emphasis on syndromal nature
1989	Musalek (A)	Dermatozoenwahn	First interdisciplinary outpatient clinic; only SPECT study
1991–1995	Trabert (D)	Dermatozoenwahn	Only epidemiological study (D), meta-analysis of 1223 cases

(continued)

Table 2 (continued)

Year of publication	Author's name	Name used for delusional infestation	New concepts
1994	Srinivasan et al. (IND)	Delusional parasitosis	Open study questioning superiority of pimozide
1995	Gallucci and beard (US)	Delusions of parasitosis	First case report on response to second-generation antipsychotic risperidone ($n = 1$)
2007	Lepping et al. (UK)	Delusional parasitosis	Systematic review of antipsychotics in DI (first), supports efficacy of antipsychotics
2008	Freudenmann and Lepping (D/UK)	Delusional parasitosis	Meta-analysis of cases treated with second-generation antipsychotics (first)
	Huber et al. (I)	Delusional parasitosis	Structural MRI study (first)
2009	Freudenmann and Lepping (D/UK)	Delusional infestation	Name change to broaden the scope of disorder
2018	Lepping et al. (UK)	Delusional infestation	Double-delusional infestation concept introduced, large study among veterinarians
2020 2018	Lepping et al. (UK) Romanov et al. (RUS)	Delusional infestation	Longer duration of untreated psychosis is associated with worse outcome

Key: F France, *D* Germany, *USA* United States, *FIN* Finland, *S* Sweden, *UK* United Kingdom, *CAN* Canada, *H* Hungary, *A* Austria, *IND* India, *I* Italy, *RUS* Russia, *ETC* electroconvulsive therapy, *SPECT* single photon emission computed tomography, *MRI*

In the early 2000's, a new name was added to the list. Driven by Ms. Mary Leitao, it gained considerable popularity through the internet and has been described as an "internet-transmitted disease" [69, 70]. This most recent addition to the DI lexicon is Morgellons Disease (MD). The situation-dependent and constantly changing symptoms described by sufferers of MD are not known and supported by the scientific community, and its protagonists vociferously refute that it is yet another identifier for DI, so how did MD become established in the DI lexicon?

In the early 2000's, Ms. Leitao started to notice red, black, blue, and white man-made fibers, "bundles of fibers," in her young son's skin which several physicians failed to identify [1]. Leitao believed that these were caused by an unknown disease. After reading "Letter to a friend upon occasion of the death of his intimate friend" written by Sir Thomas Browne (published posthumously in 1690), who had used the identifier "Morgellons" in his letter, she decided to name her son's condition "Morgellons Disease". Subsequently, Ms. Leitao founded the Morgellons Research Foundation, which aggressively lobbied Congress to have the Centers for Disease Control and Prevention (CDC) study this "unknown" disease. US Senator Barbara

Boxer, among others, supported her. Following a period of reluctance, the CDC acquiesced [71–75]. Following this, a subsequent study published in 2012 refuted her assertions [75–78]. The CDC found no causative agent. Leitao and her foundation then switched to Lyme disease, *Borrelia burgdorferi*, as the causative agent [79]. Stassa Edwards' writing in 2015 commented:

> "Disease has always had a hierarchy. To suffer from a rare disorder that has been definitively traced to a biological origin is preferable to being diagnosed with a psychiatric disorder. One patient suffers heroically, the other is simply insane. That's likely why Leitao didn't just want Morgellons recognized as a disease, but rather a disease with a biological origin, a disease like Lyme disease caused by an infectious agent [78]."

By trying to prove that MD was a true disease with a biological origin as Edwards suggested and since the CDC had refuted their original assertions, choosing Lyme disease continues to validate sufferers and the advocate medical and research communities that serve them. An example of this is an online paper written by Middelveen et al. published in 2013 [79] (with a follow-up paper in 2018), where they wrote:

> "Spirochetal infection provides evidence that this infection may be a significant factor in the illness and refutes claims that MD lesions are self-inflicted and that people suffering from the disorder are delusional [79]."

Authors who publish under the leadership of the Morgellons Research Foundation are few. The most published authors in this small group are Middelveen, Savely, Leitao, and Stricker. Most "evidence" they present is speculative and unsupported and does not fall under strict research rigor. All other authors write opinion pieces, and none have considered research into MD because its supporters when challenged seek new causes for their suffering, change their position, and with that immediately undermine their credibility.

In addition to their perceived medical sensations, other self-diagnosed Morgellons sufferers reported that they were sickened by aircraft contrails, illegal immigrants, fungi, and disease-causing pathogens that were new to science, and genetically modified foods [80–82]. This patient-driven constellation of adaptable implausible symptoms with a strong psychological component leads to a differential diagnosis of DI. Morgellons should be considered by the medical community as DI under another name and treated as such [1, 83].

4   History of Morgellons Disease and Sir Thomas Browne

Some of the earliest records of unexplained dermatitis were written in the sixteenth century. In 1544, Leonellus Faventinus de Victoriis (1574, posthumous) described seeing young children with "dermal worms" which he called "Dracontia." Other authors described them as "die durzemaden" (worms that induce wasting), Cirones, Comedones, Masclous, or Masquelons [84]. The common theme always was unexplained hairs or worms in the skin seen particularly in young children, never fibers.

Sir Thomas Browne, a seventeenth-century physician and polymath, had also described seeing dermal worms and hairs in children when he visited the Languedoc region now known as Montpellier in southeast France [84]. He coined the name Morgellons disease, which is an anglicized version of the local French descriptor, Masquelons.

Browne was born in London, England, in 1605 and lived through a tumultuous century of transition from Renaissance thinking to the age of enlightenment [85–87]. His father was a wealthy merchant, and he enjoyed an elite education at Winchester boys' school in southern England, then Oxford University. In 1629, he graduated with a Master of Fine Arts and travelled to Europe to continue his education as a physician. In 1630 (age 25), he began his medical training in Montpellier, southern France, then Padua, Italy, and finally Leiden, Germany, where he graduated in 1633. At the time, these were the three foremost centers of medical education in Europe [86]. Montpellier was an extremely poor region of France where witchcraft in the context of folk medicine was commonly practiced.

Dr. Browne was a reclusive prolific writer, best known for his books *Religio Medici* and *Urn Burial*. Approximately 30 years after his stay in Montpellier, he wrote "A letter to a friend." The text drifts from subject to subject but in it is a passing sentence describing young children suffering from Masquelons (Morgellons) (Box 2).

Though Browne credited Petrus Pichotus (nickname Picotus (Box 2)) with coining the name Morgellons in his book *De Rheumatismo* (1577), both Greenhill and Kellett [84, 87] did not find any reference of it in the text. They also questioned whether Browne ever owned Pichotus's book. Yet Browne did own several other medical volumes including a seven-volume compendium by Johannes Schenck von Grafenberg (Latin: Schenckius (Box 2)) titled *Observationum medicarum rariorum, libri VII*. In it, Schenck wrote about a rare pathological dermal condition which may have been interpreted by Browne as Morgellons. Schenck used the name "Dracunculus" to describe the condition [88].

Kellett [84] wrote this about Dr. Browne:

"It is therefore eloquent of the man and his interests, and also perhaps remarkable, that writing over forty years later, he should refer to none of these things *(he's referring to the societal upheavals of the time),* but should recall that and that the single reference he gives—see Picotus de Rheumatismo—should be not only vague, as are indeed all his references, to a degree unusual at that period, but also misleading."

Kellett was skeptical about Browne's credibility. Greenhill also suggested that Browne was prompted to write about the children of Languedoc because he had just read *De Aegagropilis* by Velshius (1660) and "had good reason for believing this disease, so whimsical and fantastic, was one that would shortly awaken general interest" [87]. His motive was to either steal the thunder from Velshius or contribute to scientific discourse. He subsequently died in 1682 before the "Letter to a friend" was published.

It is clear that these early sixteenth- and seventeenth-century writers did not understand what they were observing, yet their descriptions provide clues. All of them described infants and young children with particular dermal symptoms. Faventinus de Victoriis (1574) wrote:

> "There exists in little children certain living principles having the appearance of worms, that are called by the common folk Dracontia. They settle especially in the muscular parts of the body, to wit the arms and legs—the calves especially. Occasionally they even congregate in the flanks under the skin, and sometimes occupy the whole of the back, or failing that at least the inter-scapular region … We destroy worms of this type, that are in the habit of lurking in the pores of infants and little girls [84]."

Throughout Europe, most beds or bedding was either mattress ticking filled with straw (Fig. 1) or plain straw beds (Fig. 2). Additionally, it was usual practice to lie infants on their backs. The parts of the body that would have had direct contact with the straw would have been the back, back of legs particularly the calves, and arms (the upper half of the legs in infants would have been protected by clothing). Not only would the straw be naturally irritating, but it may also have been infested with straw itch mite *Pyemotes tritici,* endemic to the region, and the children were suffering from contact dermatitis.

Schenck wrote this observation:

> "… Worms or, as others will have it, Hairs, which are wont to infest the muscles of the arms, calves, and back in infants and children, and which are unknown to the old authorities … [88]."

Schenck died in 1598 and was published posthumously.

Schenck described how affected children wasted away while unknowingly he simultaneously described scabies, dermal infection, as "extreme warmth" and platelet-fibrin plugs as "worms." Additionally, he described more severe reactions;

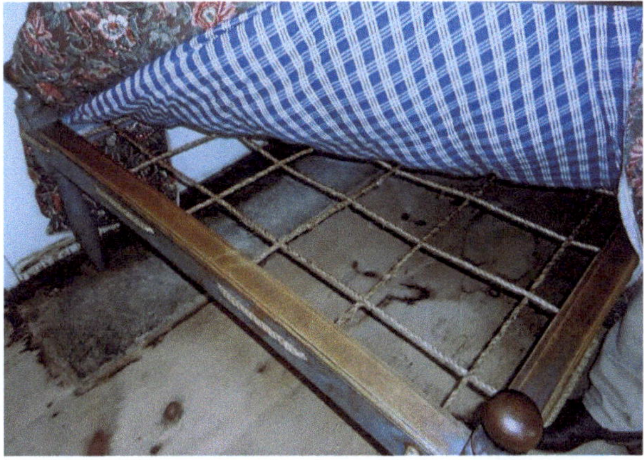

Fig. 1 Colonial bed. Sturbridge Village, Massachusetts, USA

Fig. 2 Straw bed. Hjerl
Hede Museum Village,
Denmark

"Headache, anorexia, nausea, vomiting, mild diarrhea, and joint pains may occur …" [88]. He lacked an understanding of what he was observing, yet his descriptions were accurate if his writing is not read literally. Following is an example:

> "There is a type of intercutaneous worm which is wont very frequently to infest infants under six months and not infrequently also children of two years or of about that age. They are born, in preference to all other places, in the muscles of the arms, legs and back, and arise from an excrementory humour which is contained within the pores of the body and is common at that age. This, because of the repression of transpiration and dispersal, undergoes putrefactive changes and becomes alive and, in proportion to the number of receptacles of the pores, is converted into worms, which have a shape not at all unlike those that are born in putrefying cheese*, but very much smaller. They never creep entirely out from the pores**, but protrude their little heads, which are distinguished as so many black points***. How should they not then be most troublesome, for by exciting a sensation of extreme warmth and, at the same time, of itching, they bring in their train insomnia and restlessness. Where they are packed together in large quantities and are increasing, there they plunder away from living flesh, in the same way as do pediculi the nourishing humours, taking for themselves that which should have been tender bodies. Because of this, little children pass rapidly into wasting and extreme emaciation … Our German people refer to these as Mitesser (trans. blackheads) and die zehrende Wurm (trans. the devouring worm) (*sic*), from the fact that they seize for themselves and consume the food of the infants whom they infect. The Norumbergians call them die durzemaden (trans. a local dialect term meaning worms that induce wasting) (*sic*) or, as you might put it, the worms that induce wasting [88]."

*Possibly cheese mites such as *Tyrophagus casei*.
**Possibly scabies mites.
***This may be dirt or comedo from the Latin word "*comedere*," meaning "to eat up."

Historically, comedo described parasitic worms. In current medicine, the worm-like appearance may be platelet-fibrin plugs of the skin.

Box 2. "Letter to a Friend upon Occasion of the Death of his Intimate Friend" by Dr. Thomas Browne (1674) and Illustrated by Ettmuller (1682)

"Though the beard be only made a distinction of sex, and sign of masculine heat by Ulmus, * yet the precocity and early growth thereof in him, was not to be liked in reference unto long life. Lewis, that virtuous but unfortunate king of Hungary, who lost his life at the battle of Mohacz, was said to be born without a skin, to have bearded at fifteen, and to have shown some grey hairs about twenty; from whence the diviners conjectured that he would be spoiled of his kingdom, and have but a short life; but hairs make fallible predictions, and many temples early grey have outlived the psalmist's period.+ Hairs which have most amused me have not been in the face or head, but on the back, and not in men but children, as I long ago observed in that endemial distemper of children in Languedoc, called the Morgellons#, wherein they critically break out with harsh hairs on their backs, which takes off the unquiet symptoms of the disease, and delivers them from coughs and convulsions."

Ulmus de usu barbæ humanæ. (Elm).

+The life of man is threescore and ten.

#See *Picotus de Rheumatismo.*[1]

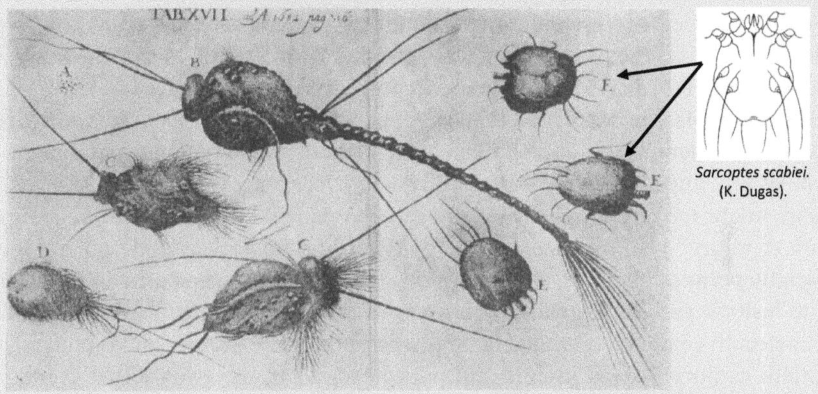

Sarcoptes scabiei.
(K. Dugas).

Drawings by Ettmuller (Acta eruditorum. Lipsiae, Grossius. 1682: p. 316).

A. "Parasite under the naked eye."

B, C, and D. "Under the micrscope, B alone being a perfect specimen."

E. "Are figures of the Cirones or *sarcoptes hominis*." [84][1]Morgellons was not mentioned in this particular book. In 1584, Petrus Pichotus authored De Rheumatismo (British Museum).

The four inaccurately illustrated specimens (left) possess heads, antennae, ocelli, brushlike setae, no legs, and no arthropod body parts. These are not known to science and may be representations of pantry pests or carpet beetle larvae. These illustrations are fanciful. The three mite illustrations on the right are suggestive of scabies mites.—Dr. Ridge.

As a conclusion, Languedoc (Montpellier) was an impoverished region of France. Possibly, scabies, straw itch mite, and/or contact dermatitis from sleeping on straw beds were likely causes of described dermatitis in young children from the region. Most written observations of that time reported dermatitis on the back of the legs, particularly the calves, arms, and backs of young children. This strongly suggests back-sleeping. There was no mention of dermatitis on the face, chest, belly, and front of legs. Browne provided an anglicized name, Morgellons, to explain some form of contact dermatitis caused by likely one or more species of mite and/or the harsh texture of the bedding itself.

Issues with the Name "Delusional Infestation"

It is very common, perhaps even axiomatic, that DI patients vigorously reject any reference to mental illness as delusion. The word "delusion" to them is highly charged. Among professionals, the term "delusional infestation" is helpful. Physicians and other medical and nonmedical professionals use identifiers to help in differential diagnosis and communication, and "delusional infestation" is currently the best option. Most patients do not agree. They can tolerate the word "infestation" but not "delusion" [70]. Patients interpret these descriptors as accusatory and/or dismissive; they are emotionally thinking in narrow terms while physicians are looking all over the place for answers. Neither word embraces any one thing and therein lie patients' anxieties and fears because of the unknown.

DI is a complex disorder with both psychiatric (actual or not) and somatic illness working in tandem. Thus, euphemisms were used to shield patients and make a DI diagnosis less impactful. This form of physician deception was felt necessary when honesty might lead to a negative response from patients by adding in their minds, "insult to injury" [70]. Altered/code/jargon wording for DI has undergone a series of socially permissible name changes (Table 2). It appears to be particularly responsive to historic periods and cultural norms, which includes how it was reported. In the nineteenth century, common euphemisms for DI were scabies and typhus. In the twentieth century, it was insects and parasites, and in the twenty-first century, a selection of names such as insects, viruses, bacteria, fungi, worms, and fibers and threads were used. The fiber and thread ideation was further euphemistically called "Morgellons Disease" [76, 89, 90].

The two euphemistic words for DI in current use are Morgellons Disease and Ekbom, but anecdotally "scabies" has also been commonly reported as an identifier. They perpetuate because it is the language that avoids upset. For example, Morgellons Disease was promoted by Mary Leitao to mean something very different from what it is. Like Ekbom [90–93], Morgellons does not possess a psychiatric connotation. Murase et al. commented on this:

> "To the layperson, the information on this Web site (*referencing the Morgellons Research Foundation*) is deceptive, particularly to someone who suffers from delusions of parasitosis. However, because the term 'Morgellons disease' does not have the word 'delusions' embedded in the term, it is a useful way to communicate with patients regarding their disease [93]."

Treating patients becomes difficult when euphemisms conceal the name of the true illness. It may lead to confusion, particularly when the literature is populated with a spectrum of terms referring to the same disorder. Applying the correct medically accepted name to a differential diagnosis is important, especially when engaging in consultations and recordkeeping. It is advisable to use the currently accepted name of "Delusional Infestation" for this reason. DI is undercounted because of the overuse of euphemisms creating a misperception that it is a rare disorder. Though diagnosis of DI is seen in negative terms by patients, it expedites physicians to their destination and accelerates appropriate care for these patients. The term "Delusional Infestation" clearly does not "tiptoe" around issues and, while blunt, ultimately works to get the job done. Ultimately, the question is which is better, relieving the patient of discomfort or preserving their feelings.

In Defense of the Name Delusional Infestation. With patience and compassion, the term delusional infestation can be accepted by patients if they are introduced to the diagnosis through developed trust. Following is a letter written by Peter Lepping, Stephen L. Walker, and Anthony P. Bewley (2020) [94] in response to an editorial written by Matthew Grant [76], which explains this.

Letter to the Editor
Why Delusional Infestation Should Remain Delusional Infestation.
   "Dear Sir,
   We read with interest the editorial by Matthew Grant about delusional infestation (DI) [70]. We have been involved in the treatment of DI for some years. We lead specialist clinics [95, 96]and have been involved in the recent development of national guidelines through the British Association of Dermatologists (BAD) [3]. We agree with Grant's summary of the difficulties that arise when treating patients with delusional beliefs. We also agree that significant risks can arise in patients with DI, both to themselves and to others through attempts to get rid of the alleged pathogens. However, we do not agree with the conclusion that the name DI should be changed, and it was interesting to note that the editorial did not suggest any alternatives. The complexity of the treatment challenges are somewhat missed in the editorial. Two of us have been party to advocating a change in the nomenclature to DI from previous names because they did not encompass varying changes in alleged pathogens [16]. Our reasons are as follows:
   First, various changes to the terminology have been attempted without improving the inherent difficulties that arise when treating patients with mono-delusional beliefs. Delusional infestation is a delusional disorder, and a change of terminology will not alter this. It is important for research and practical purposes to be as accurate as possible with a description of the disease.
   Second, most experts agree that clinicians should use symptom description for DI and similar illnesses at the starting point of discussions with patients and, thus, initially avoid the term delusional [97]. The new BAD national guidelines for DI incorporate this approach [16]. As patients usually lack capacity to make treatment decisions because of their intense delusional belief, it is ethically justified to have a gradual approach to diagnosis disclosure. This is usually in the patient's best interest and common practice with similar illnesses with high levels of disease

burden and lack of insight. Sensitivity is needed to create engagement and rapport with the patient, which facilitates effective treatment, that is, antipsychotic medication.

We do not agree that a change to the terminology would be helpful or improve patient engagement. A Cochrane systematic review of the treatment of primary DI [19] is available, and the BAD guidelines are imminent [16]. These guidelines are evidence based, to the degree that evidence currently exists. In specialist clinics and other settings, where these guidelines have been adhered to, outcomes have been promising and very much better than outcomes in standard settings.

We agree with Grant that psychiatric referrals alone are usually ineffective. We have recently shown that a longer duration of untreated illness is associated with worse outcomes [10]. All practitioners who see affected individuals should focus on early intervention that requires early engagement and rapport with the patient to facilitate meaningful and effective treatment [63]. We support any attempts to improve this challenging condition."

Peter Lepping
Wrexham Maelor Hospital Psychiatric Liaison Team,
Betsi Cadwaladr, University Health Board,
Heddfan Psychiatric Unit, Wrexham, United Kingdom,
Centre for Mental Health and Society, Bangor University, Bangor, United Kingdom,
Mysore Medical College and Research Institute, Mysore, India
Stephen L. Walker
Hospital for Tropical Diseases and Department of Dermatology, University College, London Hospitals NHS Foundation Trust, London, United Kingdom,
Department of Clinical Research, London School of Hygiene and Tropical Medicine, London, United Kingdom.
Anthony P. Bewley
Department of Dermatology, Barts Health NHS Trust, London, United Kingdom,
Queen Mary University of London, London, United Kingdom

John M. Bostwick supports their position by writing:

"Others have written of the fundamental need for 'establishment of trust' with DI patients."

Ahmad and Ramsay wrote,
"Simply acknowledging the misery open(s) therapeutic windows and encourage(s) him to take his treatment … Having typically seen many disbelieving or frankly dismissive physicians prior to this latest consultation, patients may be defensive, frustrated, and angry [98]."

Koblenzer [99] offers simple advice:

"Acknowledge to the patient that what the patient describes is exactly what that patient is experiencing. The hoary old saw goes that the point of a first date is to get a second one. This is as true for DI patients as it is for potential romantic partners."

5 Perception and the Internet

"There are two ways to be fooled. One is to believe what isn't true; the other is to refuse to believe what is true—Soren Kierkegaard."

Humans have an endless capacity for self-deception and magical imagination. Patients who suffer from DI are no exception. Self-deception, assumptions, and magical thinking often work against differential diagnoses because they either mask true symptoms or are the symptoms. Identifying the differences between the two is challenging.

Many patients perceive their parasites as actively mobile. They describe them moving freely in water, in and out of body cavities, through the skin and inanimate objects, and flying. They are often described as small white living intelligent "creatures" that can change color, often to black. Some are gelatinous. Yet these parasites always disappear when examined under magnification.

When patients relate their perceptions to people they trust who may not understand them and who respond negatively, it leads to intense frustration and unhappiness. If there are unguarded words spoken by physicians, friends, acquaintances, or family members suggesting that these patients are "mad, nuts, or crazy," they withdraw and become guarded. This perceived failure and sometimes rejection by people they trust send many vulnerable patients to the internet to seek answers and comfort.

As of this publication, two to three billion people worldwide daily use online social networks and information sites. Many social networks and search engine algorithms promote feedback loops, which those who are psychiatrically healthy easily manage. It is disastrous for DI patients because depression, insecurity, and anxiety influence thinking. Internet algorithms are engineered to track mouse clicks and time spent on particular sites to build user profiles. They prioritize, optimize, and amplify content tailored to increase engagement. In other words, they gather and steer material towards the user's interests. It promotes a laser focus on a single ideation (often populated with misinformation), and vulnerable DI patients easily fall into the trap. Patients get more and more depressed while being less and less able to disengage from the continuous feed of material the algorithms provide. It leads to harmful paranoid thinking and emotional addiction.

Social Media: Patients often lose focus. It is described as the "switch cost effect." The human brain evolved to perform single-focus tasks. Social media is designed to attract attention, and users become constantly distracted as each platform tries to outcompete with the others. The "attention engineering" of social media based on principals of casino gambling is potent. Users constantly flip from one social media interest to the next, sometimes spending moments in each at great cost. Surroundings become ignored, and the awareness of time, concentration, and focus are all compromised. Users become "users," and their lack of focus, exhaustion, and vicarious living pose serious challenges with DI treatment.

Fig. 3 **Algorithms.** Illustration by Katherine Dugas

Assume that *all* DI patients have been on the internet seeking help and information to support and share their theories of infestations and all were steered to agenda-filled websites and social networks for validation even though most deny it [69, 70]. Assume that these patients are emotionally addicted and undermined. They will challenge anything a physician might say if it contradicts their beliefs because initially learned information - though often incorrect - will be defended. It is a known phenomenon that initially learned information whether correct or incorrect is often defended in the face of expertise, because the emotional investment and inertia that go with it are slow to change [100]. However, on an intellectual level, change is possible. The intellect can override emotion, but it takes time. In the presence of a physician, many patients often gain an understanding of their disorder at an intellectual level, but once they leave a physician's office, they often revert to established thinking and habits. Closely timed, repeated visits are suggested in these cases to support patients during treatment.

To conclude, social media provides a temporary "emotional" boost by promoting a sense of pleasure with rewards and brand loyalty but leaves a desire for more (Fig. 3). While in moderation, there is little harm, but in excess and in the hands of vulnerable users, it is like "junk food." Social media websites are heavily marketed, and content is highly processed.

6   Selected Papers by Topic, *Sensu Lato*

Since the beginning of this century, DI has drawn increased interest from research and medical communities. As a result, a series of significant papers have been published that clarify the condition and identify its issues. Most of the papers are available on open-access websites. Full citations are found in references. Usually, a list such as this is relegated to an appendix or treated as a supplemental reference. In this case, we have inserted it early in the book to assist physicians to access important peer-reviewed material by topic.

General topics are listed in the left column with relevant papers on the right (Table 3).

Table 3 Selected papers by topic, *sensu lato*

General topic	Articles with content relevant to the topic
History of DI and related disorders	Ekbom KA. 2003 (trans. 1938 Yorston). The pre-senile delusion of infestation [93] Freudenmann RW, Lepping P. 2009. Delusional infestation [1] Hylwa SA. & Ronkainen SD. 2018. Delusional infestation versus Morgellons disease [76] Middelveen MJ. et al. 2018. History of Morgellons disease: From delusion to definition [80] Savely VR. et al. 2006. The mystery of Morgellons disease. Infection or delusion? [77] Schulte B. 2008. Figments of the imagination? [71] Traver JR. 1951. Unusual scalp dermatitis in humans caused by the mite Dermatophagoides (Acarine: Epidermoptidae) [8]
What is in a name?	Bewley AP, Lepping P. Freundenmann RW, Taylor R. 2010. Delusional parasitosis: Time to call it delusional infestation [16] Freudenmann RW, et al. 2012. Delusional infestation and the specimen sign: a European multi center study in 148 consecutive cases [101] Freudenmann RW, Lepping P. 2009. Delusional infestation [1] Hinkle NC. 2010. Ekbom syndrome: The challenge of "invisible bug" infestations [90] Hylwa SA. Ronkainen SD. 2018. Delusional infestation versus Morgellons disease [76] Lepping P. 2020. Why delusional infestation should remain delusional infestation. [102] Middelveen MJ, et al. 2018. History of Morgellons disease: From delusion to definition. [80] Murase JE, Wu JJ, Koo J. 2006. Morgellons disease: a rapport-enhancing term for delusions of parasitosis [93] Söderfeldt Y, Groß D. 2014. Information, consent, and treatment of patients with Morgellons disease: An ethical perspective [103]

(continued)

Table 3 (continued)

General topic	Articles with content relevant to the topic
Summary of evidence-based data on the topic	Bailey CH. et al. 2014. A population-based study of the incidence of delusional infestation in Olmsted County, Minnesota, 1976–2010 [38] Foster AA. et al. 2012. Delusional infestation: Clinical presentation in 147 patients seen at Mayo Clinic [104] Freudenmann RW, et al. 2012. Delusional infestation and the specimen sign: a European multi-center study in 148 consecutive cases [101] Hylwa SA. Ronkainen SD. 2018. Delusional infestation versus Morgellons disease [76] Lepping P. et al. 2010. Delusional infestation in dermatology in the UK: Prevalence, treatment strategies, and feasibility of a randomized controlled trial [105] Pearson ML. et al. 2012. Clinical, epidemiologic, histopathologic, and molecular features of an unexplained dermopathy [75] Todd S. et al. 2019. Delusional infestation managed in a combined tropical medicine and psychiatry clinic [95]
Is Morgellons a form of DI? What are the arguments for and against the fiber hypothesis?	Bransfield RC, Friedman KJ. 2019. Differentiating psychosomatic, somatopsychic, multisystem illnesses, and medical uncertainty. (review) [106] Elkan D. 2007. Morgellons disease: The itch that won't be scratched. [103] Freudenmann RW, Lepping P. 2009. Delusional infestation [1] Harvey WT, Bransfield RC, Mercer DE, et al. 2009. Morgellons disease, illuminating an undefined illness: a case series [107] Hylwa SA. Ronkainen SD. 2018. Delusional infestation versus Morgellons disease [76] Koblenzer CS. 2006. The challenge of Morgellons disease [108] Lai J, Xu Z, Zu Y, Hu S. 2018. Reframing delusional infestation: Perspectives on unresolved puzzles [109] Lepping P, Freudenmann RW. 2012. Delusional infestation in childhood, adolescence, and adulthood [61] Marris E. 2006. Mysterious 'Morgellons disease' prompts US investigation [74] Middelveen MJ, Stricker RB. 2011. Filament formation associated with spirochetal infection: a comparative approach to Morgellons disease [110] Middelveen MJ, Stricker RB. 2016. Morgellons disease: a filamentous borrelial dermatitis [111] Middelveen MJ, Burugu D. et al. 2013. Association of spirochetal infection with Morgellons disease [79] Middelveen MJ. et al. 2018. History of Morgellons disease: From delusion to definition [80] Murase JE, Wu JJ, Koo J. 2006. Morgellons disease: a rapport-enhancing term for delusions of parasitosis [93] Pearson ML Selby JV, Katz KA, et al. 2012. Clinical, epidemiologic, histopathologic, and molecular features of an unexplained dermopathy [75] Schulte B. 2008. Figments of the imagination? [71] Simpson L, Baier M. 2009. Disorder or delusion? Living with Morgellons disease [89] Söderfeldt Y, Groß D. 2014. Information, consent, and treatment of patients with Morgellons disease: An ethical perspective [103]

Table 3 (continued)

General topic	Articles with content relevant to the topic
Client-patient behaviors and beliefs, etc.	
Patient demographics. Psycho-epidemiology of patients	Beach SD, Kroshinsky D, Kontos N. 2014. Case 37-2014: a 35-year-old woman with suspected mite infestation [112] Bewley AP, Lepping P. Freundenmann RW, Taylor R. 2010. Delusional parasitosis: Time to call it delusional infestation [16] Donabedian H. 2007. Delusions of parasitosis [113] Foster AA. et al. 2012. Delusional infestation: Clinical presentation in 147 patients seen at Mayo Clinic [104] Freudenmann RW, Lepping P. 2008. Second-generation antipsychotics in primary and secondary delusional parasitosis [114] Freudenmann RW, Kölle M, et al. 2010. Delusional parasitosis & the matchbox sign revisited: The international perspective [114] Freudenmann RW, Lepping P. 2009. Delusional infestation [1] Geary MJ, Russell RC, et al. 2020. 30 years of samples submitted to an Australian medical entomology department [115] Harvey WT, Bransfield RC, Mercer DE, et al. 2009. Morgellons disease, illuminating an undefined illness: a case series [107] Hinkle NC. 2010. Ekbom syndrome: The challenge of "invisible bug" infestations [90] Koblenzer CS. 2006. The challenge of Morgellons disease [99] Kohorst JJ, Bailey CH, et al. 2018. Prevalence of delusional infestation—a population-based study [116] Lee CS. 2008. Delusions of parasitosis [117] Lepping P, Noorthoorn EO, et al. 2017. An international study of the prevalence of substance use in patients with delusional infestation 2017 [2] Lepping P, Aboalkaz S, Squire SB, et al. 2020. Later age of onset and longer duration of untreated psychosis are associated with poor outcomes in delusional infestation [118] Lynch PJ. 1993. Delusions of parasitosis [119] Martins AC, Mendes CP, Nico MM. 2016. Delusional infestation: a case series from a university dermatology center in São Paulo, Brazil [120] Middelveen MJ, et al. 2018. History of Morgellons disease: From delusion to definition [80] Mohandas P, Bewley A, Taylor R. 2013. Dermatitis artefacta and artefactual skin disease: The need for a psychodermatology multidisciplinary team to treat a difficult condition [121] Nguyen CM, Danesh M, Beroukhim K, Sorenson E, Leon A, Koo J. 2015. Psychodermatology: a review [122] Pearson ML. et al. 2012. Clinical, epidemiologic, histopathologic, and molecular features of an unexplained dermopathy [75] Rinshniw M, Lepping P, Freudenmann RW. 2014. Delusional infestation by proxy-what should veterinarians do? [64] Savely VR. et al. 2006. The mystery of Morgellons disease. Infection or delusion? [77] Schulte B. 2008. Figments of the imagination? [71] Todd S. et al. 2019. Delusional infestation managed in a combined tropical medicine and psychiatry clinic [95]

(continued)

Table 3 (continued)

General topic	Articles with content relevant to the topic
The internet and DI, for better or worse	Beach SD, Kroshinsky D, Kontos N. 2014. Case 37-2014: a 35-year-old woman with suspected mite infestation [112] Elkan D. 2007. Morgellons disease: The itch that won't be scratched [103] Freudenmann RW, Kölle M, et al. 2010. Delusional parasitosis & the matchbox sign revisited: The international perspective [114] Freudenmann RW, Lepping P. 2009. Delusional infestation [1] Harvey WT, Bransfield RC, Mercer DE, et al. 2009. Morgellons disease, illuminating an undefined illness: a case series [107] Hylwa SA, Ronkainen SD. 2018. Delusional infestation versus Morgellons disease [76] Koblenzer CS. 2006. The challenge of Morgellons disease [99] Lepping P, Freudenmann RW. 2012. Delusional infestation in childhood, adolescence, and adulthood [61] Murase JE, Wu JJ, Koo J. 2006. Morgellons disease: a rapport-enhancing term for delusions of parasitosis [93] Schulte B. 2008. Figments of the imagination? [71] Söderfeldt Y, Groß D. 2014. Information, consent, and treatment of patients with Morgellons disease: An ethical perspective [103] Misery L. 2013. Morgellon's syndrome: a disease transmitted via the media [123] Lustig A, Mackay S, Strauss, J. 2009. Morgellons disease as internet meme (letter) [124]
Common beliefs: The infesting organism has not yet been identified by science. Many perceived organisms biologically implausible, proteus-like, chimera of arthropods, nematode, fungus, and protozoan	Altschuler DZ, et al. 2004. Collembola (springtails) (Arthropoda: Hexapoda: Entognatha) found in skin scrapings from individuals diagnosed with delusory parasitosis [125] Christiansen KA, Bernard EC. 2008. Critique of the article Collembola (springtails) (Arthropoda: Hexapoda: Entognatha) found in skin scrapings from individuals diagnosed with delusory parasitosis [126] Donabedian H. 2007. Delusions of parasitosis [113] Ekbom KA. 2003 (trans. 1938 Yorston). The pre-senile delusion of infestation [92] Elkan D. 2007. Morgellons disease: The itch that won't be scratched [103] Freudenmann RW, Lepping P. 2009 delusional infestation [1] Harvey WT, Bransfield RC, Mercer DE, et al. 2009. Morgellons disease, illuminating an undefined illness: a case series [107] Hinkle NC. 2010. Ekbom syndrome: The challenge of "invisible bug" infestations [90] Hylwa SA, Bury JE, Davis MDP, et al. 2011. Delusional infestation, including delusions of parasitosis: Results of histologic examination of skin biopsy and patient-provided skin specimens [127] Lai J, Xu Z, Zu Y, Hu S. 2018. Reframing delusional infestation: Perspectives on unresolved puzzles [109] Lepping P. 2017. Eine seltene Störung mit hohem Leidensdruck. *Trans.* A rare disorder with high levels of suffering [128] Middelveen MJ, Stricker RB. 2016. 2016. Morgellons disease: a filamentous borrelial dermatitis [111] Middelveen MJ. et al. 2018. History of Morgellons disease: From delusion to definition [80] Murase JE, Wu JJ, Koo J. 2006. Morgellons disease: a rapport-enhancing term for delusions of parasitosis [93] Savely VR. et al. 2006. The mystery of Morgellons disease. Infection or delusion? [77] Schulte B. 2008. Figments of the imagination? [71] Shelomi M. 2013. Evidence of photo manipulation in delusional parasitosis paper [129]

Table 3 (continued)

General topic	Articles with content relevant to the topic
Why are many of the beliefs and behaviors stereotypical?	Elkan D. 2007. Morgellons disease: The itch that won't be scratched [103] Freudenmann RW, Lepping P. 2009. Delusional infestation [1] Hinkle NC. 2010. Ekbom syndrome: The challenge of "invisible bug" infestations [90] Koblenzer CS. 2006. The challenge of Morgellons disease [99] Lai J, Xu Z, Zu Y, Hu S. 2018. Reframing delusional infestation: Perspectives on unresolved puzzles [109] Lepping P, Freudenmann RW. 2012. Delusional infestation in childhood, adolescence, and adulthood [61] Pearson ML. et al. 2012. Clinical, epidemiologic, histopathologic, and molecular features of an unexplained dermopathy [75] Simpson E, Baier M. 2009. Disorder or delusion? Living with Morgellons disease [89]

Front line: Public health/extension, etc.

What is the role of "your neighborhood entomologist/ parasitologist"? (or vet!)	Christiansen KA, Bernard EC. 2008. Critique of the article Collembola (springtails) (Arthropoda: Hexapoda: Entognatha) found in skin scrapings from individuals diagnosed with delusory parasitosis [126] Geary MJ, Russell RC, et al. 2020. 30 years of samples submitted to an Australian medical entomology department [115] Hinkle NC. 2010. Ekbom syndrome: The challenge of "invisible bug" infestations [90] Lepping P, Rinshniw M, Freudenmann RW. 2015. Frequency of delusional infestation by proxy and double delusional infestation in veterinary practice: Observational study [130] Rinshniw M, Lepping P, Freudenmann RW. 2014. Delusional infestation by proxy-what should veterinarians do? [64]
The role of housing inspectors, exterminators, and landlords	Geary MJ, Russell RC, et al. 2020. 30 years of samples submitted to an Australian medical entomology department [115] Hinkle NC. 2010. Ekbom syndrome: The challenge of "invisible bug" infestations [90] Lepping P, Freudenmann RW. 2012. Delusional infestation in childhood, adolescence, and adulthood [61] Potter MF, Koehler PG. 2013. Invisible itches: Insect and non-insect causes [131]
What is the role of a state/county public health lab?	Lepping P, Baker C, Freudenmann RW. 2010. Delusional infestation in dermatology in the UK: Prevalence, treatment strategies, and feasibility of a randomized controlled trial [105]

(continued)

Table 3 (continued)

General topic	Articles with content relevant to the topic
Medicine at large and research	
Common experiences of most patients	Beach SD, Kroshinsky D, Kontos N. 2014. Case 37-2014: a 35-year-old woman with suspected mite infestation [112] Donabedian H. 2007. Delusions of parasitosis [113] Ekbom KA. 2003 (trans. 1938 Yorston). The pre-senile delusion of infestation [92] Foster AA, Hylwa SA, Bury JE, et al. 2012. Delusional infestation: Clinical presentation in 147 patients seen at Mayo Clinic [104] Freudenmann RW, Lepping P, Huber M, et al. 2012. Delusional infestation and the specimen sign: a European multicenter study in 148 consecutive cases [101] Freudenmann RW, Lepping P. 2009. Delusional infestation [1] Harvey WT, Bransfield RC, Mercer DE, et al. 2009. Morgellons disease, illuminating an undefined illness: a case series [107] Hinkle NC. 2010. Ekbom syndrome: The challenge of "invisible bug" infestations [90] Hylwa SA, Bury JE, Davis MDP, et al. 2011. Delusional infestation, including delusions of parasitosis: Results of histologic examination of skin biopsy and patient-provided skin specimens [127] Bewley A, Lepping P, Taylor RE. (Eds.) delusional infestation: Psychodermatology in clinical practice (Lepping; Chap. 12). [12] Lepping P, Huber M, Freudenmann RW. 2015. How to approach delusional infestation [63] Lynch PJ. 1993. Delusions of parasitosis [119] Middelveen MJ. et al. 2018. History of Morgellons disease: From delusion to definition [80] Bransfield RC, Friedman KJ. 2019. Differentiating psychosomatic, somatopsychic, multisystem illnesses, and medical uncertainty (review) [106] Pearson ML. et al. 2012. Clinical, epidemiologic, histopathologic, and molecular features of an unexplained dermopathy [75] Schulte B. 2008. Figments of the imagination? [71] Simpson E, Baier M. Disorder, or delusion? Living with Morgellons disease [89]

Table 3 (continued)

General topic	Articles with content relevant to the topic
Who "owns" these patients? Is there a clinical specialty that should manage them? A single specialty? A multispecialty clinic? Should all dermatology, infectious disease, psychiatry, family medicine, and internal medicine physicians provide care? Does any specialty deserve a pass that permits them to NOT see these patients?	Beach SD, Kroshinsky D, Kontos N. 2014. Case 37-2014: a 35-year-old woman with suspected mite infestation [112] Bostwick JM. 2011. Taming hornets: The therapeutic relationship in successful treatment of delusional infestation [108] Fabbro S, Aultman JM, Mostow EN. 2013. Delusions of parasitosis: Ethical and clinical considerations [132] Foster AA, Hylwa SA, Bury JE, et al. 2012. Delusional infestation: Clinical presentation in 147 patients seen at Mayo Clinic [104] Freudenmann RW, Lepping P. 2008. Second-generation antipsychotics in primary and secondary delusional parasitosis outcome and efficacy [114] Freudenmann RW, Lepping P. 2009. Delusional infestation [1] Lepping P, Baker C, Freudenmann RW. 2010. Delusional infestation in dermatology in the UK: Prevalence, treatment strategies, and feasibility of a randomized controlled trail [105] Healy R, Taylor R, Dhoat S, Leschynska E, Bewley AP. Management of patients with delusional parasitosis in a joint dermatology/ liaison psychiatry clinic. Br J Dermatol. 2009;161:197–9. https://doi.org/10.1111/j.1365-2133.2009.09183.x. [158] Hinkle NC. 2010. Ekbom syndrome: The challenge of "invisible bug" infestations [90] Hussain K, Gkini MA, Taylor R, Shinhmar S, Bewley A. 2018. A patient with delusional infestation by proxy: Issues for vulnerable adults [66] Hylwa SA, Ronkainen SD. 2018. Delusional infestation versus Morgellons disease [76] Maher S, Hallahan B, Flaherty G. 2017. Itching for a diagnosis—A travel medicine perspective on delusional infestation [133] Mohandas P, Bewley A, Taylor R. 2013. Dermatitis artefacta and artefactual skin disease: The need for a psychodermatology multidisciplinary team to treat a difficult condition [121] Rinshniw M, Lepping P, Freudenmann RW. 2014. Delusional infestation by proxy-what should veterinarians do? [64] Sambhi R, Lepping P. 2010. Psychiatric treatments in dermatology: An update [134] Tey HL, Wallengren J, Yosipovitch G. 2013. Psychosomatic factors in pruritus [21] Todd S. et al. 2019. Delusional infestation managed in a combined tropical medicine and psychiatry clinic [95] Wong S, Bewley A. 2011. Patients with delusional infestation (delusional parasitosis) often require prolonged treatment as recurrence of symptoms after cessation of treatment is common: An observational study [135]
What about "alternative practitioners" who want to give parenteral antibiotics for months or use completely untested/ unapproved therapies?	Brett AS, McCullough BL. 2012. Addressing requests by patients for non-beneficial interventions [136] Lantos PM, Shapiro ED, Auwaerter PG, et al. 2015. Unorthodox alternative therapies marketed to treat Lyme disease [137] Middelveen MJ. et al. 2018. History of Morgellons disease: From delusion to definition [80] Schulte B. 2008. Figments of the imagination? [71] Simpson L, Baier M. 2009. Disorder or delusion? Living with Morgellons disease [89] Steere AC. Arvikar SA. 2015. Editorial commentary. What constitutes appropriate treatment of post-Lyme disease symptoms and other pain and fatigue syndromes? [138]

(continued)

Table 3 (continued)

General topic	Articles with content relevant to the topic
How does the clinician decide DI is a correct dx?	Beach SD, Kroshinsky D, Kontos N. 2014. Case 37-2014: a 35-year-old woman with suspected mite infestation [112] Bransfield RC, Friedman KJ. 2019. Differentiating psychosomatic, somatopsychic, multisystem illnesses, and medical uncertainty (review) [106] Bury JE, Bostwisk JM. 2010. Iatrogenic delusional parasitosis: a case of physician-patient folie à deux [139] Cerda A. De La, Reichenberg J, Magid M. 2012. Successful treatment of patients previously labeled as having "delusions of parasitosis" with antidepressant therapy [140] Donabedian H. 2007. Delusions of parasitosis [113] Freudenmann RW, Lepping P. 2009. Delusional infestation [1] Hinkle NC. 2010. Ekbom syndrome: The challenge of "invisible bug" infestations [90] Hylwa SA, Ronkainen SD. 2018. Delusional infestation versus Morgellons disease [76] Hylwa SA, Bury JE, Davis MDP, et al. 2011. Delusional infestation, including delusions of parasitosis: Results of histologic examination of skin biopsy and patient-provided skin specimens [127] Bewley A, Lepping P, Taylor RE. (Eds.) delusional infestation: Psychodermatology in clinical practice (Lepping; Chap. 12) [12] Lepping P, Freudenmann RW. 2012. Delusional infestation in childhood, adolescence, and adulthood. [61] Lepping P, Huber M, Freudenmann RW. 2015. How to approach delusional infestation [63] Lynch PJ. 1993. Delusions of parasitosis [119] Maher S, Hallahan B, Flaherty G. 2017. Itching for a diagnosis—a travel medicine perspective on delusional infestation [133] Middelveen MJ, Stricker RB. 2016. 2016. Morgellons disease: a filamentous borrelial dermatitis [111] Middelveen MJ. et al. 2018. History of Morgellons disease: From delusion to definition [80] Reich A, Kwiatkowska D, Pacan P. 2019. Delusions of parasitosis: An update [141] Rinshniw M, Lepping P, Freudenmann RW. 2014. Delusional infestation by proxy-what should veterinarians do? [64] Schulte B. 2008. Figments of the imagination? [71] Todd S. et al. 2019. Delusional infestation managed in a combined tropical medicine and psychiatry clinic [95]

Table 3 (continued)

General topic	Articles with content relevant to the topic
How does one decide which therapies to use?	Beach SD, Kroshinsky D, Kontos N. 2014. Case 37-2014: a 35-year-old woman with suspected mite infestation [112] Brett AS, McCullough LB. 2012. Addressing requests by patients for non-beneficial interventions [136] Donabedian H. 2007. Delusions of parasitosis [113] Fabbro S, Aultman JM, Mostow EN. 2013. Delusions of parasitosis: Ethical and clinical considerations [132] Freudenmann RW, Lepping P. 2008. Second-generation antipsychotics in primary and secondary delusional parasitosis outcome and efficacy [114] Freudenmann RW, Lepping P. 2009. Delusional infestation [1] Hinkle NC. 2010. Ekbom syndrome: The challenge of "invisible bug" infestations [90] Koblenzer CS. 2010. The current management of delusional parasitosis and dermatitis artefacta [142] Lee CS. 2008. Delusions of parasitosis [117] Bewley A, Lepping P, Taylor RE. (Eds.) delusional infestation: Psychodermatology in clinical practice (Lepping; Chap. 12) [12] Lepping P, Freudenmann RW. 2012. Delusional infestation in childhood, adolescence, and adulthood [61] Lepping P, Huber M, Freudenmann RW. 2015. How to approach delusional infestation [63] Lepping P, Russell I, Freudenmann RW. 2007. Antipsychotic treatment of primary delusional parasitosis: Systematic review [143] Lynch PJ. 1993. Delusions of parasitosis [119] Mohandas P, Bewley A, Taylor R. 2013. Dermatitis artefacta and artefactual skin disease: The need for a psychodermatology multidisciplinary team to treat a difficult condition [121] Rinshniw M, Lepping P, Freudenmann RW. 2014. Delusional infestation by proxy-what should veterinarians do? [64]
What does one write in the chart?	Freudenmann RW, Lepping P. 2009. Delusional infestation [1] Murase JE, Wu JJ, Koo J. 2006. Morgellons disease: a rapport-enhancing term for delusions of parasitosis [93]

(continued)

Table 3 (continued)

General topic	Articles with content relevant to the topic
How does one broach the topic of DI as the possible dx?	Bury JE, Bostwisk JM. 2010. Iatrogenic delusional parasitosis: a case of physician-patient folie à deux [139] Cook J, Lewandowsky S, et al. 2011 (revised 2020). The debunking handbook [100] Donabedian H. 2007. Delusions of parasitosis [113] Ekbom KA. 2003 (trans. 1938 Yorston). The pre-senile delusion of infestation [92] Fabbro S, Aultman JM, Mostow EN. 2013. Delusions of parasitosis: Ethical and clinical considerations [132] Freudenmann RW, Lepping P. 2009. Delusional infestation [1] Heller MM, Murase JE, Koo J. 2011. Time and effort to establish therapeutic rapport with delusional patients [144] Koblenzer CS. 2006. The challenge of Morgellons disease [99] Koblenzer CS. 2010. The current management of delusional parasitosis and dermatitis artefacta [142] Lepping P. 2020. Why delusional infestation should remain delusional infestation [102] Bewley A, Lepping P, Taylor RE. (Eds.) delusional infestation: Psychodermatology in clinical practice (Lepping; Chap. 12) [12] Lepping P, Freudenmann RW. 2012. Delusional infestation in childhood, adolescence, and adulthood [61] Lepping P, Huber M, Freudenmann RW. 2015. How to approach delusional infestation [63] Lynch PJ. 1983. Delusions of parasitosis [119] Magid M, Reichenberg JS. 2019. Treating the difficult patient: Ten pearls to reduce resentment and regain control of the doctor-patient visit [145] Mohandas P, Bewley A, Taylor R. 2013. Dermatitis artefacta and artefactual skin disease: The need for a psychodermatology multidisciplinary team to treat a difficult condition [121] Murase JE, Wu JJ, Koo J. 2006. Morgellons disease: a rapport-enhancing term for delusions of parasitosis [93] Schulte B. 2008. Figments of the imagination? [71] Simpson L, Baier M. 2009. Disorder or delusion? Living with Morgellons disease [89] Söderfeldt Y, Groß D. 2014. Information, consent, and treatment of patients with Morgellons disease: An ethical perspective [103] Tey HL, Wallengren J, Yosipovitch G. 2013. Psychosomatic factors in pruritus [21] Todd S. et al. 2019. Delusional infestation managed in a combined tropical medicine and psychiatry clinic [95]

Table 3 (continued)

General topic	Articles with content relevant to the topic
Different types of therapies: Pharmacologic to treat psychosis Pharmacologic to treat infestation Placebo Behavioral/CBT	Beach SD, Kroshinsky D, Kontos N. 2014. Case 37-2014: a 35-year-old woman with suspected mite infestation [112] Bewley AP, Lepping P, Freudenmann RW, Taylor R, 2010. Delusional parasitosis: Time to call it delusional infestation [16] Brett AS, McCullough LB. 2012. Addressing requests by patients for non-beneficial interventions [136] Cerda A. De La, Reichenberg J, Magid M. 2012. Successful treatment of patients previously labeled as having "delusions of parasitosis" with antidepressant therapy [140] Lepping P, Gil-Candon RM, Freudenmann RW. 2005. Delusional parasitosis treated with amisulpride [146] Donabedian H. 2007. Delusions of parasitosis [113] Fabbro S, Aultman JM, Mostow EN. 2013. Delusions of parasitosis: Ethical and clinical considerations [132] Freudenmann RW, Lepping P. 2008. Second-generation antipsychotics in primary and secondary delusional parasitosis outcome and efficacy [114] Freudenmann RW, Schönfeldt-Lecuona C, Lepping P. 2007. Primary delusional parasitosis treated with olanzapine [57] Freudenmann RW, Kühnlein P, Lepping P, Schönfeldt-Lecuona C. 2008. Secondary delusional parasitosis treated with paliperidone [56] Freudenmann RW, Lepping P. 2009. Delusional infestation [1] Hinkle NC. 2010. Ekbom syndrome: The challenge of "invisible bug" infestations [90] Koblenzer CS. 2006. The challenge of Morgellons disease [99] Koblenzer CS. 2010. The current management of delusional parasitosis and dermatitis artefacta [142] Kölle M, Lepping P, Kassubek J, Schönfeldt-Lecuona C, Freudenmann RW. 2010. Delusional infestation induced by piribedil add-on in Parkinson's disease [147] Lee CS. 2008. Delusions of parasitosis [117] Bewley A, Lepping P, Taylor RE. (Eds.) delusional infestation: Psychodermatology in clinical practice (Lepping; Chap. 12) [12] Lepping P, Freudenmann RW. 2008. Delusional parasitosis: a new pathway for diagnosis [148] Lepping P, Freudenmann RW. 2012. Delusional infestation in childhood, adolescence, and adulthood [61] Lepping P, Huber M, Freudenmann RW. 2015. How to approach delusional infestation [63] Lepping P, Baker C, Freudenmann RW. 2010. Delusional infestation in dermatology in the UK: Prevalence, treatment strategies, and feasibility of a randomized controlled trial [105] Lepping P, Russell I, Freudenmann RW. 2007. Antipsychotic treatment of primary delusional parasitosis: Systematic review [143] Lynch PJ. 1983. Delusions of parasitosis [119] Middelveen MJ. et al. 2018. History of Morgellons disease: From delusion to definition [80] Mohandas P, Bewley A, Taylor R. 2013. Dermatitis artefacta and artefactual skin disease: The need for a psychodermatology multidisciplinary team to treat a difficult condition [121] Reich A, Kwiatkowska D, Pacan P. 2019. Delusions of parasitosis: An update [141] Sambhi R, Lepping P. 2010. Psychiatric treatments in dermatology: An update [134] Simpson L, Baier M. 2009. Disorder or delusion? Living with Morgellons disease [89] Wong S, Bewley A. 2011. Patients with delusional infestation (delusional parasitosis) often require prolonged treatment as recurrence of symptoms after cessation of treatment is common: An observational study [135]

(continued)

Table 3 (continued)

General topic	Articles with content relevant to the topic
Patients who self-medicate. How to limit harm? Can there be benefits?	Ekbom KA. 2003 (trans. 1938 Yorston). The pre-senile delusion of infestation [92] Freudenmann RW, Lepping P. 2009. Delusional infestation. [1] Geary, M. J et al. 2020. 30 years of samples submitted to an Australian medical entomology department [115] Hinkle NC. 2010. Ekbom syndrome: The challenge of "invisible bug" infestations [90] Hylwa SA, Bury JE, Davis MDP, et al. 2011. Delusional infestation, including delusions of parasitosis: Results of histologic examination of skin biopsy and patient-provided skin specimens [127] Lai J, Xu Z, Zu Y, Hu S. 2018. Reframing delusional infestation: Perspectives on unresolved puzzles [109] Bewley A, Lepping P, Taylor RE. (Eds.) delusional infestation: Psychodermatology in clinical practice (Lepping; Chap. 12) [12] Lepping P, Freudenmann RW. 2012. Delusional infestation in childhood, adolescence, and adulthood. [61] Lepping P, Huber M, Freudenmann RW. 2015. How to approach delusional infestation [63] Lepping P, Rishniw M, Freudenmann RW. 2015. Frequency of delusional infestation by proxy and double delusional infestation in veterinary practice: Observational study [130] Lynch PJ. 1983. Delusions of Parasitosis [119] Simpson L, Baier M. 2009. Disorder or delusion? Living with Morgellons disease [89]
Specimens and the role of the clinical laboratory	Donabedian H. 2007. Delusions of parasitosis [113] Elkan D. 2007. Morgellons disease: The itch that won't be scratched [103] Freudenmann RW, Lepping P, Huber M, et al. 2012. Delusional infestation and the specimen sign: a European multicenter study in 148 consecutive cases [101] Freudenmann RW, Lepping P. 2009. Delusional infestation [1] Geary MJ, Russell RC, et al. 2020. 30 years of samples submitted to an Australian medical entomology department [115] Heller MM, Murase JE, Koo J. 2011. Time and effort to establish therapeutic rapport with delusional patients [144] Hinkle NC. 2010. Ekbom syndrome: The challenge of "invisible bug" infestations [90] Hylwa SA, Bury JE, Davis MDP, et al. 2011. Delusional infestation, including delusions of parasitosis: Results of histologic examination of skin biopsy and patient-provided skin specimens [127] Hylwa SA, Ronkainen SD. 2018. Delusional infestation versus Morgellons disease [76] Bewley A, Lepping P, Taylor RE. (Eds.) delusional infestation: Psychodermatology in clinical practice (Lepping; Chap. 12) [12] Lepping P, Huber M, Freudenmann RW. 2015. How to approach delusional infestation [63] Middelveen MJ, Stricker RB. 2011. Filament formation associated with spirochetal infection: a comparative approach to Morgellons disease [110] Middelveen MJ, Stricker RB. 2016. 2016. Morgellons disease: a filamentous borrelial dermatitis [111] Middelveen MJ. et al. 2018. History of Morgellons disease: From delusion to definition [80] Pearson ML. et al. 2012. Clinical, epidemiologic, histopathologic, and molecular features of an unexplained dermopathy [75] Schulte B. 2008. Figments of the imagination? [71] Todd S. et al. 2019. Delusional infestation managed in a combined tropical medicine and psychiatry clinic [95]

Table 3 (continued)

General topic	Articles with content relevant to the topic
What research is needed? Is there a role for collaboration among clinicians, scientists, public health officers, and exterminators ad infinitum?	Grant JE, Odlaug BL, Won Kim S. N-Acetylcysteine, a glutamate modulator, in the treatment of trichotillomania: A double-blind, placebo-controlled study [43] Bewley AP, Lepping P, Freudenmann RW, Taylor R. 2010. Delusional parasitosis: Time to call it delusional infestation. Br J Dermatol 163: 1–2 [16] Freudenmann RW, Lepping P. 2008. Second-generation antipsychotics in primary and secondary delusional parasitosis outcome and efficacy [114] Freudenmann RW, Lepping P. 2009. Delusional infestation [1] Lai J, Xu Z, Zu Y, Hu S. 2018. Reframing delusional infestation: Perspectives on unresolved puzzles [109] Lepping P, Baker C, Freudenmann RW. 2010. Delusional infestation in dermatology in the UK: Prevalence, treatment strategies, and feasibility of a randomized controlled trial [105] Middelveen MJ. et al. 2018. History of Morgellons disease: From delusion to definition [80] Pearson ML. et al. 2012. Clinical, epidemiologic, histopathologic, and molecular features of an unexplained dermopathy [75] Todd S. et al. 2019. Delusional infestation managed in a combined tropical medicine and psychiatry clinic [95]
What are we missing here? Have we missed something diagnostically?	Bransfield RC, Friedman KJ. 2019. Differentiating psychosomatic, somatopsychic, multisystem illnesses, and medical uncertainty (review) [106] Elkan D. 2007. Morgellons disease: The itch that won't be scratched [103] Freudenmann RW, Lepping P. 2009. Delusional infestation [1] Lai J, Xu Z, Zu Y, Hu S. 2018. Reframing delusional infestation: Perspectives on unresolved puzzles [109] Middelveen MJ, Stricker RB. 2011. Filament formation associated with spirochetal infection: a comparative approach to Morgellons disease [110] Middelveen MJ. et al. 2018. History of Morgellons disease: From delusion to definition [80] Savely VR. et al. 2006. The mystery of Morgellons disease. Infection or delusion? [77] Schulte B. 2008. Figments of the imagination? [71] Simpson L, Baier M. 2009. Disorder or delusion? Living with Morgellons disease [89] Ahmed A, Affleck AG, Angus J, et al. 2022 British Association of Dermatologists guidelines for the management of adults with delusional infestation [3]

(continued)

Table 3 (continued)

General topic	Articles with content relevant to the topic
Medicine (dermatology)	
The initial dermatology exam. I. Attempts to identify possible infestations	Beach SD, Kroshinsky D, Kontos N. 2014. Case 37-2014: a 35-year-old woman with suspected mite infestation [112] Bury JE, Bostwisk JM. 2010. Iatrogenic delusional parasitosis: a case of physician-patient folie à deux [139] Donabedian H. 2007. Delusions of parasitosis [113] Fabbro S, Aultman JM, Mostow EN. 2013. Delusions of parasitosis: Ethical and clinical considerations [132] Foster AA, Hylwa SA, Bury JE, et al. 2012. Delusional infestations: Clinical presentations in 147 patients seen at Mayo Clinic [104] Freudenmann RW, Lepping P. 2009. Delusional infestation. Geary MJ, Russell RC, et al. 2020. 30 years of samples submitted to an Australian medical entomology department [115] Hinkle NC. 2010. Ekbom syndrome: The challenge of "invisible bug" infestations [90] Hylwa SA, Bury JE, Davis MDP, et al. 2011. Delusional infestation, including delusions of parasitosis: Results of histologic examination of skin biopsy and patient-provided skin specimens [127] Lai J, Xu Z, Zu Y, Hu S. 2018. Reframing delusional infestation: Perspectives on unresolved puzzles [109] Lee CS. 2008. Delusions of parasitosis [117] Bewley A, Lepping P, Taylor RE. (Eds.) delusional infestation: Psychodermatology in clinical practice (Lepping; Chap. 12) [12] Lynch PJ. 1983. Delusions of parasitosis [119] Middelveen MJ. et al. 2018. History of Morgellons disease: From delusion to definition [80] Potter MF, Koehler PG. 2013. Invisible itches: Insect and non-insect causes [131] Schulte B. 2008. Figments of the imagination? [71]

Table 3 (continued)

General topic	Articles with content relevant to the topic
The initial dermatology exam. II. Attempts to identify other causes of formication	Beach SD, Kroshinsky D, Kontos N. 2014. Case 37-2014: a 35-year-old woman with suspected mite infestation [112] Bransfield RC, Friedman KJ. 2019. Differentiating psychosomatic, somatopsychic, multisystem illnesses, and medical uncertainty (review) [106] Bury JE, Bostwisk JM. 2010. Iatrogenic delusional parasitosis: a case of physician-patient folie à deux [139] Foster AA, Hylwa SA, Bury JE, et al. 2012. Delusional infestations: Clinical presentations in 147 patients seen at Mayo Clinic [104] Fowler E, Maderal A, Yosipovitch G. 2019. Treatment-induced delusions of infestation associated with increased brain dopamine levels [149] Freudenmann RW, Lepping P. 2009. Delusional infestation [1] Geary MJ, Russell RC, et al. 2020. 30 years of samples submitted to an Australian medical entomology department [115] Gordon-Elliot JS, Muskin PR. 2013. Managing the patient with psychiatric issues in dermatologic practice [150] Heller MM, Wong JW, Lee ES, et al. 2013. Delusional infestations: Clinical presentation, diagnosis, and treatment (review) [151] Hinkle NC. 2010. Ekbom syndrome: The challenge of "invisible bug" infestations [90] Hylwa SA, Bury JE, Davis MDP, et al. 2011. Delusional infestation, including delusions of parasitosis: Results of histologic examination of skin biopsy and patient-provided skin specimens [127] Lee CS. 2008. Delusions of parasitosis [117] Bewley A, Lepping P, Taylor RE. (Eds.) delusional infestation: Psychodermatology in clinical practice (Lepping; Chap. 12) [12] Lepping P, Huber M, Freudenmann RW. 2015. How to approach delusional infestation[63] Lynch PJ. 1983. Delusions of parasitosis [119] Middelveen MJ. et al. 2018. History of Morgellons disease: From delusion to definition [80] Nguyen CM, Danesh M, Beroukhim K, Sorenson E, Leon A, Koo J. 2015. Psychodermatology: a review [122] Schulte B. 2008. Figments of the imagination? [71]
Medicine (psychiatry)	
What is a delusion? What is monosymptomatic psychosis? "Why won't the patient listen to reason?"	Bilowski, J. 2009. Dermatologic manifestations of psychiatric disease [41] Bransfield RC, Friedman KJ. 2019. Differentiating psychosomatic, somatopsychic, multisystem illnesses, and medical uncertainty. (review) [106] Freudenmann RW, Lepping P. 2009. Delusional infestation [1] Huber MR. et al. 2018. Regional gray matter volume and structural network strength in somatic vs. non-somatic delusional disorders [24] Lee CS. 2008. Delusions of parasitosis [117] Lepping P, Freudenmann RW. 2012. Delusional infestation in childhood, adolescence, and adulthood [61] Lynch PJ. 1983. Delusions of Parasitosis [119] Maher S, Hallahan B, Flaherty G. 2017. Itching for a diagnosis—a travel medicine perspective on delusional infestation [133] Martins AC, Mendes CP, Nico MM. 2016. Delusional infestation: a case series from a university dermatology center in São Paulo, Brazil [120] Sambhi R, Lepping P. 2010. Psychiatric treatments in dermatology: An update [134] Simpson L, Baier M. 2009. Disorder or delusion? Living with Morgellons disease [89]

(continued)

Table 3 (continued)

General topic	Articles with content relevant to the topic
Is there a neuropsychiatric basis for DI? Some sort of genetic predisposition with an environmental trigger that generates a particular PANDAS-like syndrome?	Alraies MC, Keller E, Shaheen K, Alyraiyes AH. 2011. The clinical picture. A 40-year-old woman with excoriated skin lesions [152]
	Beach SD, Kroshinsky D, Kontos N. 2014. Case 37-2014: a 35-year-old woman with suspected mite infestation [112]
	Bilowski, J. 2009. Dermatologic manifestations of psychiatric disease [41]
	Bury JE, Bostwisk JM. 2010. Iatrogenic delusional parasitosis: a case of physician-patient folie à deux [139]
	Cerda A. De La, Reichenberg J, Magid M. 2012. Successful treatment of patients previously labeled as having "delusions of parasitosis" with antidepressant therapy [140]
	Lepping P, Gil-Candon R, Freudenmann RW. 2005. Delusional parasitosis treated with amisulpride [146]
	Freudenmann RW, Lepping P. 2009. Delusional infestation [1]
	Heller MM, Wong JW, Lee ES, et al. 2013. Delusional infestations: Clinical presentation, diagnosis, and treatment (review) [151]
	Hinkle NC. 2010. Ekbom syndrome: The challenge of "invisible bug" infestations [90]
	Lee CS. 2008. Delusions of parasitosis [117]
	Bewley A, Lepping P, Taylor RE. (Eds.) delusional infestation: Psychodermatology in clinical practice (Lepping; Chap. 12) [12]
	Lepping P, Freudenmann RW. 2012. Delusional infestation in childhood, adolescence, and adulthood [61]
	Middelveen MJ, Stricker RB. 2016. 2016. Morgellons disease: a filamentous borrelial dermatitis [111]
	Mohandas P, Bewley A, Taylor R. 2013. Dermatitis artefacta and artefactual skin disease: The need for a psychodermatology multidisciplinary team to treat a difficult condition [121]
	Nguyen CM, Danesh M, Beroukhim K, Sorenson E, Leon A, Koo J. 2015. Psychodermatology: a review [122]
	Reich A, Kwiatkowska D, Pacan P. 2019. Delusions of parasitosis: An update [134]
	Tey HL, Wallengren J, Yosipovitch G. 2013. Psychosomatic factors in pruritus [21]

Table 3 (continued)

General topic	Articles with content relevant to the topic
Asking a patient to try an antipsychotic medication	Beach SD, Kroshinsky D, Kontos N. 2014. Case 37-2014: a 35-year-old woman with suspected mite infestation [112] Bostwick JM. 2011. Taming hornets: The therapeutic relationship in successful treatment of delusional infestation [108] Donabedian H. 2007. Delusions of parasitosis [113] Fabbro S, Aultman JM, Mostow EN. 2013. Delusions of parasitosis: Ethical and clinical considerations [132] Freudenmann RW, Lepping P. 2009. Delusional infestation [1] Heller MM, Wong JW, Lee ES, et al. 2013. Delusional infestations: Clinical presentation, diagnosis, and treatment (review) [151] Koblenzer CS. 2006. The challenge of Morgellons disease [99] Koblenzer CS. 2010. The current management of delusional parasitosis and dermatitis artefacta [142] Lee CS. 2008. Delusions of parasitosis [117] Bewley A, Lepping P, Taylor RE. (Eds.) delusional infestation: Psychodermatology in clinical practice (Lepping; Chap. 12) [12] Lepping P, Freudenmann RW. 2012. Delusional infestation in childhood, adolescence, and adulthood [61] Lepping P, Huber M, Freudenmann RW. 2015. How to approach delusional infestation [63] Lynch PJ. 1983. Delusions of parasitosis [119] Martins AC, Mendes CP, Nico MM. 2016. Delusional infestation: a case series from a university dermatology center in São Paulo, Brazil [120] Murase JE, Wu JJ, Koo J. 2006. Morgellons disease: a rapport-enhancing term for delusions of parasitosis [93] Nguyen CM., Danesh M, Beroukhim K, Sorenson E, Leon A, Koo J. 2015. Psychodermatology: a review [122] Söderfeldt Y, Groß D. 2014. Information, consent, and treatment of patients with Morgellons disease: An ethical perspective [103] Wong S, Bewley A. 2011. Patients with delusional infestation (delusional parasitosis) often require prolonged treatment as recurrence of symptoms after cessation of treatment is common: An observational study [135]

(continued)

Table 3 (continued)

General topic	Articles with content relevant to the topic
Medicine (neurology)	
Neurologic disorders. Can neurologic problems create sensations of infestation (formication)?	Grant JE, Odlaug BL, Won Kim S. N-Acetylcysteine, a glutamate modulator, in the treatment of trichotillomania: A double-blind, placebo-controlled study [43]
	Beach SD, Kroshinsky D, Kontos N. 2014. Case 37-2014: a 35-year-old woman with suspected mite infestation [112]
	Fowler E, Maderal A, Yosipovitch G. 2019. Treatment-induced delusions of infestation associated with increased brain dopamine levels [149]
	Freudenmann RW, Lepping P. 2009. Delusional infestation [1]
	Fried RG. 2012. Take your "pick". Could neurotic excoriations be a Forme Fruste of Tourette's disorder [153]
	González-Rodriguez A, Molina-Andreu O, Penadés R, Catalán R, Bernardo M. 2015. Structural and functional neuroimaging findings in delusional disorder diagnostic and therapeutic implications [13]
	Huber M, Karner M, Kirchler E, Lepping P, Freudenmann RW. 2008. Striatal lesions in delusional parasitosis revealed by magnetic resonance imaging [23]
	Huber MR. et al. 2018. Regional gray matter volume and structural network strength in somatic vs. non-somatic delusional disorders [24]
	Hylwa SA, Bury JE, Davis MDP, et al. 2011. Delusional infestation, including delusions of parasitosis: Results of histologic examination of skin biopsy and patient-provided skin specimens [127]
	Koblenzer CS. 2006. The challenge of Morgellons disease [99]
	Kölle M, Lepping P, Kassubek J, Schönfeldt-Lecuona C, Freudenmann RW. 2019. Delusional infestation induced by piribedil add-on in Parkinson's disease [147]
	Ojeda-López C, et al. 2015. Delusional Parasitosis as a treatment complication of Parkinson disease. Psychosomatics [154]
	Swick BL, HW. Walling. 2005. Drug-induced delusions of parasitosis during treatment of Parkinson's disease [155]
	Lai J, Xu Z, Zu Y, Hu S. 2018. Reframing delusional infestation: Perspectives on unresolved puzzles [109]
	Bewley A, Lepping P, Taylor RE. (Eds.) delusional infestation: Psychodermatology in clinical practice (Lepping; Chap. 12) [12]
	Marschall MA, Dolezal RF, et al. 1991. Chronic wounds and delusions of parasitosis in the drug abuser [156]
	Middelveen MJ, Stricker RB. 2016. Morgellons disease: a filamentous borrelial dermatitis [111]
	Nguyen CM, Danesh M, Beroukhim K, Sorenson E, Leon A, Koo J. 2015. Psychodermatology: a review [122]
	Reich A, Kwiatkowska D, Pacan P. 2019. Delusions of parasitosis: An update [141]
	Berceli D, et al. effects of self-induced unclassified therapeutic tremors on quality of life among non-professional caregivers: a pilot study [48]

Table 3 (continued)

General topic	Articles with content relevant to the topic
Relationships	
Building relationships. The consequences of confrontation	Bostwick JM. 2011. Taming hornets: The therapeutic relationship in successful treatment of delusional infestation [108] Bury JE, Bostwisk JM. 2010. Iatrogenic delusional parasitosis: a case of physician-patient folie à deux [139] Cook J, Lewandowsky S, et al 2011 (revised 2020). The debunking handbook [100] Donabedian H. 2007. Delusions of parasitosis [113] Fabbro S, Aultman JM, Mostow EN. 2013. Delusions of parasitosis: Ethical and clinical considerations [132] Freudenmann RW, Lepping P. 2009. Delusional infestation [1] Gordon-Elliot JS, Muskin PR. 2013. Managing the patient with psychiatric issues in dermatologic practice [150] Heller MM, Murase JE, Koo J. 2011. Time and effort to establish therapeutic rapport with delusional patients [144] Heller MM, Wong JW, Lee ES, et al. 2013. Delusional infestations: Clinical presentation, diagnosis, and treatment (review) [151] Hinkle NC. 2010. Ekbom syndrome: The challenge of "invisible bug" infestations [90] Koblenzer CS. 2006. The challenge of Morgellons disease [99] Koblenzer CS. 2010. The current management of delusional parasitosis and dermatitis artefacta [142] Lee CS. 2008. Delusions of parasitosis [117] Bewley A, Lepping P, Taylor RE. (Eds.) delusional infestation: Psychodermatology in clinical practice (Lepping; Chap. 12) [12] Lepping P, Freudenmann RW. 2012. Delusional infestation in childhood, adolescence, and adulthood [61] Lepping P, Huber M, Freudenmann RW. 2015. How to approach delusional infestation [63] Magid M, Reichenberg JS. 2019. Treating the difficult patient: Ten pearls to reduce resentment and regain control of the doctor-patient visit [145] Mohandas P, Bewley A, Taylor R. 2013. Dermatitis artefacta and artefactual skin disease: The need for a psychodermatology multidisciplinary team to treat a difficult condition [121] Murase JE, Wu JJ, Koo J. 2006. Morgellons disease: a rapport-enhancing term for delusions of parasitosis [93] Rinshniw M, Lepping P, Freudenmann RW. 2014. Delusional infestation by proxy-what should veterinarians do? [64] Schulte B. 2008. Figments of the imagination? [71] Söderfeldt Y, Groß D. 2014. Information, consent, and treatment of patients with Morgellons disease: An ethical perspective [103] Tey HL, Wallengren J, Yosipovitch G. 2013. Psychosomatic factors in pruritus [21] Todd S. et al. 2019. Delusional infestation managed in a combined tropical medicine and psychiatry clinic [95]

(continued)

Table 3 (continued)

General topic	Articles with content relevant to the topic
Dealing with family members who: a. Believe that the patient is infested and possibly serve as enablers (folie à deux/trois, etc.) b. Think that the patient is nuts c. Do not know what to think but are worried about the mental and physical health of the loved one	Donabedian H. 2007. Delusions of parasitosis [113] Hinkle NC. 2010. Ekbom syndrome—the challenge of "invisible bug" infestations [90] Foster AA, Hylwa SA, Bury JE, et al. 2012. Delusional infestations: Clinical presentations in 147 patients seen at Mayo Clinic [104] Freudenmann RW, Kühnlein P, Lepping P, Schönfeldt-Lecuona C. 2008. Secondary delusional parasitosis treated with paliperidone [56] Freudenmann RW, Lepping P. 2009. Delusional infestation [1] Geary MJ, Russell RC, et al. 2020. 30 years of samples submitted to an Australian medical entomology department [115] Hussain K, Gkini MA, Taylor R, Shinhmar S, Bewley A. 2018. Patient with delusional infestation by proxy: Issues for vulnerable adults [66] Bewley A, Lepping P, Taylor RE. (Eds.) delusional infestation: Psychodermatology in clinical practice (Lepping; Chap. 12) [12] Lepping P, Freudenmann RW. 2012. Delusional infestation in childhood, adolescence, and adulthood [61] Lepping P, Rishniw M, Freudenmann RW. 2015. Frequency of delusional infestation by proxy and double delusional infestation in veterinary practice: Observational study [130] Lynch PJ. 1983. Delusions of parasitosis [119] Murase JE, Wu JJ, Koo J. 2006. Morgellons disease: a rapport-enhancing term for delusions of parasitosis [93] Rinshniw M, Lepping P, Freudenmann RW. 2014. Delusional infestation by proxy-what should veterinarians do? [64] Schulte B. 2008. Figments of the imagination? [71] Schwartz E, Witztum E, Mumcuoglu KY. 2001. Travel as a trigger for shared delusional parasitosis [157]
Support groups (groups that endorse the belief in an infestation vs. groups that help patients adjust to the bewildering panoply of physical emotional distress)	Elkan D. 2007. Morgellons disease: The itch that won't be scratched [103] Freudenmann RW, Lepping P. 2009. Delusional infestation [1] Middelveen MJ. et al. 2018. History of Morgellons disease: From delusion to definition [80] Schulte B. 2008. Figments of the imagination? [71] Simpson L, Baier M. 2009. Disorder or delusion? Living with Morgellons disease [89]
How to address the inexorable decline to a patient's personal life, family life, employment, and bank account?	Beach SD, Kroshinsky D, Kontos N. 2014. Case 37-2014: a 35-year-old woman with suspected mite infestation [112] Bransfield RC, Friedman KJ. 2019. Differentiating psychosomatic, somatopsychic, multisystem illnesses, and medical uncertainty (review) [106] Hinkle NC. 2010. Ekbom syndrome: The challenge of "invisible bug" infestations [90] Schulte B. 2008. Figments of the imagination? [71] Schwartz E, Witztum E, Mumcuoglu KY. 2001. Travel as a trigger for shared delusional parasitosis [157] Simpson L, Baier M. 2009. Disorder or delusion? Living with Morgellons disease [89] Tey HL, Wallengren J, Yosipovitch G. 2013. Psychosomatic factors in pruritus [21] Elkan D. 2007. Morgellons disease: The itch that won't be scratched [103] Foster AA, Hylwa SA, Bury JE, et al. 2012. Delusional infestations: Clinical presentations in 147 patients seen at Mayo Clinic [104] Lepping P, Freudenmann RW. 2012. Delusional infestation in childhood, adolescence, and adulthood [61] Lynch PJ. 1983. Delusions of parasitosis [119]
Should one have family members in the room when discussing the dx plan? Should one ask the patient for permission to talk with family?	Bostwick JM. 2011. Taming hornets: The therapeutic relationship in successful treatment of delusional infestation [108]

References

1. Freudenmann RW, Lepping P. Delusional Infestation. Clin Microbiol Rev. 2009;22:690–732.
2. Lepping P, Noorthoorn EO, Kemperman PMJH, Harth W, Reichenberg JS, Squire SB, Shinhmar S, Freudenmann RW, Bewley A. An international study of the prevalence of substance use in patients with delusional infestation. J Am Acad Dermatol. 2017;77:778–9.
3. Ahmed A, Affleck AG, Angus J, et al. British Association of Dermatologists guidelines for the management of adults with delusional infestation. Brit J Dermatol. 2022;187:472–80.
4. Bailey CH, Anderson LK, Lowe GC, Pittelkow MR, Bostwick JM, Davis DPM. A population-based study of the incidence of delusional infestation in Olmsted County, Minnesota, 1976-2010. Br J Dermatol. 2014;170:1130–5. https://doi.org/10.1111/bjd/12848.
5. ICD-11. International Classification of Diseases. 11th Ed. The World Health Organization. 2022. https://icd.who.int/en
6. Hurko O, Provost TT. Neurology and the skin. J Neruol Neurosurg Psychiatry. 1989;66:417–30.
7. Ramirez-Bermudez J, Espinola-Nadurille M, Loza-Taylor N. Delusional parasitosis in neurological patients. Gen Hosp Psychiatry. 2010;32:294–8.
8. Brown GE, Sorenson E, Malakouti M, Leon A, Reichenberg JS, Magid M, Howard JL, Murase JE, Koo JYM. The spectrum of ideation in patients with symptoms of infestation: from overvalued ideas to the terminal delusional state. J Clin Exp Dermatol Res. 2014;5:1–2.
9. Hinkle NC. Delusory Parasitosis. Amer Ent. 2000;46:17–25.
10. Traver JR. Unusual scalp dermatitis in humans caused by the mite, Dermatophagoides (Acarine: Epidermoptidae). Proc Entomol Soc Wash. 1951;53:1–25.
11. Trabert W. 100 years of delusional parasitosis. Met-analysis of 1,223 case report. Psychopathology. 1985;20:238–46.
12. Romanov DV, Lepping P, Bewley A, Huber M, Freudenmann RW, Lvov A, et al. Longer duration of untreated psychosis is associated with poorer outcomes for patients with delusional infestation. Acta Derm Venereol. 2018;98:848–54. https://doi.org/10.2340/00015555-2888.
13. Freudenmann RW, Lepping P. Second-generation antipsychotics in primary and secondary delusional parasitosis outcome and efficacy. J Clin Psychopharmacol. 2008;28:500–8.
14. Bewley A, Lepping P, Taylor RE, editors. Psychodermatology in clinical practice (Lepping; chap. 12; delusional infestation, page 133). Springer; 2021. p. 441.
15. González-Rodriguez A, Molina-Andreu O, Penadés R, Catalán R, Bernardo M. Structural and functional neuroimaging findings in delusional disorder diagnostic and therapeutic implications. Open Psychiatry J. 2015;2015:17–25.
16. Chute CG, et al. Revision Steering Group. World Health Organization Joint Task Force. The International Classification of Diseases (ICD). 11th ed. American Psychiatric Association; 2018.
17. Kupfer DJ, First MB, Reiger DA, editors. A research agenda for DSM-V. American Psychiatric Association; 2002. p. 332.
18. Bewley AP, Lepping P, Freudenmann RW, Taylor R. Delusional parasitosis: time to call it delusional infestation [published correction appears in Br J Derm. 2010 Oct; 163 (4):898. Freudenmann, R W [corrected to Freudenmann, R W]]. Br J Dermatol. 2010;163:1–2. https://doi.org/10.1111/j.1365-2133.2010.09841.x.
19. Regier DA, Kuhl EA, Kupfer DJ. The DSM-5: classification and criteria changes. World Psychiatry. 2013;12:92–8. https://doi.org/10.1002/wps.20050.
20. Bourgeois JA, Bienenfeld D, et al. Delusional disorder. Medscape 2017.
21. Assalman I, Ahmed A, Alhajjar R, Bewley AP, Taylor R. Treatments for primary delusional infestation. Cochrane Database Syst Rev. 2019;12:CD011326. https://doi.org/10.1002/14651858.CD011326.pub2.
22. Krämer J, et al. Abnormal cerebellar volume in somatic vs. non-somatic delusional disorders. Cerebellum Ataxias. 2020;7:8.
23. Tey HL, Wallengren J, Yosipovitch G. Psychosomatic factors in pruritus. Clin Dermatol. 2013;31:31–40. https://doi.org/10.1016/j.clindermatol.2011.11.004.

24. Corlett PR, Taylor JR, Wang XJ, Fletcher PC, Krystal JH. Toward a neurobiology of delusions. Prog Neurobiol. 2010;92:345–69.
25. Huber MR, Karner M, Kirchler E, Lepping P, Freudenmann RW. Striatal lesions in delusional parasitosis revealed by magnetic resonance imaging. Prog Neuropsychopharmacol Biol Psychiatry. 2008;32:1967–71.
26. Huber MR, Wolf C, Lepping P, Kirchler E, Karner M, Sambataro F, Herrnberger B, Corlett PR, Freudenmann RW. Regional gray matter volume and structural network strength in somatic vs. non-somatic delusional disorders. Prog Neuropsychopharmacol Biol Psychiatry. 2018;82:115–22.
27. Flann S, Shotbolt J, Kessel B, Vekaria D, Taylor R, Bewley A, Pembroke A. Three cases of delusional parasitosis caused by dopamine agonists. Clin Exp Dermatol. 2010;35:740–2. https://doi.org/10.1111/j.1365-2230.2010.03810.x.
28. Altunay IK, Ates B, Mercan S, Demirci GT, Kayaoglu S. Variable clinical presentations of secondary delusional infestation: an experience of six cases from a psychodermatology clinic. Int J Psychiatry Med. 2012;44:335–50.
29. Reich A, Ständer S, Szepietowski J. Drug-induced pruritus: a review. Acta Derm Venereol. 2009;89:236–44.
30. Fleury V, Wayte J, Kiley M. Topiramate-induced delusional parasitosis. J Clin Neurosci. 2008;15:597–8.
31. Krauseneck T, Soyka M. Delusional parasitosis associated with Pemoline. Psychopathology. 2005;38:103–4.
32. Steinert T, Studemund H. Acute delusional parasitosis under treatment with ciprofloxacin. Pharmacopsychiatry. 2006;39:159–60.
33. Robaeys G, DeBie J, Ranst MV, Buntinx F. An extremely rare case of delusional parasitosis in a chronic hepatitis C patient during pegylated interferon alpha-2b and ribavirin treatment. World J Gastroentrol. 2007;13:2379–80.
34. Marshall CL, Ellis C, Williams V, Taylor RE, Bewley AP. Iatrogenic delusional infestation: an observational study. Br J Dermatol. 2016;175:800–2. https://doi.org/10.1111/bjd.14558.
35. Huber M, Kirchler E, Karner M, Pycha R. Delusional parasitosis and the dopamine transporter. A new insight of etiology? Med Hypotheses. 2007;68:1351–8.
36. Mowla A, Asadipooya K. Delusional parasitosis following heroin withdrawal: a case report. Am J Addict. 2009;18:334–5. https://doi.org/10.1080/10550490902925888.
37. Lopez RP, Rachael T, Leight S, Smalligan RD. Gabapentin-induced delusions of parasitosis. South Med J. 2010;103:711–2.
38. Maleki K, Weisshaar E. Arzneimittelinduzierter pruritus. Hautarzt. 2014;65:436–42. https://doi.org/10.1007/s00105-013-2700-4.
39. Kogame T, Kamitani T, Yamazaki H, Ogawa Y, Fukuhura S, Kabashima K, Yamamoto Y. Longitudinal association between polypharmacy and development of pruritus: a nationwide cohort study in a Japanese population. J Eur Acad Dermatol Venereol. 2021;35:2059–66.
40. Schmoll D. Sekundärer SD. Dermatozoenwahn bei Wegener Granulomatose. (delusion of parasitosis due to Wegener's granulomatosis). Fortschr Neurol Psychiatr. 2009;79:234–7.
41. Bilowski J. Dermatologic manifestations of psychiatric disease. Pract Dermatol. 2009;0709. Psy. feature:33–8.
42. Grant JE, Odlaug BL, Won KS. N-Acetylcysteine, a glutamate modulator, in the treatment of trichotillomania: a double-blind, placebo-controlled study. Arch Gen Psychiatry. 2009;66:756–63. https://doi.org/10.1001/archgenpsychiatry.2009.60.
43. Oaklander AL. Shingles, postherpetic neuralgia and postherpetic itch. Int J Pain Med Pal Care. 2002;2:1–10.
44. Alexander JOD. Arthropods and human skin. Berlin, Heidelberg. (London): Springer-Verlag; 1984. p. 436. https://doi.org/10.1007/978-1-4471-1356-0_1.
45. Oaklander AL, Bowsher D, Galer B, Haanpää M, Jensen MP. Herpes zoster itch: preliminary epidemiologic data. J Pain. 2003;4:338–43.

46. Narumoto J, Ueda H, Tsuchida H, et al. Regional cerebral blood flow changes in a patient with delusional parasitosis before and after successful treatment with risperidone: a case report. Prog Neuro-Psychopharmacol Biol Psychiatry. 2006;30:737–40.
47. Akahane T, Hayashi H, Suzuki H, et al. Extremely grotesque somatic delusions in a patient of delusional disorder and its response to risperidone treatment. Gen Hosp Psychiatry. 2009;31:185–6.
48. Berceli D, Salmon M, Bonifas R, Ndefo N. Effects of self-induced unclassified therapeutic tremors on quality of life among non-professional caregivers: a pilot study. Glob Adv Health Med. 2014;3:45–8.
49. Nagaratnam N, O'Neile L. Delusional parasitosis following occipito-temporal cerebral infraction. Gen Hosp Psychiatry. 2000;22:129–32.
50. Floris G, Cannas A, Melis M, Solla P, Marrosu MG. Pathological gambling, delusional parasitosis and adipsia as a post-haemorrhagic syndrome: a case report. Neurocase. 2008;14:385–9.
51. Moryś JM, Jeżewska M, Koreniewski K. Neuropsychiatric manifestations of some tropical diseases. Int Marit Health. 2015;66:30–5.
52. JAMA. Vitamin B12—Revisited. Important Physician Update. JAMA. 2005; 293; 1082-1088 Mar, 2, 2005/Postgraduate Med. Vol. 110/No.1/July 2001/Mayo Interpretive Handbook. 2005/Metro—Nov. 2005/PYA.
53. Schröder SD, Griebl SW. Zinc deficiency-associated dermatitis. N Engl J Med. 2020;383:e103. https://doi.org/10.1056/NEJMicm2003516.
54. Oaklander A, Seo WK, Park MH. Neuropathic pruritus following Wallenberg syndrome. Neurology. 2009;73:1605–6.
55. Robaeys G, De Bie J, Van Ranst M, Buntinx F. An extremely rare case of delusional parasitosis in a chronic hepatitis C patient during pegylated interferon alpha-2b and ribavirin treatment. World J Gastroenterol. 2007;13:2379–80. https://doi.org/10.3748/wjg.v13.i16.2379.
56. Freudenmann RW, Kühnlein P, Lepping P, Schönfeldt-Lecuona C. Secondary delusional parasitosis treated with paliperidone. Clin Exp Dermatol. 2008;34:375–7.
57. Freudenmann RW, Schönfeldt-Lecuona C, Lepping P. Primary delusional parasitosis treated with olanzapine. Int Psychogeriatr. 2007:1–9. https://doi.org/10.1017/S1041610207004814.
58. O'Keeffe ST. Delusions of infestation during delirium in a patient with restless legs syndrome. J Geriatr Psychiatry Neurol. 1985;8:120–2.
59. Chang MW. Trichotillomania: N-Acetylcysteine shows promise medicis, the Derm Co 2009; 17 : p 89.
60. Yang EJ, Beck KM, Koo J. Folie à famille: a systematic review of shared delusional infestation. J Am Acad Dermatol. 2019;81:1211–5. https://doi.org/10.1016/j.jaad.2019.04.023.
61. Lepping P, Freudenmann RW. Chapter 10: delusional infestation. In: Childhood, adolescence, and adulthood. Pediatric Psychodermatology: a clinical manual of child and adolescent Psychocutaneous disorders; 2012. p. 211–36. ISBN 978-3-11-027393-9.
62. Papoiu AD, Wang H, Coghill RC, et al. Contagious itch in humans. A study of visual transmission of itch in atopic dermatitis and healthy subject. Br J Dermatol. 2011;164:1298–303.
63. Lepping P, Huber M, Freudenmann RW. How to approach delusional infestation. BMJ. 2015;350:6. https://doi.org/10.1136/bmj.h1328.
64. Rishniw M, Lepping P, Freudenmann RW. Delusional infestation by proxy-what should veterinarians do? Can Vet J. 2014;55:887–91.
65. Sarkar S, Ghosal MK, Bandyopadhyay D, Ghosh SK, Guha P. Delusional parasitosis by proxy. Clin Exp Dermatol. 2009;34:487–8. https://doi.org/10.1111/j.1365-2230.2009. 03555.x.
66. Hussain K, Gkini MA, Taylor R, Shinhmar S, Bewley A. A patient with delusional infestation by proxy: issues for vulnerable adults. Dermatol Ther. 2018;31:1–3. https://doi.org/10.1111/dth/12724.
67. Fisher JD. Emergency department presentation of 'delusional parasitosis by proxy'. Delusional parent, injured child. Am J Emerg Med. 2019;37:1806.e1–2.
68. Tapp T, Mofid M. Take the compendium challenge. Parasit The compendium; 1998. p. 433–9.

69. Manschreck TC. Delusional disorder and shared psychotic disorder. In: Comp. Textbook of psych. seventh ed. Baltimore: Williams &Wilkins; 2000. p. 1243–64.
70. Grant M. Between a rock and a hard place: balancing physician deception and patient self-harm in the management of delusional infestation. Am J Trop Med Hyg. 2020;102:3–4. (84)
71. Schulte B. Figments of the imagination? Washington Post; 2008. article/2008/01/16/AR2008022603134_pf.html:1-10
72. CDC web archive. Press briefing transcripts (subject; the CDC Morgellons disease study), vol. cdc.gov/media/transcripts/2008/t080116.htm. Moderator Dave Daigle with Dr. Michele Pearson; 2008. p. 2–6.
73. Dermatology Times. CDC sends Morgellons investigators to California. An MJH life sciences brand. Dermatology Times; 2006. p. 1–3.
74. Marris E. Mysterious "Morgellons disease" prompts US investigation. Nat Med. 2006;12:982.
75. Pearson ML, Selby JV, Katz KA, et al. Clinical, epidemiologic, histopathologic, and molecular features of an unexplained dermopathy. PLoS One. 2012;7:e29908.
76. Hylwa SA, Ronkainen SD. Delusional infestation versus Morgellons disease. Clin Dermatol. 2018;36:714–8. https://doi.org/10.1016/j.clindermatol.2018.08.007.
77. Savely VR, Leitao MM, Stricker RB. The mystery of Morgellons disease infection or delusion? Am J Clin Dermatol. 2006;7:1–5.
78. Edwards S. Real delusions of an unreal disease: a history of Morgellons. Jezabelcom; 2015.
79. Middelveen MJ, Burugu D, Poruri A, et al. Association of spirochetal infection with Morgellons disease. F1000Res. last updated: 2014. 2013;2:25.
80. Middelveen M, Fesler MC, Stricker RB. History of Morgellons disease: from delusion to definition. Clin Cosmet Investig Dermatol. 2018;11:71–90.
81. Molyneux J. AKA 'Morgellons'. Am J Nurs. 2008;108:25–6. https://doi.org/10.1097/01.NAJ.0000317988.01798.2f.
82. Jamison L, editor. The devil's bait. Symptoms, signs, and the riddle of Morgellons. Harpers Magazine; 2013. p. 64–70.
83. Söderfeldt Y, Groß D. Information, consent, and treatment of patients with Morgellons disease: an ethical perspective. Am J Clin Dermatol. 2014;15:71–6.
84. Kellett CE. Sir Thomas Browne and the disease called the Morgellons. Ann Med Hist. 1935;7:467–79.
85. Preston C. Sir Thomas Browne: selected writings. Carcanet Press; 2006. p. 168.
86. Hughes JT. The medical education of sir Thomas Browne, a seventeenth-century student at Montpellier, Padua, and Leiden. J Med Biogr. 2001;9:70–6.
87. Greenhill WA, editor. Sir Thomas Browne's Religio Medici: letter to a friend &c. and Christian morals/Thomas Browne. London: Macmillan and company Limited; 1926. p. 382.
88. Schenck von Grafenberg J. Observationum medicarum rariorum libri VII. Sumptibus Joannis Beyeri; 1665. p. 918.
89. Simpson L, Baier M. Disorder, or delusion? Living with Morgellons disease. J Psychosoc Nurs Ment Health Serv. 2009;47:36–41.
90. Hinkle NC. Ekbom syndrome: the challenge of "invisible bug" infestations. Annu Rev Entomol. 2010;55:77–94.
91. Hinkle NC. Ekbom syndrome: a delusional condition of "bugs in the skin". Curr Psychiatry Rep. 2011;13:178–86. https://doi.org/10.1007/s11920-011-0188-0.
92. Ekbom KA. The pre-senile delusion of infestation. (translation by Yorston et al. 1938. Der praesenile dermatozoenwahm, Acta. Psychiatry. Et Neur. Scand. 13: 227-259.). Hist Psychiatry. 2003;14:232–56.
93. Murase JE, Wu JJ, Koo J. Morgellons disease: a rapport-enhancing term for delusions of parasitosis. J Am Acad Dermatol. 2006.; 55:913–4. https://doi.org/10.1016/j.jaad.2005.10.044.
94. Lepping P, Walker SL, Bewley AP. Why delusional infestation should remain delusional infestation. Letter to the Editor. Creative Commons Attribution. 2020; (CC-BY).
95. Todd S, Squire SB, Bartlett R, Lepping P. Delusional infestation managed in a combined tropical medicine and psychiatry clinic. Trans R Soc Trop Med Hyg. 2019;113:18–23. (95)

96. Healy R, Taylor R, Dhoat S, Leschynska E, Bewley AP. Management of patients with delusional parasitosis in a joint dermatology/liaison psychiatry clinic. Br J Dermaol. 2009;161:197–9. (96)

97. Moriarty N, Alam M, Kalus A, O'Connor K. Current understanding, and approach to delusional infestation. Am J Med. 2019;132:1401–9. (99)

98. Ahmad K, Ramsey B. Delusional parasitosis; lessons learnt. Acta Derm Venereol. 2009;89:165–8.

99. Koblenzer CS. The challenge of Morgellons disease. J Am Acad Dermatol. 2006;55:920–2.

100. Cook J, Lewandowsky S, et al. The debunking handbook. 2011 and revised 2020. https://sks.to/db2020: doi:https://doi.org/10.17910/b7.1182.

101. Freudenmann RW, Lepping P, Huber M, Dieckmann S, Bauer-Dubau K, Ignatius R, Misery L, Schollhammer M, Harth W, Taylor RE, Bewley AP. Delusional infestation and the specimen sign: a European multicenter study in 148 consecutive cases. Br J Dermatol. 2012;167:247–51.

102. Lepping P. Why delusional infestation should remain delusional infestation. Am J Trop Med Hyg. 2020;103:700. https://doi.org/10.4269/ajtmh.19-0927.

103. Elkan D. Morgellons disease—the itch that won't be scratched. Newscientist/Health; 2007.

104. Foster AA, Hylwa SA, Bury JE, Davis MD, Pittelkow MR, Bostwick JM. Delusional infestation: clinical presentation in 147 patients seen at Mayo Clinic. J Am Acad Dermatol. 2012;67:673.e1–10. https://doi.org/10.1016/j.jaad.2011.12.012.

105. Lepping P, Baker C, Freudenmann RW. Delusional infestation in dermatology in the UK: prevalence, treatment strategies, and feasibility of a randomized controlled trail. Clin Exp Dermatol. 2010;35:1–4. https://doi.org/10.1111/j.1365-2230.2010.03782.x.

106. Bransfield RC, Friedman KJ. Differentiating psychosomatic, somatopsychic, multisystem illness and medical uncertainty (review). Healthcare. 2019;7:1–28.

107. Harvey WT, Bransfield RC, Mercer DE, et al. Morgellons disease, illuminating an undefined illness: a case series. J Med Case Rep. 2009;3:1–8. https://doi.org/10.4076/1752-1947-3-8243.

108. Bostwick JM. Taming hornets: the therapeutic relationship in successful treatment of delusional infestation. Gen Hosp Psychiatry. 2011;33:533–4.

109. Lai J, Xu Z, Xu Y, Hu S. Reframing delusional infestation: perspectives on unresolved puzzles. Psychol Res Behav Manag. 2018;11:425–32.

110. Middelveen MJ, Stricker RB. Filament formation associated with spirochetal infection: a comparative approach to Morgellons disease. Clin Cosmet Investig Dermatol. 2011;4:167–77.

111. Middelveen MJ, Stricker RB. Morgellons disease: a filamentous borrelial dermatitis. Int J Gen Med. 2016;9:349–54.

112. Beach SR, Kroshinsky D, Kontos N. Case 37-2014: a 35-year-old woman with suspected mite infestation. N Engl J Med. 2014;371:2115–23.

113. Donabedian H. Delusions of Parasitosis. Clin Infect Dis. 2007;45:131–4.

114. Freudenmann RW, Kölle M, Schönfeldt-Lecuona C, Dieckmann S, Harth W, Lepping P. Delusional parasitosis and the matchbox sign revisited: the international perspective. Acta Derm-Venereol. 2010;90:517–9. https://doi.org/10.2340/00015555-0909.

115. Geary MJ, Russell RC, Moerkerken L, Hassan A, Doggett SL. 30 years of samples submitted to an Australian medical entomology department. Aust Ent. 2020;60:172–97. https://doi.org/10.1111/aen.12480.

116. Kohorst JJ, Bailey CH, Andersen LK, Pittelkow MR, Davis MDP. Prevalence of delusional infestation—a population-based study. JAMA Dermatol. 2018;154:615–7. https://doi.org/10.1001/jamadermatol.2018.0004. https://jamanetwork.com/journals/jamadermatology/fullarticle/2676941

117. Lee CS. Delusions of parasitosis. Dermatol Ther. 2008;21:2–7.

118. Lepping P, Aboalkaz S, Squire SB, et al. Later age of onset and longer duration of untreated psychosis are associated with poor outcomes in delusional infestation. Acta Derm Venereol. 2020;100:1–2. https://doi.org/10.2340/00015555-3625.

119. Lynch PJ. Delusions of parasitosis. Semin Dermatol. 1993;12:39–45.

120. Martins AC, Mendes CP, Nico MM. Delusional infestation: a case series from a university dermatology center in São Paulo. Brazil Int J Dermatol. 2016;55:864–8. https://doi.org/10.1111/ijd.13004.

121. Mohandas P, Bewley A, Taylor R. Dermatitis artefacta and artefactual skin disease: the need for a psychodermatology multidisciplinary team to treat a difficult condition. Bri J Dermatol. 2013;169:600–6.

122. Nguyen CM, Danesh M, Beroukhim K, Sorenson E, Leon A, Koo J. Psychodermatology: a review. Pract Dermatol. 2015:49–54.

123. Misery L. Morgellons syndrome: a disease transmitted via the media. Ann Dermatol Venereol. 2013;140:59–62.

124. Lustig A, Mackay S, Strauss J. Morgellons disease as internet meme (letter). Psychosomatics. 2009;50:90.

125. Altschuler DZ, Crutcher M, Dulceanu N, Cervantes BA, Terinte C, Sorkin LN. Collembola (springtails) (Arthropoda: Hexapoda: Entognatha) found in scrapings from individuals diagnosed with delusory parasitosis. JNY Entomol Soc. 2004;112:87–95.

126. Christiansen KA, Bernard EC, et al. Ent News. 2008;119:537–40.

127. Hylwa SA, Bury JE, Davis MDP, Pittelkow M, Bostwick JM. Delusional infestation, including delusions of parasitosis: results of histologic examination of skin biopsy and patient-provided skin specimens. Arch Dermatol. 2011;147:1041–5.

128. Lepping P. Eine seltene Störung mit hohem Leidensdruck. CliniCum derma; 2017. p. 3/17.

129. Shelomi M. Evidence of photo manipulation in delusional parasitosis paper. J Parasitol. 2013;99:583–858.

130. Lepping P, Rishniw M, Freudenmann RW. Frequency of delusional infestation by proxy and double delusional infestation in veterinary practice: observational study. Br J Psychiatry. 2015;206:160–3.

131. Potter MF, Koehler PG. Invisible itches: insect and non-insect causes. UF/IFSA pub; 2022. p. 1–6.

132. Fabbro S, Aultman JM, Mostow EN. Delusions of parasitosis: ethical and clinical considerations. J Am Acad Dermatol. 2013;69:156–9.

133. Maher S, Hallahan B, Flaherty G. Itching for a diagnosis—a travel medicine perspective on delusional infestation. Travel Med Infect Dis. 2017;18:70–2. https://doi.org/10.1016/j.tmaid.2017.05.011.

134. Sambhi R, Lepping P. Psychiatric treatment in dermatology: an update (review article). Clin Exp Dermatol. 2010;35:120–5.

135. Wong S, Bewley A. Patients with delusional infestation (delusional parasitosis) often require prolonged treatment as recurrence of symptoms after cessation of treatment is common: an observational study. Br J Dermatol. 2011;165:893–6. https://doi.org/10.1111/j.13652133.2011.10426.x.

136. Brett AS, McCullough LB. Addressing requests by patients for nonbeneficial interventions. JAMA. 2012;307:149–50.

137. Lantos PM, Shapiro ED, Auwaerter PG, et al. Unorthodox alternative therapies marketed to treat Lyme disease. Clin Infect Dis. 2015;60:1776–82.

138. Steere AC, Arvikar SA. Editorial commentary what constitutes appropriate treatment of post-Lyme disease symptoms and other pain and fatigue syndromes? Clin Infect Dis. 2015;60:1783–5.

139. Bury JE, Bostwisk JM. Iatrogenic delusional parasitosis: a case of physician-patient folie à duex. Gen Hosp Psychiatry. 2010;32:210–2.

140. De La CA, Reichenberg J, Magid M. Successful treatment of patients previously labeled as having "delusions of parasitosis" with antidepressant therapy. J Drug Derm. 2012;11:1506–7.

141. Reich A, Kwiatkowska D, Pacan P. Delusions of parasitosis: an update. Dermatol Ther. 2019;9:631–8.

142. Koblenzer CS. The current management of delusional parasitosis and dermatitis artefacta. Skin Therapy Lett. 2010:1–7. www.skintherapyletter.com/2010/15.9/1.html.

143. Lepping P, Russell I. Freudenmann RW. Antipsychotic treatment of primary delusional parasitosis: systematic review. Br J Psychiatry. 2007191:198-205. doi:https://doi.org/10.1192/bjp.bp.106.029660.

144. Heller MM, Murase JE, Koo JYM. Time and effort to establish therapeutic rapport with delusional patients. Arch Dermatol. 2011;147:p1046.
145. Magid M, Reichenberg JS. Treating the difficult patient: ten pearls to reduce resentment and regain control of the doctor-patient visit. Skinmed. 2019;17:11–4.
146. Lepping P, Gil-Candon R, Freudenmann RW. Delusional parasitosis treated with amisulpride. Prog Neurol Psychiatry. 2005;9:12–6.
147. Kölle M, Lepping P, Kassubek J, Schönfeldt-Lecuona C, Freudenmann RW. Delusional infestation induced by piribedil add-on in Parkinson's disease. Pharm-Psychiatry. 2010;43:240–2. https://doi.org/10.1055/s-0030-1261881.
148. Lepping P, Freudenmann RW. Delusional parasitosis: a new pathway for diagnosis. Clin Exp Dermatol. 2008;33:113–7.
149. Fowler E, Maderal A, Yosipovitch G. Treatment-induced delusions of infestation associated with increased brain dopamine levels. Acta Derm Venereol. 2019;99:327–8. https://doi.org/1 0.2340/00015555-3093.
150. Gordon-Elliott JS, Muskin PR. Managing the patient with psychiatric issues in dermatologic practice. Clin Dermatol. 2013;31:3–10.
151. Heller MM, Wong JW, Lee ES, et al. Delusional infestations: clinical presentation, diagnosis, and treatment. Int J Dermatol. 2013;52:775–83. https://doi.org/10.1111/ijd.12067.
152. Alraies MC, Keller E, Shaheen K, Alyraiyes AH. The clinical picture. A 40-year-old woman with excoriated skin lesions. Cleveland Clin. J Med. 2011;78:161–3.
153. Fried RG. Take your "pick": could neurotic excoriations be a forme fruste of Tourette's disorder? Dermatol. 2012;20:1–5.
154. Ojeda-López C, Aguilar-Venegas LC, Tapia-Orozco M, Cervantes-Arriaga A, Rodríguez-Violante M. Delusional Parasitosis as a treatment complication of Parkinson disease. Psychosomatics. 2015;56:696–9. https://doi.org/10.1016/j.psym.2015.07.004.
155. Swick BL, Walling HW. Drug-induced delusions of parasitosis during treatment of Parkinson's disease. J Am Aca Derm. 2005;53:1086–7.
156. Marschall MA, Dolezal RF. Cohen M, Marschall S F. Chronic wounds, and delusions of parasitosis in the drug abuser. Plast Reconstr Surg 1991; 88 : 328-330.
157. Schwartz E, Witztum E, Mumcuoglu KY. Travel as a trigger for shared delusional parasitosis. J Travel Med. 2001;8:26–8.
158. Healy R, Taylor R, Dhoat S, Leschynska E, Bewley AP. Management of patients with delusional parasitosis in a joint dermatology/ liaison psychiatry clinic. Br J Dermatol. 2009;161:197–9. https://doi.org/10.1111/j.1365-2133.2009.09183.x.

Part II
Diagnosis and Management

The Interview Processes

Gale E. Ridge

DI patients are not routine and should be approached holistically. Rarely is there a DI patient with a single problem; there are often many contributing factors, each with a variable history ranging from predispositions to trauma. Thus, it is important to understand where the patient is in the progress of their illness. As illustrated in the flowchart, Fig. 5, a physician faces several decision points when progressing toward a differential diagnosis. Although there are many similarities among DI patients, each patient is slightly different. They possess a complex mix of somatic and psychiatric conditions, and untangling these can be challenging.

In the interview process, always be pleasant and welcoming. Avoid using any negative language or cute euphemisms. To quote Mark Twain,

> "The difference between the right word and the almost right word is the difference between lightning and the lightning bug,
> … be very careful what you say."

It is important to know how to interact and connect with DI patients to build early trust. Many patients in the formative period of DI ideation genuinely want answers, but for any number of reasons, their physicians failed to provide them. It is often a matter of goals. Physicians and patients may not agree on root causes; they agree to disagree, but if their goals are the same, to stop "the pain," then cures may be found through trial and error with different approaches to care. Patients and physicians often miss this. Frustrated, many patients start to research possible causes by themselves while continuing to seek help. Their research may either lead to medical diagnosis, treatment, and resolution [1] or move in a direction of "commitment" where they begin to believe that they are truly parasitized. They have been left to diagnose themselves, and in their minds fill a perceived void with information their

G. E. Ridge (✉)
Department of Entomology, The Connecticut Agricultural Experiment Station,
New Haven, CT, USA
e-mail: gale.ridge@ct.gov

G. E. Ridge (ed.), *The Physician's Guide to Delusional Infestation*,
https://doi.org/10.1007/978-3-031-47032-5_2

physicians were not able to provide. Additionally, a genuine diagnosis can go wrong, and patients reject results, remaining committed to their beliefs. It is at this point many DI patients start to migrate from physician to physician described as "doctor shopping/hopping," seeking affirmation. The longer a patient remains stuck in DI ideation while dismissing genuine medical results, the more trapped they become.

Empathy: With unhappy stressed patients, metaphorically speaking, "crashing through the front door" may not be a good idea, especially with high-functioning patients. Choosing slower, more considered side entrances such as windows and side or backdoors for alternative entrances often yields better results. Use phrases such as "It sounds as if you would like to be understood" rather than "What do you mean?" Check for depression and possible suicidal ideation, particularly in long-term cases. Some more desperate patients may say, "I think I'm going to die" or "I feel like dying." This should be taken seriously.

Never try to calm patients by *telling* them to "calm down." This can upset them more because it is perceived to be a lack of understanding. You may not agree with what they say, but treat patients with kindness because they are sensitive to their situation and may quickly become unsettled.

If patients' behaviors and descriptions suggest histories of doctor shopping/hopping (Box 1), it is advisable to take note of this and allow the patients to continue to describe their experiences. Do not deny these patients their beliefs. Work with what they are saying rather than react to what they are saying. Look more at their actions rather than their words. Reflect to patients what they are thinking so they can hear themselves. Do not interrupt their streams of thought. It is more important for the physician to listen and gather information. Be with each patient as an individual; it is not about "you" (the physician), and remain reassuring. Talk with patients, not at them. Convey that you genuinely want to help. Be authentic, be yourself, and pay attention to details.

Box 1. Doctor Shopping/Hopping.
In the United States, doctor shopping/hopping is common with DI patients. This is when patients migrate through multiple medical professionals [2]. Other groups include pest management professionals/operators, university extension services, veterinarians, epidemiologists, entomologists, social services, and health departments. It is driven by several factors. These include:

Long waits to get appointments, failed appointments, incompatible office hours, and inconvenience of access to clinics or physician's offices,

Insufficient time for adequate communication between patients and providers,

Undesirable physician/clinician attributes as perceived by the patient,

A lack of understanding and patience with medically unexplained patient descriptions and symptoms by medical providers,

Symptom persistence and patient disbelief in diagnoses,

Unresolved somatic discomfort and obsessive preoccupation with it along with psychiatric and/or emotional dysfunction including the delusion.

United States vs. European approaches to DI care.
The medical system in the United States is vulnerable to doctor shopping/hopping by patients. Patients often become their own advocates in search of care following disappointing encounters with medical professionals. They start to go from physician to physician looking for answers. Choice of care evolves to align with a developed belief system. It is self-selective. Issues remain unresolved, and patients drift away from mainstream medicine toward fringe influencers where they languish.

In comparison, doctor shopping/hopping in Europe is rare because of the way European medical systems and services are set up. A patient visits their primary care physician who assigns them to a specific individual specialist such as a dermatologist or neurologist. The patient then remains with the specialist who organizes care by consulting and including other medical experts and professionals.

Ask patients what they have already tried. By doing this, it again conveys acknowledgment of issues, and you are taking them seriously. The goal is not to undo the delusion but to understand it and address it. Watch for exceptions, and presume nothing; occasionally, there are true infestations such as filarial worms, and that evidence may be subtle. Ask appropriate questions to get as much initial information as possible from the patient (Box 2).

Box 2. Interviewing.
Always take detailed notes because clues and patterns may become apparent. Do not react or respond to the patient's narratives and answers. Remain calm and inviting. Always consider the backfire effect. Following are suggested questions and objectives that may be helpful during the interview process.

The way a question is phrased can either open a patient up or unintentionally close them down. For example, questions might be asked in one of the two ways:

"Are you longing to be understood?" or

"It sounds as if you would like to be understood?"

The first question is more directed <u>at</u> the patient using the word "you," placed earlier in the sentence, while the second is more reflective aligning more <u>with</u> the patient and gentler in tone. Additionally, observational comments such as:

"It must be hard for you when people are not taking you seriously." or

"I can imagine it is very hard for you when people are not taking you seriously,"

are helpful.

> Additionally, encourage patients to change their language use. This causes a frameshift in thinking. With dermatology patients, suggest that they use the short terms "pinching" or "pricking" rather than "bite" or "biting" when they talk about their sensations. The jury is out as to what is happening, and by removing these inflammatory words from conversations, it begins to lessen the perception of being attacked by an unknown parasite/pathogen while opening space for the consideration of other options.

Keep conversations focused because some patients can drift from idea to idea. If you feel a patient is increasing their resistance, change direction. Tailor conversations to respect patients' thinking. It is often helpful to direct conversations away from the patient to a third-person space. Encourage patients to ask questions. Often, if patients are given the opportunity to have their questions answered, they may be more inclined to compare new information with their own preexisting beliefs and they may start to listen.

Storytelling: Normalize DI patient's reactions and experiences by describing suitable cases. Tell stories. It pushes back against the patient's sense of isolation. Seeing they are not alone provides comfort.

Joining with patients: Patients may have doctor shopped/hopped before seeing their current medical provider. Therefore, it is important to quickly develop a relationship. Frame the new relationship as a cooperative "journey" the physician and patient are on together.

Clarify professional level of expertise while expressing a willingness to help. Emphasize that the patient has responsibilities to the physician in keeping appointments and recording experiences as much as the physician has responsibilities to finding a cure.

Humans are highly tuned to read changes in emotional expression, which can be as quick as 1/40th of a second. Many who have endured DI for an extended period will be highly reactive to facial expression. If a physician "tunes out," it will immediately be detected. Be "present," and do not allow distractions. Tell colleagues and staff not to interrupt interviews unless prearranged.

Figure out what is reality and what is not to minimize conflict. Patients can express anger directed at their physicians, but they are not angry at them but frustrated at the situation (Fig. 4). To counter this, physicians might redirect the anger and rally around a "common cause" by joining with the patients in their anger. Redirect the conversation and partner with these patients. For example, write on a piece of paper one of the patient's grievances or issues, then move to sit next to the patient. Put the paper on a table in front you both, and discuss the problem together. This is one technique of joining.

Language: Cruelty and callousness toward patients through language are often reported. Below is a table from the Texas Council for Developmental Disabilities that has examples of what to say and what not to say to patients. Language and the use of it are integral in the healing process of DI. If it is misspoken or interpreted by patients negatively, the opportunity for a cure may be lost (Table 1).

Table 1 Phrases physicians should or should not say in the presence of DI patients.

Examples of what you should say	Examples of what you should not say
Person diagnosed with a mental health disorder	Crazy, nuts, nut case, insane, psycho, mentally ill, emotionally disturbed, demented, raving
Person diagnosed with a cognitive disability or with an intellectual and developmental disability	Retard, dumb, dumb-ass, slow, idiot, moron, mentally retarded, impediment
Person with a physical disability	A cripple, a gimp
People without disabilities	Normal, healthy, whole, or typical people
People with disabilities	The handicapped, disabled
Person who has or been diagnosed with.....	Person afflicted with, suffers from, a victim of.....
Person of short stature, little person	A dwarf, midget, pigmy
Take a deep breath	Calm down

Patients can be deceptive. Some patients test. By establishing a good rapport early, it may avoid this as well as subsequent confusion and distrust. Encourage respectfulness from patients by saying:

> "I give you the benefit of doubt, you give me the benefit of doubt, but we need to work together to find the answers together."

Psychogenic and somatic itch: The cingulate cortex is important in processing sensations of itch as well as modulating emotion and cognition that includes reward anticipation. Thus, itch can become rewarding through scratching, and these two often form a vicious cycle. Some patients become highly focused on their pruritus to the exclusion of all else. The downstream consequences are anxiety, depression, loss of focus, and with that sometimes employment, ability to learn, and sleep. These result in dissatisfaction and a lower quality of life. Pruritic sensations, whether psychogenic or somatic in origin, are not independent of one another. They often coexist [3].

Personality traits: Certain personality traits lend themselves to the development of DI. One trait is increased conscientiousness. This includes perfectionism and strict/compulsive behaviors (Obsessive Compulsive Disorder (OCD)) such as excessive tidiness and cleanliness. Patients also exhibit selflessness and focus on the perceived needs and social signals from relevant others. They are intuitive toward the welfare of others and lose awareness of their own needs. Self-isolation to protect others from perceived parasites/pathogens is common. Their negative focus makes it difficult for them to appreciate any personal success [4].

1 Medical Office Staff and Delusional Infestation

It is not uncommon for patients to phone interview physician practices. The phone interview saves time, assesses the suitability of a physician or nonmedical professional, and allows conversation cutoff if dissatisfied. It is much harder to do this in person. Medical office staff who answer phones need to have training in identifying potential DI callers. When working with suspected DI callers, avoid being drawn into long conversations. Be understanding, but do not hesitate to interrupt if they go into lengthy descriptions. Listen for repeated use of the irregular verb "bite" and the pronouns "they" and "them." Take notes. NEVER speculate if asked questions by a suspected DI caller; remain neutral on any subject that might be brought up.

Questions to ask: Listen for the overuse of words such as "they," "them," and "biting." Patients have difficulty describing their persecutors and sensations, so resort to using these descriptors. They may also use the phrase, "it sounds crazy but …" or "I'm not crazy …." This may suggest experience from being told they were. Following are some sample questions that might be asked:

"Date of birth?"

Older patients are more prone to DI.

"When did you start to feel these sensations?"

This provides a timeline, conveys interest, and introduces alternative language into conversation that avoids the use of the word "bite."

"Have you seen other physicians or nonmedical professionals about this complaint?"

This checks for doctor shopping/hopping. Nonmedical professionals may include pest management professionals/operators, university extension services, health departments, veterinarians, and/or entomologists.

"Do you live alone?"

Isolated patients are at higher risk for DI. Additionally, conscientious DI patients often isolate to protect others, breaking important relationship bonds.

"What is your recent travel history? Have you eaten any unusual or raw foods?"

This may point to more unusual sources of discomfort such as tropical arthropod parasitism or consumption of filarial worms.

"Have you been on the internet?"

This suggests that patients may have become caught up in algorithmic steering and confirmation bias, which feeds DI pathologies.

"Have you tried over-the-counter remedies?"

This may suggest tendency to self-harm whether intentional or not.

"Is there someone you know and trust who can accompany you to appointments?"

The role of a second person whom patients trust is to provide emotional support, be a memory repository (many patients lack first memories and have difficulties accepting physician's advice while struggling with DI), and act as surrogates for physicians in patient's homes (emotional support humans (ESHs)). On the other hand, they may contribute to perpetuating the belief system.

"What do you want by seeing the physician?"

This may provide a momentary stop point of self-reflection for a patient. The question inserts the concept of a goal, which prompts a patient to react either positively or negatively to the question. How they react may provide additional information for the physician, particularly in how they plan to work with a patient.

2 List of What to Look for While Interviewing DI Patients

Look for DI sign: Doctor shopping/hopping, skin sores, and believed fibers/parasites under the skin or in the body.

Patient's age: Older more isolated patients are at higher risk.

Race and ethnicity. These can be DI influencers. Ask about Hispanic ancestry. Hispanics with Caribbean heritage who are over 60 are more susceptible to vitamin B12 deficiency. This can cause formication sensations [5].

Diet. Diet will influence a sense of well-being. It plays a significant role in DI. The alimentary canal (the second brain) is an influencer of moods and psychological well-being. What is being eaten plays an important role. The consumption of junk foods or suspected eating disorder behaviors should be considered.

Eyesight: Patients may have poor eyesight and cannot see detail, misinterpreting objects.

Body weight: Ask about weight issues. Patients can gain or lose weight during DI.

Skin type: DI seems to be more prevalent in patients with Fitzpatrick skin types I, II, and III [6].

Companion animals: Do patients own and/or co-sleep with pets, other animals, or have pet birds? This checks for flea or mite species that are present in domestic animals.

Children: Do patients have children? If so, how old are they? Both pets and particularly young children may need protection.

Travel history: If a patient has traveled, check for possible identified human disease-causing parasites, fungi, or pathogens found in the region of the world they had visited. Question about unusual foods eaten. Look for obligatory myiasis and/or facultative or accidental myiasis picked up from travel.

Work: Question about patient's profession and avocation.

Military service: Patients may have been exposed to toxic chemicals while in military service, which may have caused long-term harm.

Acquaintances: Ask for a list of close friends, relevant others, colleagues, and/or family. They may become important in assisting recovery through reinforcement and emotional support. On the other hand, friends, family or others may abandon, attack, or even enable patients, which works against effective treatment.

Events prior to patient visit(s): Ask about the sequence of events prior to and after disease onset. Look for any traumatic events which might trigger DI ideation such as injuries or death of loved ones.

Patient efforts: What have patients done so far to address their problems? Are they self-medicating?

Pain: Some patients are more tolerant of pain than others. Sensitivity to pain should be checked.

Quality of life: Assess quality-of-life issues (accommodation, food, relationships, pets, transportation, etc.).

Hygiene history: Excessive self and residential cleaning, including pesticide treatments, should be checked for.

Resources. Identify patient's resources such as financial security, accommodation, and level of education.

Reflective questioning: Pay attention to how patients process information. Use "reflective questioning." It garners cooperation and enables patients to generate new patterns of cognition and behavior. Examples of questioning are the following:

"How is it that we find ourselves together today?"
"How do you feel about …?"
"Who else is worried about this?"
"What else do you think you could have done?"

Notice if patients are hyperaware of facial expressions and reading into the interviewer's reactions. This indicates a strong negative history. Listen for bizarre and unbelievable accounts.

How do patients frame their thoughts? This may indicate the level of guardedness.

Listen to what is being said and how it is being said. Tune into what is *not* being said.

Stories from patients are often rambling and lack sense. Biologically implausible descriptions and explanations indicate a possible detachment from reality around the DI ideation. If patients are allowed to talk about other experiences or subjects, their conversations often become normal.

Patient desperation: When patients express desperation, offer encouragement. Many have been through a great deal and are at the end of their rope. These patients need reassurance and comfort, and this is a moment to provide it. This often provides emotional relief for these patients, promotes cooperation and trust, and allows physicians time they may not have to figure out a differential diagnosis.

Internalizing experience: Patients can internalize "experience thinking" because a previous physician found convincing evidence of infestation and had treated for it. This may cause a placebo effect with remission, but following several weeks, the fixed delusions return. These patients then rationalize and believe that their infestations are resistant to treatment, so when subsequent physicians cannot find evidence, "they" (the parasites) either have morphed into new or different life-forms or are "hiding." It is common for these patients to entertain the idea that their parasites "are new to science."

Withheld information: Patients often withhold information either deliberately or from fear and anxiety. Their previous experiences with other physicians may have been very negative, and they may not want to bias a new physician into possibly confirming a previous provider's findings. Many patients experience previous physician skepticism and feel helpless, frustrated, angry, or humiliated, resulting in withholding of vital information.

Box 3 provides a checklist of DI indicators, which might be checked off during interviews.

Box 4 provides some "Do's and Don'ts" when working on the psychological investment components of DI [7].

Box 3. Checklist for DI Conditions*

More than 7 yes checks, suspect DI.	Yes	No
Body tension. DI patients are physically tense		
Overuse of the pronouns "they," "them," and "biting"		
Excoriations, particularly on easily reached areas of the body. Injuries on the back are often crescent shape and linear		
Dermal injuries are often circular made by fingernails from picking and/or linear as several parallel lines which match spacing between fingers		
Often there is a denial of psychological disease		
More severe patients lack eye contact, preferring to look down		
Descriptions of implausible biology and behavior of perceived parasites		
Common are well-rehearsed descriptions of parasite behavior		
Confusion or switching descriptive parasite behaviors when challenged. (It is similar to fighting the Hydra of Lerna: one head is cut off, then another appears)		
Supports beliefs using others, e.g., folie à deux		
Doctor shopping/hopping		
Either not inclined to volunteer information about medication/illegal drug use or overly generous with detail. Patients often have multiple medications		
Admission of self-medication using over-the-counter drugs		
When the topic is changed, conversations become normal two-way exchanges		
When focused on the DI, conversations are patient-driven and patient-dominated. It is not unusual to listen to long monologues		
Defensive outward aggression when challenged by facts and information		
The physician may experience "the stranger effect" when patients share too much confidential information. Patients may be hesitant to share issues with family or relevant others from fear and/or negative judgment. They view the physician as a confidant		
Specimen sign, e.g., high volume of samples and images		
Is the patient listening but not hearing?		
Paranoia. The sense of being persecuted		
Increased age and isolation are common indicators. Isolation can be a preexisting risk factor for DI [7]. Patients are generally older than 35 years. Highly traumatized patients, those with psychiatric disorders, or drug abusers present younger		
High use of the internet where algorithms steer patients to selected information based on their search criteria and browsing history		
Diet. Check for weight gain or loss and quality of food consumed		
Other observations		
*Checklist based on Ely et al. [3] and the author's experiences.		

Box 4. Suggested "Do's and Don'ts" When Working on the Psychological and Emotional Investment Components of DI. Amended from Freudenmann, and Lepping [7]

- Take time; take a careful history, including trips to tropical resorts.
- Perform the diagnostic investigations needed (even if you are sure that the patient has no infection).
- Examine all specimens carefully.
- Acknowledge the patient's suffering. Show empathy and offer to help to reduce distress.
- Paraphrase the symptoms ("you are itching," "the sensations," "the crawling," etc.) instead of reinforcing or questioning them.
- Indicate that you are familiar with the problem and that you were able to help other patients with similar problems.
- Answer that you did not find any pathogens so far, but you believe the patient's symptoms.
- You may want to consider giving etiological options such as overactivity in the nervous system, neuronal memory after a previous infestation, or abnormal neuron-adaptive processes in the brain.
- Try to introduce antipsychotics as the only substances helpful against these processes, as suggested by current research. You need to point out that in higher doses, these medications can be used against psychosis, but at low doses they can relieve discomfort. Patients will google the medication. In the United States, pimozide does not have a license for psychosis, which may be helpful. In the United States, amisulpride is not available (as of 2023).
- Use the names "unexplained dermopathy" or "Concern for Infestation (CI)" if a patient asks for a diagnosis.
- Introduce antipsychotics as helpful against the patient's distress and itching (there are antihistaminic components in many antipsychotics).
- Do not try to convince the patient or question the patient's beliefs. Equally, do not agree with the patient's delusion.
- Do not attempt immediate psychiatric referral or try to establish psycho-pharmacological therapy too soon.
- Try and avoid words like "delusion(al)," "psychotic," "psychological," and "psychiatric" at an early stage.
- Do not use phrases like "calm down, be happy it's not infectious, and it is only psychogenic." This can upset the patient because of the inference of a psychiatric problem which has been voiced.
- Do not simply prescribe an antipsychotic because different approaches are needed according to the type of DI.
- Do not prescribe antibiotics or any other anti-infective without a real infection further reinforcing the delusion.
- Do not overlook frank aggression against other healthcare professionals and yourself.
- Do not forget to ask patients with despair and signs of manifest depression about suicidal ideation and to evaluate any risk to others.

3   Family, Caregivers, Friends, and Associates and Their Important Roles in Patient Care

Who patients know has the potential of playing an important role in their care. It is strongly suggested that patients have someone they trust accompany them to appointments. A second supportive person, the "emotional support human," in the room with the patient and physician is very helpful. They often clarify facts, dates, events, and experiences. They may provide additional information and be calming. It also allows physicians space beyond one-to-one interactions through clarification of issues that may get overlooked. Additionally, when patients are home, emotional support humans can remind them of what was said, of information given, and of any physician instructions. They serve as information repositories when patients may not be equipped to remember what they were told, particularly if they are experiencing the five stages of grief. They also often ground patients when anxieties spiral out of control as well as monitor behaviors. Emotional support humans can be very potent allies.

Problem family members, caregivers, and associates: Families, friends, co-workers, relevant others, employers, service staff (nurses, home help, cleaning staff, etc.), unenlightened professionals, social groups, or organizations may perpetuate the idea of an infestation in the minds of patients. Often those who are emotionally invested, well-meaning, or misguided will unwittingly reinforce a patient's faulted belief systems. It can manifest as sympathetic physical symptoms (folie à deux) as a form of emotional support. It leads to chronic investment and protracted periods of suffering for patients. These associates often undermine the work of physicians and need to be identified and invited to reconsider.

When a family member, friend, colleague, or confidant are seen as possibly working against a patient's best interests whether deliberate or not, take them aside and ask:

"What do you think is going on, what are your thoughts?"

This invites the skeptic to be more participatory and, if approached tactfully, permits them to speak without judgment. This may help in the progression of diagnosis and subsequent patient care.

4   Family Members, Caregivers, and Associates Who Seek Help

"My family and I are at a loss for what to do. We know there are no bugs, but he thinks there are. He gets furious with us if we tell him there's nothing there and he gets all depressed and then more determined than ever to prove it to his doctors. Since he is blind and can't drive, my mother has to miss so many days of work to bring him to all these appointments to 'prove' them wrong, as I'm sure you can assume, nothing ever gets accomplished, and he goes home more mad and more determined to prepare for the next doctor." J. S.

The greatest struggle families, caregivers, and associates have when aware that a person they know is suffering from unexplained DI is to find help. This account illustrates the frustration and often powerlessness those with bonds to DI patients experience. Not only are they witnessing a loved one slowly self-destruct, but they also feel powerless to do anything about it. For caring family and friends, this is very distressing.

Isolation: To become isolated is part of the DI ideation. The first reason for isolation is out of concern for those the patient cares about. The patient wants to protect loved ones from their perceived parasites or pathogens. This is noble, but as they withdraw from those whom they care about, patients lose bonded connections. There are variations of this thinking, but the theme remains the same: a move to isolate in order to protect. Second, the reverse may happen. Significant others may withdraw from a patient out of concern of themselves becoming infested. Additionally, a socially aggressive or neurotic patient who is difficult and/or unpleasant to be around due to their DI behaviors may be abandoned by their significant others.

To counter this, ignore all parties' beliefs. Explain that no infestations have been found, though patients may believe that they are infested and encourage family, friends, and associates to maintain contact. Encourage the pursuit of normal social activities. This is therapy through presence. Though there will be resistance by patients and possibly relevant others, persistence is necessary. Keep patients engaged. By maintaining contact, the perceived "belligerent support" felt by patients will evolve to appreciation, even welcomed as they start to heal.

Enabling: Enabling can be inadvertent, well-meaning, deliberate, or grounded in fear. Enablers can counteract professional help, and their behavior may be very harmful if not dangerous for DI patients. They allow and support destructive patterns of behavior. Enabling is often a misguided attempt to "help." Some features of enabling are as follows:

Enablers may be codependent and promote protectionism and denial. They:

- May ignore deleterious behaviors by patients, such as social withdrawal, lying, repetitive self-mutilation, or excessive bathing and cleaning.
- May "clean up" after patients, such as paying the bills on their behalf.
- May attempt to rationalize by hiding issues or blaming others.
- Make allowances for patient's loss of responsibility and control.
- May feel resentment toward patients. This occurs when enablers continue to help patients while inwardly feeling more and more contempt for them. They may increasingly become more and more exhausted by the situation.
- May keep quiet about inner doubts for fear of confrontation and a desire to keep patients safe and happy. These actions are governed by a fear which they may not be aware of.
- May prioritize the needs of patients over their own.
- May not follow through any ultimatums or rules set by them.

Enabler antidote: If enablers are identified, seek support for them as well. Get them to understand the situation. Advise them to approach established organizations such as Al-Anon or Nar-Anon for guidance on how to appropriately support DI patients while achieving more self-autonomy without guilt. In the right context, encourage enablers to say "No" to DI patient requests. Help them establish healthy boundaries. Enablers, though they may remain supportive, need to make it clear to patients that they are no longer going to tolerate their DI-related behaviors. Enablers need to remain consistent. DI patients' behaviors are akin to addiction and will often induce guilt. DI sufferers can react vociferously to enablers' changed attitudes, so enablers need to be prepared for this.

In severe cases, particularly with self-mutilation where family, friends, or associates are unable or unwilling to assist, stage an intervention. Speak to health departments or similar entities that possess legal powers to see if there are programs in place where patients may be involuntarily assessed for suspected yet undiagnosed medical and/or psychiatric complications.

Backfire-effect: Some people possess a tendency to question their understanding of a subject if there is new evidence that does not agree with original information. This is beneficial thinking. They adjust their thinking and reconsider why they may be wrong. It is a mechanism of self-reflection and adaptation, which is useful particularly in changing social, environmental, and cultural situations. To "change one's mind" when encountering new evidence permits flexibility and development.

For others, adjusting to new evidence through consideration and a change in thinking is more challenging. They possess a tendency to reject evidence that does not align with prior information and subsequent beliefs driven often by insecurity and anxiety. They fall into the "backfire-effect." The backfire-effect is a type of cognitive dissonance that is selective and biased to support faulty beliefs and information. These are defended despite evidence to the contrary. Many patients become guarded for contradiction. They are not new to this experience and are primed for conflict. For some, they use the backfire-effect to deepen their convictions. Their emotions override their intellectual capacity.

The backfire-effect can be strong and damaging. Those who are deeply invested seek help mostly for validation. They trigger easily and are ready for contradiction, instantly doubling down. They usually cut off contact with parting shots of recrimination, blame, and a declaration they are right. They withdraw and become lost to care.

Case study of enabling and subsequent cure: In 2018, a young mother contacted one of the authors about her uncle. The uncle was legally blind and had severe rheumatoid and psoriatic arthritis. He lived with her mother and grandmother. He took high doses of morphine with a history of steroid use. For 6 years, he had complained of bugs feeding on him and had seen numerous physicians and hospital emergency departments about it.

The grandmother was the enabler:

> "My poor grandma is 82 and she refuses to give up on her son but every morning she's afraid of what she's going to find when she goes to check on him."

> "My grandmother has a co-dependency relationship going on with him and she was brain-washed to believe they may actually be real." J. S.

The uncle's behavior created intolerable stress in the family, and finally the grandmother asked J. S. to find an entomologist. Following several exchanges, they visited one of the authors.

The 2 hour interview and follow-ups resulted in the grandmother realizing her son was mentally ill. Sitting and listening to his descriptions during the interviews gave her the space to clearly see his condition. Yet at the time, she felt powerless. It did nothing for the uncle except make him angry. He began to seriously threaten suicide. Eventually, he was admitted to a psychiatric ward and detoxed off the morphine:

> "While he was in the psych unit, my grandma was able to get a break from him and really got to see that she was keeping him sick. She cleaned his bedroom and she found two large bottles of random pills and she now sees that he's a drug addict." J. S.

In addition, J. S.'s mother "flipped out on her" (the grandmother) and said that she must not be doing her job, because the bugs have been the topic of every doctor appointment he has had for the last 6 years and that he is lying to her, but all she needs to do is call his primary doctor and he would confirm that he had delusional parasitosis …:

> "… My grandma got my uncle back on the phone and BLASTED him! She told him enough was enough. Everyone is going out of their way to try to help him, and he refuses to accept the help. She told him that he's putting too much stress on everyone in the family, and he will not be allowed home unless he agrees to in patient first! That was the first time in my whole life that she ever stood firm and didn't enable him." J. S.

Following treatment, the uncle recovered.

5   Gaslighting*

Gaslighting by family and friends: Gaslighting was first brought to public notice by the 1938 thriller play "Gaslight" (changed later to "Angel Street"), written by Patrick Hamilton. The premise of the story was that a husband tried to convince his wife she was going insane in order to control her. Personal effects were moved, home gas lights dimmed, and footsteps heard. All these the wife experienced while her husband the perpetrator continually told her that it was a figment of her delusional mind.

Since then, the definition of gaslighting has broadened to include motivations other than control, e.g. indecision, denial, fatigue, well-meaning intentions, stress, fear, and the desire to avoid inconvenience. In the context of DI, it remains a form of psychological manipulation in which the "gaslighter" influences the DI patient's thinking by sowing self-doubt. The twist is that instead of an effort to try to convince a DI patient they are ill and require treatment to recover, negative responses by those associated with these patients, including some physicians, has the *opposite* effect and patients are "gaslighted" into denial.

An example of this is:

"Stop thinking bugs are on you, nothing's there, it's all in your head."

This form of gaslighting may be expressed without malice but can progress to impatience and intolerance. It also causes patients to begin to doubt themselves and increase their insecurities.

Gaslighting by patients: DI patients can themselves gaslight, particularly with new physicians or experts who do not know them. Truths may be withheld or slanted; examples include lies regarding nondisclosure of recreational drug use [8] and previous treatment attempts. Opinions may be frequent, and assertions of a strong knowledge of unusual parasite biologies and behaviors described. Invested patients are quick to test if new physicians are compliant by using earnest confident speech and aggressive body language. Previous encounters with physicians have not given them the answers they want. Patients may also enlist compliant relatives and/ or associates to support them. What they are doing is to deceive for personal advantage, even though they are the ones that need the help.

*Please note that "gaslighting" is viewed as a colloquial term by the American Psychological Association. They write:

HYPERLINK "https://dictionary.apa.org/gaslight" Gaslight

Verb. to manipulate another person into doubting his or her perceptions, experiences, or understanding of events. The term once referred to manipulation so extreme as to induce mental illness or to justify commitment of the gaslighted person to a psychiatric institution but is now used more generally. It is usually considered a colloquialism, though occasionally it is seen in clinical literature, referring, for example, to the manipulative tactics associated with antisocial personality disorder.—**gaslighted**. *Adjective.* [from *Gaslight*, a 1938 stage play and two later film adaptations (1940, 1944) in which a wife is nearly driven to insanity by the deceptions of her husband] [9].

6 Munchausen Syndrome by Proxy (MSP) and Munchausen Syndrome (MS)

Munchausen syndrome by proxy (MSP) is a psychological disorder where a person in the role of caregiver of a child, animal, older and/or infirm adult exaggerates or invents symptoms of their victim to gain attention. Behavior and actions may include exaggerated or invented symptoms, lying about symptoms, medical test altering such as urine sample contamination, medical record falsification, or induction of illness through deliberate acts of harm such as starvation, promoting infection, or direct physical acts of injury from dermal excoriation to poisoning. The

MSP patient often exhibits intense love and concern towards their victim and is cooperative with medical staff and authorities. Diagnosis of MSP is extremely difficult as well as treatment because the condition is built on lying and manipulation and often these patients have difficulties telling fact from fiction.

Munchausen syndrome (MS) is a factitious disorder where a person deliberately fakes, exaggerates, self-induces symptoms or injuries to garner attention. Behaviors are similar to MSP, but are directed at the self. Again, diagnosis is difficult because of lying and manipulation by the patient and prognosis is poor. This is due to a readiness to doctor shop/hop when not "satisfied" disrupting medical records and a strong resistance to a Munchausen syndrome diagnosis. Munchausen syndrome is often the result of undiagnosed underlying psychological issues.

7 Sampling by Patients and Professionals, and Useful Equipment

Invite patients to observe and participate in their own analyses. This allows them to be involved and active in the investigation process rather than passively waiting for results.

If possible, have available a stereomicroscope hooked up to a monitor so that patients can easily view samples. Allow them to select what they think are their most egregious specimens for viewing, and ask them to describe what they see. It often promotes curiosity and questioning, builds rapport, and reassures patients that their physicians are taking the time for a thorough diagnosis. However, patients can become upset at not seeing what they saw at home. They may make excuses such as their samples escaped in transit, they mysteriously disappeared, or they "changed." Samples that are not obvious should be sent to microbiology, entomology, or parasitology laboratories for additional analysis. For some patients, describing their symptoms and specimens as a medical mystery is beneficial.

Encourage patients to collect specimens in a more consistent manner. The following are two easy sampling techniques patients might use, which reduce cross contamination and provide clean specimens. It keeps patients busy and engaged. Do not conclude from just one sample. Accept a few samples selected by the patients, but do not allow too many (10 samples is a reasonable maximum). Know how to spot environmentally contaminated samples. Unhelpful household debris and environmental detritus that do not point toward an infestation can be problematic.

1. **Dermal sampling** (Fig. 1)
If a patient feels a pricking sensation which they describe as "biting" on the skin, they should gently tap a small piece of scotch/cellotape on the area. Lightly adhere the tape onto a piece of glass (microscope slide if provided or glass jar or cup), and label location, time, and date. Repeat 2–4 times.
Analysis: Gently peel the tape off glass, invert the tape sticky side up, and examine under a dissecting microscope.

2. **Vacuum sampling** (Fig. 2)

Place a coffee filter over the open end of a vacuum cleaner hose to cover it. Holding coffee filter firmly with one hand, turn on vacuum cleaner. Vacuum over a selected area. Debris will be caught by the filter. Turn off the vacuum to stop suction, fold the coffee filter in on itself to trap debris, put into a ziplock bag, and label with location and date. Repeat 2–3 times.

Analysis: In the laboratory, flush coffee filter samples using ethyl alcohol into a petri dish. Examine under a dissecting microscope. If there is a suspicion of mites, cover petri dish and hold for 24 h. Mites will hydrate and be easily seen. Slide mount in Hoyer's medium (oil) ventral side facing up for identification to species level (Part III, Chap. 4, Fig. 1).

Fig. 1 Scotch/sello tape sampling. Place scotch/cellotape onto pricking/biting sensation on the skin, gently lift off, place onto glass, and label location, time, and date

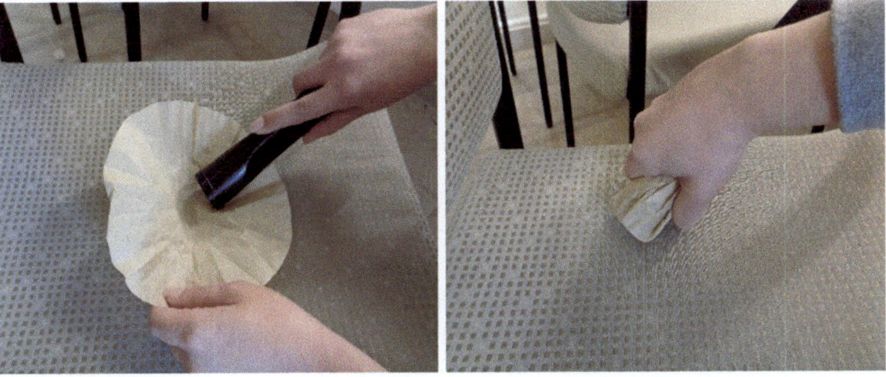

Fig. 2 Coffee filter sampling. Wrap coffee filter over the end of the vacuum cleaner hose, hold firmly, and vacuum location for specimens. When you see a disk of dirt caught by the filter, turn off vacuum cleaner, put coffee filter into a ziplock bag, seal, and label with location, time, and date

8   Information for Patients

Box 5. Scabs.
A scab is a dry rough protective crust of dried blood and serum that oozes out
of the skin when the top layer is injured. It can include skin debris.

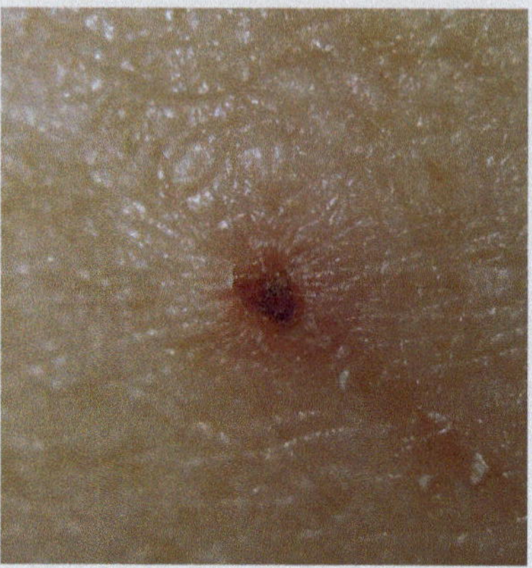

A scab is the body's response to an injury. Immediately following an injury,
special blood cells called platelets clot at the site and dry, creating a natural
bandage to exclude germs and debris. The above image shows a new injury.
The inflammation around the scab shows the skin naturally decontaminating
the wound. Many patients misunderstand what scabs are and interpret the
crystalline properties of the dried blood and sera present in scabs as eggs,
worms, or parasites.

The following section is for patients. It provides information about the skin, gastrointestinal tract, and patient instructions.

The skin and GI tract are complex organs. They often are involved in unexplained skin and/or intestinal sensations. This section helps explain how they work.

The Skin: The skin is very complex. It is the largest organ in the human body
making up 16% of overall body mass. In adults, its average weight is approximately
8 lb. (3.6 kg) with a surface area of 22 ft^2 (2.04 m^2). It is about 1/4 inch (2 mm) thick
with three layers, the epidermis, dermis, and hypodermis (subcutaneous fat layer).
It shields the body from pathogens and chemicals, sun radiation, and water. The skin
also regulates body temperature, is highly sensory with several neurological systems, and stores water, fat, and vitamin D that assists in bone calcium formation.
The skin is the body's interface with the environment. It provides information about
temperature, a spectrum of touches from soft and comfortable to itch and pain.

A relatively large region of the brain is dedicated to communication with the skin
[10, 11]. It is a wide band of cells located at the top the brain, known as the parietal lobe.

Hong Liang Tey wrote:

"Just as the eyes are the windows to our soul, the skin is a surface reflection of the inner depths of our mind. The skin and the brain are polar terminal differentiations from the same embryonic neuroectoderm, and pruritus is a symptom that demonstrates the complex yet intricate link between the two. To illustrate this point, itch can be induced simply by thinking about it [11]."

Box 6. The Itch.

Irritants on the skin trigger a fundamental human response, which in most cases is "the itch." The itch is a neurological stimulus prompting the instinct to self-protect, scratch, and groom. In 1660, Samuel Hafenreffer, a German physician, wrote a definition describing the sensations of an itch:

"An unpleasant sensation that provokes the desire to scratch [12]"

Itching is as potent as pain [13–15]. Itching is a healthy response that prompts self-grooming to remove dirt, debris, dead skin, or unwanted organisms. It originates early in human evolutionary history and is an act of self-preservation. The fingertips evolved with adapted nerve-rich pads to feel for slight skin imperfections and flattened fingernails to facilitate efficient safe removal of most irritants.

Most patients with an itch disorder pick and dig at themselves using their fingernails or a tool in response to a wide spectrum of dermal perceptions, symptoms, and/or irritants expressed as itchiness [16]. Much of this is "context dependent and that can change over time" [17]. It is human nature to "pick," and everyone does it.

The Gastrointestinal (GI) Tract: The GI tract is a long hollow tubelike organ with companion organs connected to it to assist in breaking down food for energy, growth, immune protection, and cell repair. It consists of the mouth, esophagus, stomach, small and large intestines, and anus. Companion organs are the liver, gallbladder, and pancreas.

Populating the GI tract is a microbiome consisting of nearly 40 billion bacteria and, to a lesser extent, archaea, protists, fungi, and viruses. There are up to 40 species of gut bacteria that make up to 60% of the dry mass in human feces. Their genetic diversity is 100 times more than that of the human genome. The brain is physically and biochemically connected to the GI tract principally through the vagus nerve (gut-brain axis) and gut microbe-generated chemistry. The GI tract has been described as the "second brain" [18, 19]. It is lined with a mat of highly complex important neurotransmitters, and these often determine mental health. There are over 100 million of these neurons, more than the spinal cord or the peripheral nervous system [19]. This mat of neurons controls the gut independently of the brain.

The gut microbiome manages inflammation, emotions, appetite, stress, anxiety, and depression. It is a region rich in serotonin. Up to 95% of serotonin found in the body is in the GI tract. If patients are prescribed antidepressant medications called "selective serotonin reuptake inhibitors" (SSRIs) to increase serotonin levels in the body, it can interfere with natural serotonin levels in the gut, resulting in irritable bowel syndrome. The health of serotonin levels in the gut has been linked to protection against osteoporosis and synapse formation abnormalities in autistic children. Often, these children have GI motor abnormalities and elevated gut-produced serotonin levels in their bloodstreams [20–22].

Box 7. Patient Instructions.
Instructions which can be followed to speed up a differential diagnosis and care:

1. Do not seek information on the internet.
There are strong reasons why internet research should be avoided. Modern internet searches and browsing are highly algorithm driven. Algorithms are designed to use past searches, clicks, and browsing histories to deliver a more personalized internet experience. In effect, internet searches will be biased towards what algorithms "think" users want. The more searching that is done on a particular subject, the more likely algorithms may deliver misleading, fringe, or commercially biased material.
2. Do not self-treat.
The cause of a problem must first be identified before treatment. It may be tempting to self-treat, but this can do more harm than good. Self-medicating with alternative or veterinary medications is ill advised. Self-treatment using chemicals and/or pesticides on the body, personal possessions, or living space is not advised. These can be dangerous and may cause avoidable harm.
3. Collect samples methodically to streamline the identification process.
Skin samples: <u>Do not pick</u> at or detach skin as samples. Apply a piece of clear scotch/sello tape to the affected area, remove, and stick the tape to a glass jar. Label, identifying part of the body the sample was collected from, time of day, and date.
Vacuum samples: Cover the open end of a vacuum cleaner hose with a coffee filter, and secure using a rubber band or hold in place. Vacuum surfaces of interest. Samples will be trapped by the filter. Turn off the vacuum cleaner, and put folded coffee filter into a ziplock plastic bag. Label, location, and time of sampling.
Less is more: Over-collecting and submitting many samples will greatly slow down the identification process. Limit submissions to ten or fewer samples.
4. **Moral support:** Have a family member, friend, or close associate accompany you to consultations and/or physician visits. Emotional support is important as part of the healing process.
5. **List medications, drugs, supplements, etc. currently being taken:** This assists in speeding up differential diagnosis.

Fig. 3 Specimen sign

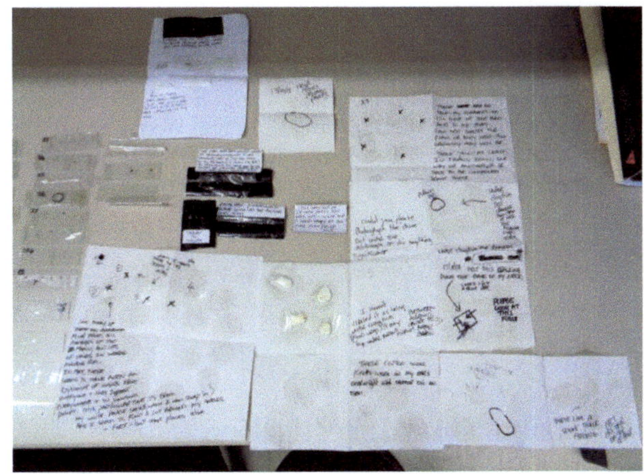

9 Specimen Sign

Specimen Sign was first described by Perrin in 1896 [23] and formalized in an editorial in the Lancet based on a report by Lyell [24]. It is a strong indicator of DI. Many patients present a high number of specimens to prove the seriousness of infestation [25]. These manifest as photographs, scotch/sello tape samples, videos, and containers of material (Fig. 3).

These are often uninvited haphazardly collected specimens. It is also not uncommon for patients in periods of anxiety to continue submitting specimens during their care period. Freudenmann and Kölle et al. [26] list materials perceived by patients as their pathogens as skin debris, dander, cloth, man-made threads, hair, dirt, dust, crusted dried blood, insect parts, flies, sand, food particles, water, and wood.

Patients may leave a medical office at ease only to return home to preexisting triggers and anxieties. The influence of an office visit lessens, and they fall back into preexisting habits. This may prompt more sampling. If this is the case, set limits on volume. If there is a high volume of specimens, ask patients to select two or three samples they are confident represent what they perceive. It avoids being buried by time-consuming specimens.

Letters: It is not uncommon for personal often handwritten letters to be included with specimens. These may provide important information. The following letter (misspellings included) illustrates a stream of thought that shows anxiety, paranoia, folie à deux, double-delusional infestation involving an animal, and arthropod behaviors that are not natural:

"To whom it may concern,

In the morning our whole body burns. I am sending these containers with a few plastic bags. With different bugs in them. They have wings. They sit in our hairs; they get through our clothes it makes holes in the cotton clothes. They bite the skin it like it eats the surface of the skin and you see the blood some I have to put alcohol on a tissues and keep sponging the blood off. It even pets in our private my husband had little bubble around his private.

Even my cat is bothered, with these bugs they try to get in his ears. These bugs were inplant outside my house. They crack the vinel Siding to put them in. We are constantly itching in are hair and body. Theirs constant movement in my hair. I put Vicks at night in my hare and eyebrows my husband to. It drives us insane. The seterpee occasionally will fall on us like it comes from the air, its hard to explain. We are elderly my husband has Demechia and prostate and he is 89 years old. And me I am 78 years old I have lime disease. We need someone to solve this. It is so strange. Exterminator couldn't identify the bugs. Thank you

Sincerely,

Anonymous."

10 Laboratory Investigation

Perform a full panel series of laboratory tests and in some cases a brain scan to look for underlying medical condition(s). Following is a list of additional laboratory tests specific for DI patients suggested by Bury and Bostwick [27] and Heller et al. [28] (Table 2).

Consider a general psychiatric assessment if the patient agrees to be tested and ask for:

Anxiety: Use the GAD-7 screening scale.

Depression: Use the PHQ-9 screening scale.

Obsessive compulsive disorder (OCD): Use the Yale-Brown Obsessive-Compulsive screening scale [29].

Table 2 Additional laboratory tests for DI patients

Complete blood count (CBC)	Serum calcium
Age-appropriate cancer screening	Serum creatinine
Albumin	Serum creatinine
Antinuclear antibody	Serum electrolytes
Blood urea nitrogen	Serum glucose
C-reactive protein	Serum Ig E (immunoglobulin E)
Erythrocyte sedimentation rate	Thyroid function tests
Folate	Total protein
Foliate	Urinalysis
Hepatitis C	Urine toxicology (look for drug use)
HIV testing	Vitamin B12 (Hispanics, particularly those with Caribbean heritage)
Iron studies	Brain scan including limbic system
Liver function tests	Rapid plasma regain test for syphilis
Phosphorus	Rheumatoid factor
Pregnancy test if of childbearing age	

A full examination of skin disorders can reveal issues [30]. ELISA testing for antibodies may reveal hidden issues.

Drugs and drug side effects: Ask about medications and other ingested products. Illegal drug use in DI patients particularly in younger men is twice as common compared to the general population [30, 31]. Lepping et al. [32] suggest that as many as 60% of DI patients may be on or have taken illegal drugs. Often, motivation for taking illicit drugs is to relieve distress. Stimulants can also be major triggers for DI [33, 34]. Additionally, consider over-the-counter medications (OTCs), internet-advised self-treatments, and online chat rooms. Ask about the use of veterinary medications if patients own pets that may include Ivermectin or Gabapentin. Following is an example of a veterinary medication overdose where a patient purchased Albendazole online [35]:

> "The patient was admitted for a syndrome consisting of alopecia, pancytopenia, and elevated transaminases. While hospitalized, he developed neutropenic fever necessitating broad-spectrum antibacterials which precipitated Clostridioides difficile colitis. The authors analyzed serial blood and urine specimens by liquid chromatography–mass spectrometry to document supratherapeutic concentrations of albendazole and its active metabolite [35]."

Many DI patients present with an extensive list of prescribed medications. The combination of elevated susceptibility to DI thinking driven by possible drug side effects should be considered. The list of medications which cause dermatologic side effects is considerable. Common side effects are pruritus, paresthesia, urticaria, rash/redness/red spots, and crawling sensations. Other side effects include dry skin, flushed skin, blisters, welts, tingling, pins and needles, numbness, psoriasis, peeling skin, and exanthema. Consider drug screening and/or proxy screening of pet medications if there is a suspicion of use.

There are numerous articles that include reviews and case reports on drug and chemical-induced dermatitis. Fifty of the most prescribed drugs in the United States are listed below. The source for Table 3 is Fuentes et al. [36], with a comparative list of data by Clincalc [37], and Drugreport [38] in Table 4. Side effects' information was obtained from the Mayo Clinic.

Table 3 Top 50 prescribed drugs in the United States, pharmacological action, and dermal side effects [36].

	Fuentes et al. [51]	Drug class	Pharmacological action	Paresthesia	Pruritus	Urticaria	Rash/redness/red spots	Crawling
1	Lisinopril	ACEi	Antihypertensive		X		X	
2	Levothyroxine	Hormones	Hypothyroidism	X	X	X	X	
3	Atorvastatin	HMG-CoA reductase inhibitor	Antihyperlipidemic		X	X	X	
4	Metformin	Biguanide	Antidiabetic				X	
5	Simvastatin	HMG-CoA reductase inhibitor	Antihyperlipidemic					
6	Omeprazole	PPI	Anti-GERD	X	X		X	
7	Amlodipine besylate	CCB	Antihypertensive	X	X	X		X
8	Metoprolol	Beta-blocker	Antihypertensive		X		X	
9	Acetaminophen; hydrocodone	Opioid	Analgesic					
10	Albuterol	Beta-2 agonist	Bronchodilator		X	X	X	
11	Hydrochlorothiazide	Diuretic	Antihypertensive		X	X	X	
12	Losartan	ARB	Antihypertensive	X	X		X	X
13	Gabapentin	Anticonvulsant	Anticonvulsant				X	
14	Sertraline	SSRI	Antidepressant		X	X	X	
15	Furosemide	Loop diuretic	Antihypertensive					
16	Acetaminophen	Analgesic	Analgesic		X	X	X	
17	Atenolol	Beta-blocker	Antihypertensive					
18	Pravastatin	HMG-CoA reductase inhibitor	Antihyperlipidemic					
19	Amoxicillin	Antibiotic	Antibiotic		X	X	X	
20	Fluoxetine	SSRI	Antidepressant		X	X	X	
21	Citalopram	SSRI	Antidepressant	X	X			
22	Trazodone	SSRI	Antidepressant	X	X		X	X
23	Alprazolam	Benzodiazepine	Antianxiety	X	X		X	X
24	Fluticasone	Steroid	Corticosteroid					

	Fuentes et al. [51]	Drug class	Pharmacological action	Paresthesia	Pruritus	Urticaria	Rash/redness/red spots	Crawling
25	Bupropion	Dopamine/norepinephrine reuptake inhibitor	Antidepressant		X	X	X	X
26	Carvedilol	Beta-blocker	Antihypertensive		X		X	
27	Potassium chloride	Electrolytes	Electrolyte supplement					
28	Tramadol	Opioid	Analgesic		X	X	X	
29	Pantoprazole	PPI	Anti-GERD					
30	Montelukast	Leukotriene receptor antagonist	Antagonist					
31	Escitalopram	SSRI	Antidepressant					
32	Prednisone	Corticosteroid	Anti-inflammatory					
33	Rosuvastatin	HMG-CoA reductase inhibitor	Antihyperlipidemic					
34	Ibuprofen	NSAID	Analgesic		X	X	X	
35	Meloxicam	NSAID	Analgesic			X	X	
36	Insulin glargine	Long-acting insulin	Antidiabetic		X		X	
37	Hydrochlorothiazide and lisinopril	ACEi/diuretic	Antihypertensive	X	X	X	X	X
38	Clonazepam	Benzodiazepine	Anticonvulsant		X		X	
39	Aspirin	Salicylate	Antiplatelet				X	
40	Clopidogrel	Antiplatelet medication	Antiplatelet		X	X	X	
41	Glipizide	Sulfonylurea	Antidiabetic	X	X		X	X
42	Warfarin	Blood thinner	Anticoagulant					
43	Cyclobenzaprine	Muscle relaxant	Analgesic		X	X	X	
44	Insulin human	Hormone	Antidiabetic		X	X	X	
45	Tamsulosin	Alpha-10-antagonist	Urinary retention			X	X	
46	Zolpidem	Sedative-hypnotics	Hypnotic	X	X			X

(continued)

Table 3 (continued)

	Fuentes et al. [51]	Drug class	Pharmacological action	Paresthesia	Pruritus	Urticaria	Rash/redness/red spots	Crawling
47	Ethinyl estradiol/ norgestimate	Hormonal contraceptive	Contraceptive					
48	Duloxetine	SNRI	Antidepressant		X	X	X	
49	Ranitidine	Histamine H2 antagonist	Antiulcerant	X	X	X	X	
50	Venlafaxine	SNRI	Antidepressant	X	X		X	

Table 4 Comparative order of the top 50 prescribed drugs in the United States [37, 38]

	Clincalc (2019)	Drugreport (2020)
1	Atorvastatin	Lisinopril
2	Levothyroxine	Atorvastatin
3	Lisinopril	Levothyroxine
4	Metformin	Metformin
5	Metoprolol	Amlodipine
6	Amlodipine	Metoprolol
7	Albuterol	Omeprazole
8	Omeprazole	Simvastatin
9	Losartan	Losartan
10	Gabapentin	Albuterol
11	Hydrochlorothiazide	Gabapentin
12	Sertraline	Hydrochlorothiazide
13	Simvastatin	Acetaminophen/hydrocodone
14	Montelukast	Sertraline
15	Acetaminophen; hydrocodone	Fluticasone
16	Pantoprazole	Montelukast
17	Furosemide	Furosemide
18	Fluticasone	Amoxicillin
19	Escitalopram	Pantoprazole
20	Fluoxetine	Escitalopram
21	Rosuvastatin	Alprazolam
22	Bupropion	Prednisone
23	Amoxicillin	Bupropion
24	Dextroamphetamine; dextroamphetamine saccharate; amphetamine; amphetamine aspartate	Pravastatin
25	Trazodone	Acetaminophen
26	Duloxetine	Citalopram
27	Prednisone	Amphetamine
28	Tamsulosin	Ibuprofen
29	Ibuprofen	Carvedilol
30	Citalopram	Trazodone
31	Meloxicam	Fluoxetine
32	Pravastatin	Tramadol
33	Carvedilol	Insulin glargine
34	Potassium chloride	Clonazepam
35	Tramadol	Tamsulosin
36	Clopidogrel	Atenolol
37	Insulin glargine	Potassium chloride
38	Aspirin	Meloxicam
39	Atenolol	Rosuvastatin
40	Venlafaxine	Clopidogrel
41	Alprazolam	Propranolol
42	Ethinyl estradiol; norethindrone	Aspirin

(continued)

Table 4 (continued)

	Clincalc (2019)	Drugreport (2020)
43	Allopurinol	Cyclobenzaprine
44	Hydrochlorothiazide; lisinopril	Hydrochlorothiazide/lisinopril
45	Cyclobenzaprine	Glipizide
46	Clonazepam	Duloxetine
47	Zolpidem	Methylphenidate
48	Azithromycin	Ranitidine
49	Oxycodone	Venlafaxine
50	Warfarin	Zolpidem

11   The Aspect of Time with Patient Care

How much time should be reserved for suspected or confirmed DI cases? Patients often present bizarre narratives and nonroutine symptoms, which impede speedy differential diagnoses. The guideline prepared by the British Association of Dermatologist (BAD), clinical standards unit, suggests allowing 45 min for initial appointments with 30 min for subsequent appointments [39]. They also suggest offering:

> "Treatment for at least 1 year after symptoms have resolved to adults with DI and restart treatment if symptoms recur following cessation of treatment [40]."

Though DI cases are time consuming, it is important in some cases to set time limits. Depending on each specific case, one of three options might be used to manage time:

1. State at the beginning of an interview there is another appointment following. If time runs out, a subsequent appointment will be made.
2. If an interview starts without a preset time limit, then once the physician feels the interview has become unproductive, say that another patient is expected for an appointment and end the visit.
3. Pre-arrange a time for a staff member to send a "page" or pretend to be "paged" to end an interview.

Setting time limits helps patients focus. Being aware of a limited time, patients will generally describe their experiences more succinctly with less "waffle." In other words, it provides a time discipline, which is helpful to both the physician and the patient.

12   Delusional Infestation: The Five Stages of Grief

Anger, denial, depression, and bargaining are experienced by many DI patients during their illness. Time plays a major role and is underappreciated for its potency. The longer patients remain attached to their beliefs, the more they will move between anger, denial, depression, and bargaining. It becomes a vicious cycle.

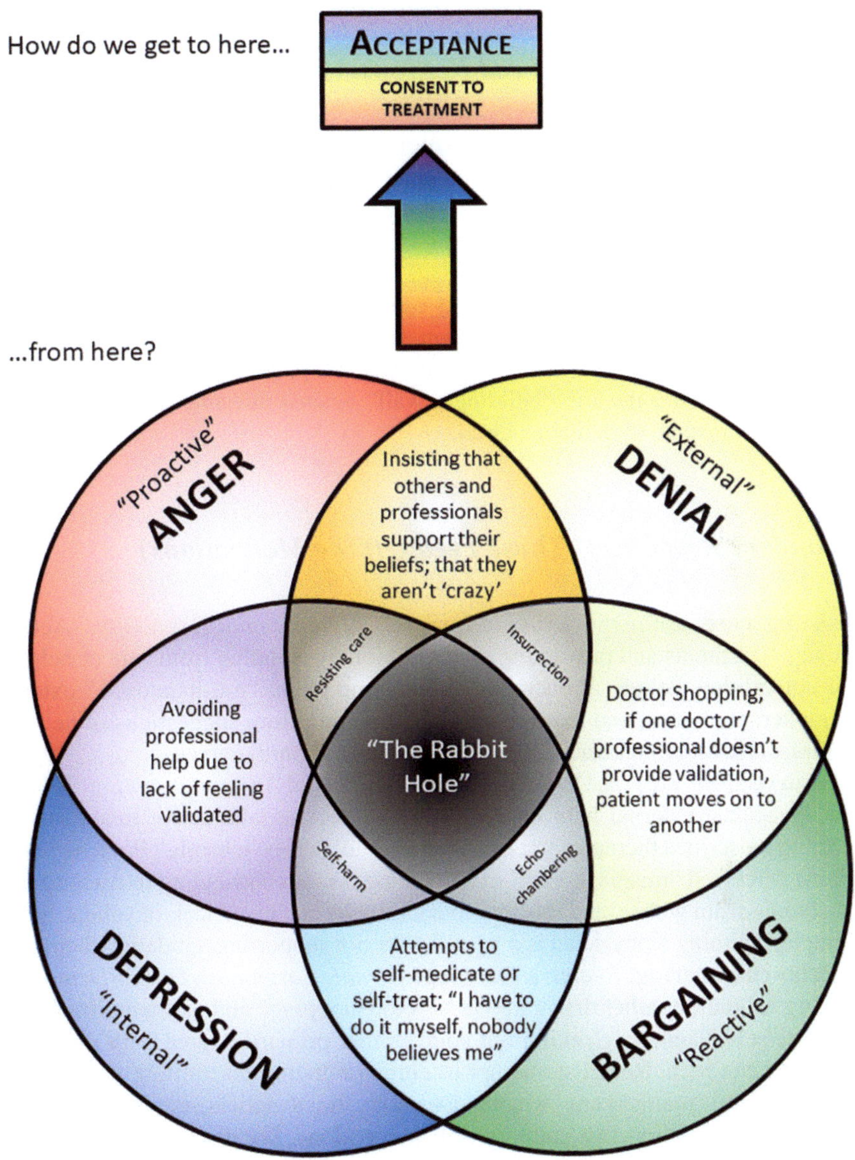

Fig. 4 Acceptance and consent to treatment. (Design K. Dugas)

Kübler-Ross identified the "state of loss" as the five stages of grief and acceptance (Fig. 4). For long-term, deeply emotionally invested patients, acceptance is nearly impossible. Patients become trapped in a cycle of thinking that becomes increasingly difficult to escape. The Venn diagram and companion explanations illustrate the complexity of actions and reactions which DI patients experience.

Explanation of the DI Venn diagram, Fig. 4.

By Katherine Dugas.

13   Explaination of the Venn Diagram

Primary Motives (These Feed Responses)

**"Anger"/Proactive*: "acting in anticipation of future problems, needs, or changes." Taking direct action against a perceived situation—opposite to bargaining/reactive.

"Bargaining"/Reactive: "readily responsive to a stimulus; occurring as a result of stress or emotional upset." Countering in response to external events—opposite to anger/proactive.

"Denial"/External: "to attribute to causes outside the self." Pushing away, "deporting" unwanted stimuli—polar opposite to depression/internal.

"Depression"/Internal: "to incorporate (values, patterns of culture, etc.) within the self." Retreating inward, "building a wall" against unwanted stimuli, "head in the sand"—opposite to denial/external.

Secondary Responses (These Feed Off the Motivations)

Resisting care: *Proactive/External/Internal; Anger/Denial/Depression*. Actively refusing treatments and preventing rehabilitation by isolating from care or help that will not validate beliefs. Proactively anticipating being "misdiagnosed," externalizing alternate issues, and internalizing by retreating from suggestions of help. For example, refusing to take/accept prescribed medication/therapies.

Insurrection: *External/Proactive/Reactive; Denial/Anger/Bargaining*. Working directly against external help or treatments by seeking alternative treatments and rejecting prescribed therapies, "taking matters into their own hands." Proactive attack against perceived "misdiagnoses," externalizing issues contributing to discomfort as not coming from within, and reacting to continuous perceived lack of validation. For example, claiming a physician is a "quack" for not supporting/validating them.

Echo-chambering: *Reactive/External/Internal; Bargaining/Denial/Depression*. Seeking validation rather than care via "doctor shopping" and/or attempting to self-medicate/self-diagnose. Seeking out sources that primarily agree with rather than those that can help. Reactive attempt to cure continuing discomfort, externalizing by refusing the "mainstream" when it does not validate beliefs, and internalizing by seeking any validation that supports self-perceptions. For example, using internet forums or social media groups dedicated to alternative therapies that are not peer reviewed/scientifically supported.

Self-harm: *Internal/Reactive/Proactive; Depression/Bargaining/Anger*. Attempts to isolate from external care by self-medicating, self-treating, or self-diagnosing resulting in self-harm.

Internalizing perceived issues, reacting to discomfort by trying to find independent solutions, and proactively attempting to force validation without physician input. For example, self-mutilation by attempting to extract perceived parasites. Applying treatments off-label leading to accidental poisoning or other forms of severe harm.

*Definitions are based on the Merriam-Webster dictionary [41].

14   Rumination and Other Responses and Conditions

Rumination: This is a process of continuous thinking about the same thoughts. These are often driven by two predispositions, anxiety and worry. Thoughts and duration of thinking are variable from casual brief consideration to chronic all-consuming thinking that uses considerable periods of time. Rumination has at least five levels from mild to psychotic and is prevalent in DI. During 2020, in the early days of the SARS-CoV-2 pandemic, there were rumors that the virus was spread by touching grocery bags. This scenario is used to illustrate degrees of response DI patients experience as they navigate uncertainty.

1. **Mild**: The "just in case" scenario. This may be mild superstition or a response to an unsubstantiated warning. Some people avoided touching grocery bags "just in case" due to the rumors, often even after they were refuted. Yet if grocery bags were accidentally touched by them, they were not overly concerned.
2. **Anxiety:** Those with anxiety do a great deal of self-talk. It is a primary psychiatric disorder and can either be episodic or present most of the time as a generalized anxiety disorder [39]. Patients worry and are predisposed to looking for something wrong. Perception of contaminated grocery bags is more intense. Overreaction and constant thinking about coming home with infected bags prompt a cascade of avoidance behaviors. Rumination is a well-established part of the anxiety.
3. **Obsession:** This is when often negative thinking continually preoccupies and intrudes on quality of life. Rumination over "contaminated grocery bags" is consuming. They will wear protective clothing and gloves and frequently wash. Obsessives need to control to feel safe. Thus, patients will look for evidence to support beliefs. In a sense, patients become lost in their obsessive anxiety and do not know what or whom to trust. Their experience is that "nothing bad has happened so what they are doing must be working." These DI patients obsessively ruminate to hold onto a level of control in their lives. When seeking help, it sets up an inner conflict where they had found a sense of safety in a self-created world, yet life has become intolerable. To surrender their belief system is terrifying. Patients want to be heard, but they are scared and so retreat to a safe place inside their ruminations.
4. **Depression**: This is a primary and common disorder. One in six people experience depression. Symptoms include loss of sleep, energy, appetite, weight, libido, concentration, memory, and psychomotor agitation. More severe depression includes hopelessness, worthlessness, and suicidal ideation. A person is considered depressed if symptoms persist for at least 2 weeks [39]. Depression is present with DI patients.
5. **Psychosis**: This is when a person has lost contact with reality. Rumination is deeply internalized, and perception distorted. Understanding what is real and what is not is difficult. Often, there are psychiatric generated delusions and hallucinations, inappropriate behavior, and sometimes nonsensical speech. Normal quality-of-life activities become impossible. Perceived contaminated grocery bags would be impossible to tolerate.

Crying: Allow patients to cry. Many patients are scared, depressed, and isolated, and the simple act of comfort using touch and giving time for a "good cry" is cathartic. It shows that the physician cares and can be an indication that the patient may be moving away from being guarded to more openness and trust. Patients who have suffered from DI for a short period often transition more easily to trust through involuntary crying, while those with long-term investment may remain more rigidly faithful to their beliefs. Tears of anger are not uncommon with patients. Crying from pain requires intervention.

Disgust: Disgust is seldom identified by patients, but it is present. It serves as a function of evolved pathogen avoidance by humans. Several instinctive responses are nausea, revulsion, gagging, and a strong desire to withdraw from a perceived pathogen. These are innate defense mechanisms to reduce the risk of being infested [42]. Itching can be incorporated into a disgust response resulting in scratching as well as self-hugging. Self-hugging results in reduced access by perceived ectoparasites as well as providing emotional comfort.

Anger: Patients can become angry when a physician's findings do not agree with their beliefs and expectations. The wider the gap between the two, the angrier a patient might become. The physician is doing their job and so should not own the patient's anger. Often, the anger though directed at the physician is in reality underlying frustration. Anger that evolves into aggression is different. If any patients are predisposed to anger, it is important to work in teams of two. This often averts angry encounters.

15   Delusional Infestation: Aspects from a Psychodermatological Perspective

We all tend to try and predict outcomes and test them against reality, and then we align our predictions with reality. In DI patients, the response is different. DI patients consider unlikely explanations to be likely ones leading to delusional formation. It is not uncommon to find people with a type of sensation that leads them to believe they have an infestation. From a psychiatric point of view, there are mono delusional disorders that do not possess some sort of somatic component [40] and where somatosensory pathways are not in the same way engaged. Additionally, there is evidence that certain neuron pathway signatures might play a role particularly in hyperactive dopaminergic network pathways [43].

It has been suggested that diagnosis should be "exclusive," which means that the disorder cannot be better explained through substance abuse or use of medicinal drugs, mental comorbidities, or undiagnosed somatic disorders [44]. It puts a delusional diagnosis squarely in the camp of a psychiatric diagnosis. In many cases, it is not so.

Self-inflicted injury: Dermatologists want to see primary lesions, prior to being altered by kicking, scratching, or rubbing. This is because secondary lesions can mask underlying skin disorders. It is not uncommon for patients with secondary lesions to deny scratching or picking and declare that these are part of the infestation. On rare occasions, pathological skin picking is life threatening [45–47].

Clusters of three or four parallel lines, all the same length, suggest active scratching. Look for blood under fingernails. Additionally, check to see if the middle of the back between the shoulder blades is clear. Most patients cannot reach this area.

Decision flowchart: As the interview(s) progresses and information about patient experiences and symptoms is gathered, certain decision points and parameters may be applied. It is noticeable that many cases, though different, possess common themes. Bifurcated decisions are possible based on this information, which may lead to quicker differential diagnoses. The following flow chart might be used to assist in differential diagnosis (Fig. 5).

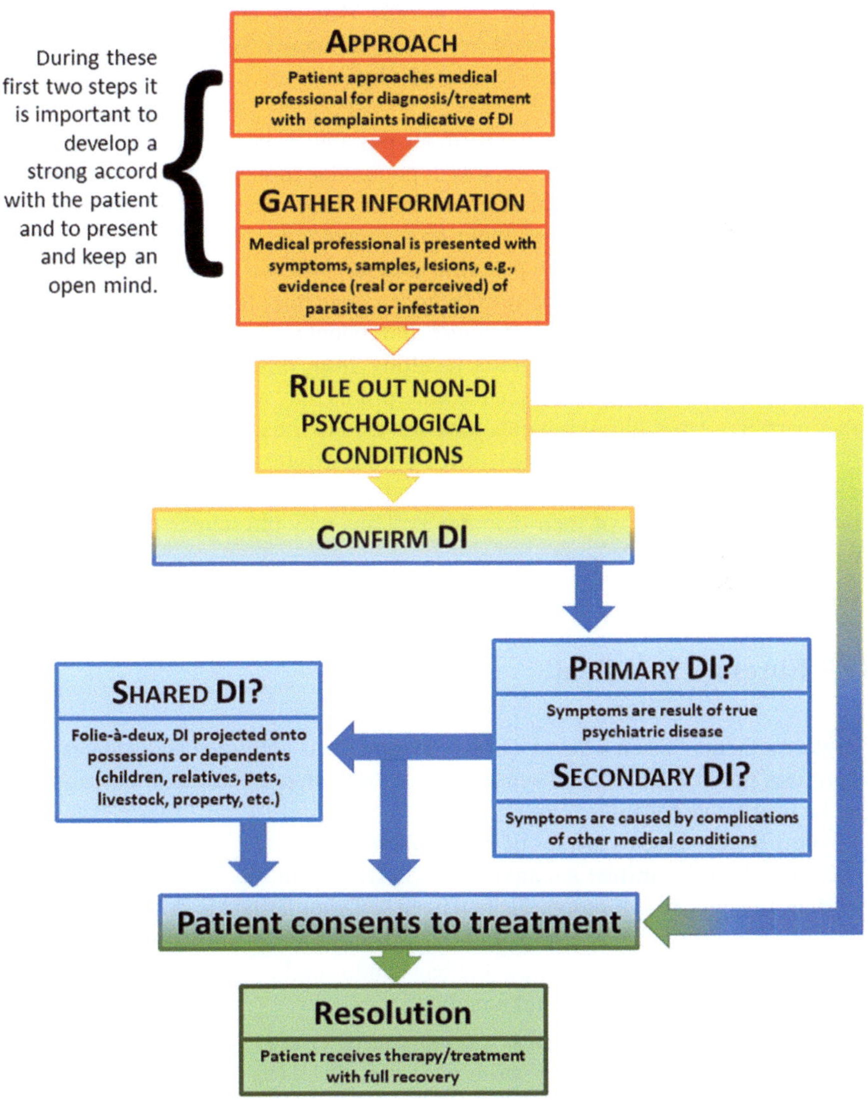

Fig. 5 Physician differential diagnosis flow chart. (Design K. Dugas)

16   International Travel

Illness following international travel particularly in the tropics is possible [48]. Some tropical pathogens and parasites cause neuropsychiatric disturbance and function loss in the brain cerebral system. This affects control for anxiety, fear, and personality. The most important of these are malaria *Plasmodium* spp., neurocysticercosis, schistosomiasis *Schistosoma* spp. (Class: Trematoda), trypanosomiasis (Family: Trypanosomatidae) (African sleeping sickness and American Chagas disease), and Dengue fever. These can cause "profound changes in the nervous system functions" and should be considered when questioning DI patients about their travel histories [49](see chapter "When Delusional Patients are Really Infested: The Exception to the Rule," by Richard Pollack, and chapter "The Role of the Clinical Parasitology Laboratory in Delusional Infestation," by Bobbi Pritt and Blaine Mathison).

Additionally, always ask patients about possible exposure to common human parasites such as ticks, mosquitoes, Ceratopogonidae (no-see-ums), lice, scabies, or bed bugs. Also consider parasitism either accidental or deliberate by parasites such as Tumbo or Mango flies *Cordylobia anthropophaga*, cheese skippers (*Piophila casei* (L.) found in decomposed pecorino cheese), bird and rodent mites, black flies (transmission of the parasitic worm *Onchocerca volvulus* which causes river blindness), human bot fly *Dermatobia hominis*, and rodent and rabbit bot flies (*Cuterebra* spp., Family: Oestridae). Dermatitis resulting from physical contact with exotic arthropods with urticating hairs such as caterpillars may also be possible. Check for other parasites especially if patients have traveled to the tropics.

Some patients identify foreign travel as the origin of their parasitism. They believe they picked up a previously "unknown organism or disease." These patients conclude that the rarity of their disorder explains why physicians cannot find their "mysterious creatures."

17   Eureka - It Is Real!

Eureka moments - when a direct cause of symptoms is identified and addressed - are liberating, provide a profound sense of relief, revitalize treatment momentum, and can be instructive for future cases.

Following are four examples of eureka moments:

Case 1: Scabies mites: An anxious OCD DI patient was being treated with risperidone to alleviate symptoms. He had been self-mutilating his skin using a needle, carefully putting what he claimed to be mites into tiny scientific ziplock bags. His physician then decided to examine the bags and found to his surprise a single scabies mite in each. His doctor exclaimed:

> "This guy was not only delusional he was a little bit OCD, and he was able to dig out with a needle each mite and put it individually into a tiny, labeled bag!"

Following appropriate scabies treatment and a period of risperidone treatment to alleviate the DI ideation and anxiety, the patient was cured.

Case 2: *Microsporum canis*: Following is a description written by Wang and Chen [50]:

"A 6-year-old boy presented to the dermatology clinic with a 3-month history of rash on his face and scalp, with associated hair loss. On physical examination, he had multiple scaly, erythematous plaques on his scalp, neck, and face as well as patches of alopecia and bilateral post auricular lymphadenopathy. Areas of bright green fluorescence were observed on his scalp on Wood's lamp examination. After preparation of a hair sample with potassium hydroxide, microscopic examination showed hyphae and spores on the outside of hair shafts, indicating an ectothrix infection. *Microsporum canis* was identified on molecular testing of scalp scrapings, and direct microscopic evaluation of fungal cultures showed septate hyphae and macroconidia (Panel B). *M. canis* is a zoophilic fungus that is often acquired from domestic animals. The patient's parents reported no pets at home but mentioned that the patient may have had exposure to animals while staying in the countryside several months earlier. Treatment with oral and topical terbinafine was initiated, and after 6 weeks of therapy, the skin lesions had resolved. The parents were contacted 3 and 6 months after the completion of therapy, and they confirmed that there had been no recurrence of the rash."

Case 3: Vitamin B12 deficiency: The following case report highlights an often-overlooked consideration of ethno-specific differential diagnoses with predisposition to vitamin deficiency [5]. Patient, Ms. A, is Hispanic. Hispanics over 60 years of age with Caribbean ancestry often have lower levels of B12 compared to other Hispanic populations. Ms. A developed DI and had used scissors and bleach to "clean her body of pests." Following is part of a letter written by her attending physicians:

"A 60-year-old Hispanic woman with a past medical history of irritable bowel syndrome (IBS), hypothyroidism, migraines, and a 22-year psychiatric history of depression and anxiety who was initially admitted for depression and suicidal ideation because of unremitting delusional parasitosis. Prior pharmacotherapy with upward titrations of mirtazapine and fluphenazine, followed-by risperidone and venlafaxine, were all unsuccessful at reducing her symptoms of delusional parasitosis. Initial intake labs revealed cobalamin (B-12) deficiency without macrocytic anemia. Four days after intramuscular (IM) cobalamin injections, the patient's sensation of formication nearly completely resolved. Symptoms of delusional parasitosis were almost completely resolved at one-month follow-up" [5].

The authors suggest that B12 should be ruled out in a differential diagnosis in patients with possible formication delusions <u>before</u> consideration of antipsychotic therapy.

Box 8. List of Actions When Working with DI Patients.
Bifurcated differential diagnosis: psychiatric and somatic:

1. Establish time limits for appointments.
2. Do not judge patients. Be kind, and give them time. Their reality is not your reality.
3. Request someone patients trust to accompany them during appointments.
4. Ask when sensations began, and allow patients to explain their circumstances and experiences. Listen for repeated use of red-flag pronouns "they" and "them." Take note of described parasite behaviors and biologies to compare against actual known parasites.

5. Order complete blood panel tests including additional tests listed in Table 1, as well as other tests judged appropriate.
6. Limit patient sampling to a few patient-selected samples.

Treatment

1. If not dealing with psychosis, address patient emotional distress with short-term small, measured doses of SGAs or similar drugs as bridge medications if needed (for more information, see Part III, Chap. 3 and Table 5).
2. Treat diagnosed somatic disease.
3. Use ESHs* to assist with emotional support, monitoring, and home care.
4. Follow up with patients and ESHs for at least 18 months to limit recidivism.

*Emotional support humans (see glossary)

Case 4: Shared DI following travel: On rare occasions, travel can trigger shared DI. In one case, five family members were triggered by a gift of yak wool sweaters (jumpers) brought back from Nepal by one of the daughters. The mother developed pruritus, then the father, and younger brother. A grandmother and cousin who lived in separate cities also developed pruritus. An older brother and the daughter were not affected. All tests of the family members and their clothing were negative for parasites. It took over 18 months for the delusion to subside through medical supervision. This was an example of folie à cinq [51].

Physicians and providers such as those who work in travel, tropical medicine clinics, and infectious disease departments should be alert to conditions and experiences that might trigger DI onset in travelers. A greater awareness provides fewer unnecessary investigations, timely treatment, and "improved clinical outcomes" [52].

18 Delusional Infestation and Veterinarians

Of the six forms of DI mentioned in this book, the two forms DI by proxy (DIP) and double DI (DDI) may be most difficult for veterinarians to manage. To review, DIP is when patients project their delusions onto animals, children, dependents, or objects. The recipients of the delusions do not or cannot share the delusions. DDI is when patients share their delusions with animals and/or children, or dependents who again do not or cannot share the delusions. Additionally, animals are particularly vulnerable, because they cannot speak for themselves, give consent, or ask for help.

There can be significant dangers when working with DI clients and animals. In their attempts to get rid of alleged infestations, it is not uncommon for clients to use pesticides and/or strong chemicals on their animals. Since animals cannot speak for themselves, the veterinarian's knowledge of the situation is totally dependent on what clients reveal. It is important not to dismiss or ignore clients' claims and thoroughly investigate. As dictated by the species that has been brought in and alleged pathogen(s), perform appropriate diagnostic tests including toxicological analyses. Also thoroughly interview the client.

Box 9. How to Determine Whether the Animal Owner Believes the Animal and Others Are Infested. Suggested Questions to ask.

1. *"Do you feel you are infested, too?"*
If they respond "yes," you can ask to see the lesions. If they show you lesions, you can ask "Have you consulted a physician or a dermatologist about these?"

(a) If they answer "no," this allows you to suggest a consultation between you and their physician to help explain the animal's "problem."
(b) If they answer "yes," you can ask what was determined by the physician to help you diagnose their animal's "problem."

If they do not show you lesions, look for evidence of self-trauma. If they respond "no," proceed to the following questions.
2. If the client has children, ask: *"Do you think your child(ren) are infested?"*
3. *"Are there other people close to you who think they are infested or whom you think are infested* (although physicians failed to find pathogens)?"

How to determine whether the animal owner likely suffers from DI by proxy (DIP) or double delusion (DDI). Suggested questions:

1.   *"Have you seen any other veterinarians about this problem with your animal?"*
If they respond "yes," you can ask what the outcomes of those visits were.
2.   *"Have you attempted any treatment of the condition, or has medication been prescribed by a previous veterinarian?"*
3.   *"Do you have any samples of the parasitic organism which can be analyzed or examined?"*
4.   *"Does your animal have any lesions (sores) you attribute to a parasitic infestation?"*

Veterinarian interviewing: Adapt the interview techniques for human patients to the veterinary practice. Always acknowledge what patients are reporting about their animal(s) and/or themselves. Agree to disagree, with the understanding that both the veterinarian and client may arrive at a point where they disagree yet can respect the other's points of view. Avoid validating the client's faulty beliefs about their animal's health by offering to prescribe either a placebo or a parasitic medication when there is no evidence of parasitism. This could reinforce a client's belief system and lead to a behavior pattern of expecting subsequent treatments in the absence of proof. Only treat what is found; if nothing is found, *do not treat.* Finally, being blunt with clients is not helpful. For example, do not say,

"The pet has nothing, settle down, it's all in your head/psychological/your fault" or *"You are wrong. I know it better than you do, I am the professional/doctor."* [53]

This will immediately sabotage any chances of protecting the safety of an animal since unhappy alienated clients will leave and likely not return. This may put animals at greater risk. If there is a conclusion that a client is suffering from either DIP or DDI, it may be necessary to use some subterfuge. Depending on the level of trust, ask to speak with the client's physician and/or family members under the pretext of alerting the physician and/or family members about personal concerns regarding the animal's "infestations" and general health. State that it would be easier to do this in person. If the client agrees, then a confidential conversation with a physician explaining the situation may open opportunities for more directed future evaluations of the client by a physician.

If you conclude there is an immediate, elevated risk to the welfare of an animal and/or children/dependent(s) and/or the client themselves, <u>do not hesitate to intervene</u>. <u>Do not "second guess" the situation.</u> Move quickly, and contact appropriate agencies, authorities, or departments who have qualified trained staff with skills for help and support.

Animals in high risk of harm: Animals that are most at risk of harm are dogs. Dogs have evolved through domestication certain behaviors to understand and interact with humans. They are acutely attentive to subtle cues and can read human emotions, gestures, and actions fluently. They acquire many skills over time as they live with their caregivers. It is not uncommon for attentive dogs to reflect behaviors of their owners, which can be particularly troublesome with DI patients. These owners perceive their pets are experiencing the same symptoms as they are when in fact the animals are simply responding to unconsciously given cues by their owners. The dogs are not sick, but their mirroring behavior may confirm in the minds of their owners a shared experience and so treatment by owners, abuse, and numerous veterinarian visits including requests to euthanize the animals are not uncommon [54]. Following is a flow chart for veterinarians suggesting decision points when working with DIP and/or DDI clients and how to appropriately respond [53] (Fig. 6).

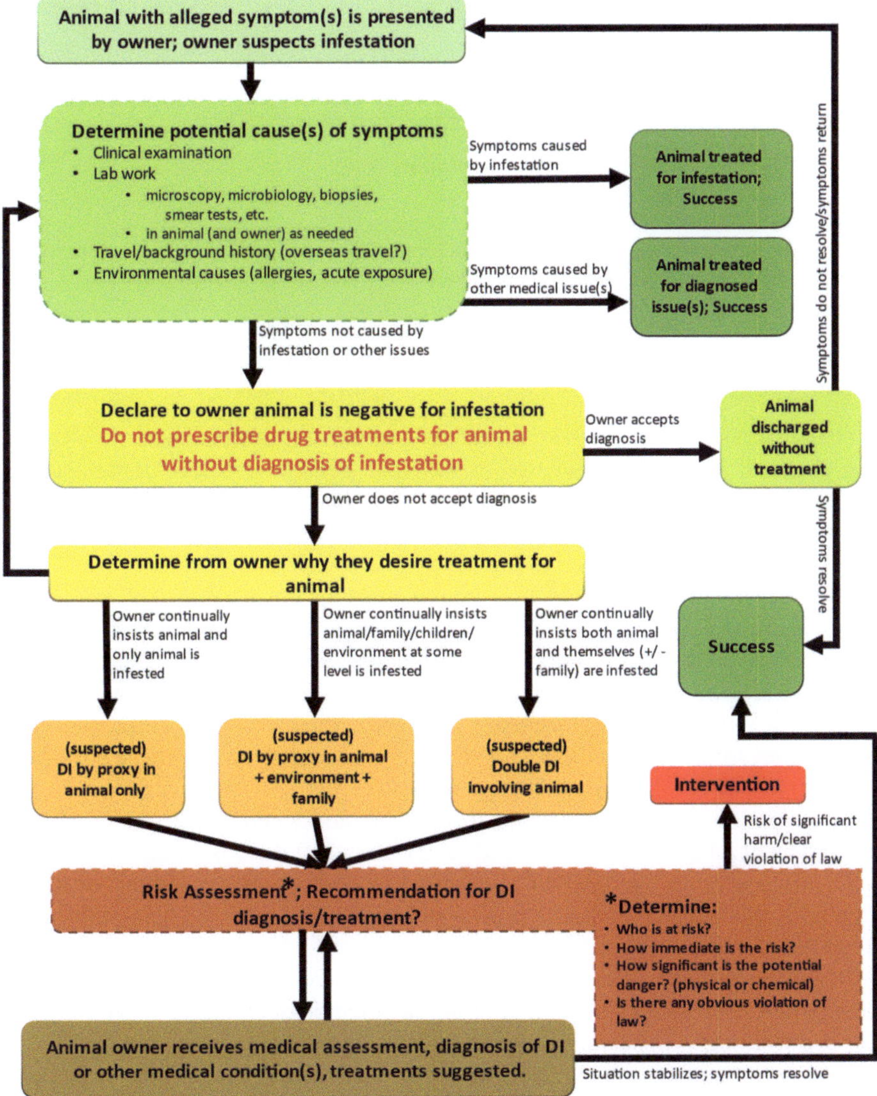

Fig. 6 Veterinarian differential diagnostic flow chart [53]. (Design K. Dugas)

19 Pest Management Professionals/Pest Control Officers

Pest management professionals (PMPs) (also called pest control officers or pest control operators) are often the first professionals patients contact during symptom onset. Additionally, they may be contacted at any time during DI experience. PMPs often face accounts by patients of bizarre parasite activity and unreasonable requests for treatment. These clients pose a unique challenge requiring a particular approach.

Initial contact: Train office/business staff to identify suspected DI callers by listening to what they say.

Speech red flags include the following:

Repeated use of particular pronouns and verbs, "They, them, bites, etc."
  Callers cannot apply a name to their perceived parasites and what they are doing, so generic pronouns and verbs are used as substitutes.

Repeated statements of being "bitten."
  Descriptions of crawling, biting, or parasites freely moving around the body or infesting objects, associates, or animals strongly suggest DI.

Unrecognizable/Bizarre descriptions of perceived parasite biology and behavior.
  DI clients are often intellectually high functioning, so their descriptions can sound very convincing. Do not be fooled.

Office/business staff should take notes and give them to a technician or collaborative entomologist with knowledge of regional pests of medical importance for review. If office/business staff are suspicious they had spoken with a DI caller, warn technicians who are scheduled to service the client. Have in place a service plan and protocol for DI cases in the event legal or regulatory challenges arise.

Service visits/calls: Always visit with *two* technicians not one. This conveys to clients that they are being taken seriously; it protects against errant behavior by clients and provides opportunities for either technician to step away from a situation and take a break. Allocate extra time for a service call, but set a time limit. DI clients particularly with initial visits are likely to overwhelm technicians with descriptions and samples to prove their case. If a service call becomes inordinately protracted, exit with an excuse about another appointment.

Interview clients before starting inspections and sample collections. Explain limitations of a PMP's duties and skills and clarify that there is no medical training or licensing given to technicians to examine or diagnose bites, lesions, or human secretions. Maintain a very strict professional position, and avoid being drawn into client beliefs and delusions.

Maintain a single focus. Investigate whether there are arthropods or other possible parasites present. When walking into a situation, technicians do not know if they are working with actual parasitism, true mental illness, depression, loneliness, anxiety, or an undiagnosed underlying medical condition with evolved emotional investment. Never allow clients to agitate for confirmation of their beliefs, and NEVER SPECULATE with these clients. If any speculation agrees with a client's belief, they will attach to it. They see a technician as the professional who just confirmed their beliefs.

Clients often self-treat using over-the-counter (OTC) pesticides as well as use nonregulated pesticides purchased online from foreign countries. One pest management professional who worked in New Hampshire, USA, reported discovering a client's bathtub full of illegal online purchased pesticides. DI clients generally "give little thought to the dangers" of these pesticides to themselves and those they associate with [40].

Take notes! It is extremely important to take notes. It provides a record which should be preserved against future service calls, need for referral, or in rare cases protection from retaliation. In the first visit, ask:

"When did you first remember feeling ..." or *"When did you start feeling ..."* and finish using the client's own words.

This questioning is important in assessing investment by the client and is a strong indicator of which direction future service interactions may take.

If the answer is greater than 6 months before the time of the service call, expect strong emotional investment with resistance to advice. These clients are using intuition to drive their beliefs, not reason. They tend to be rigid, defensive, and not willing to listen to advice and guidance. These clients will likely be challenging.

If the answer is less than 6 months before the time of the service call, emotional investment may not be as strong, and reason and advice may be more welcomed. These clients are more likely to have two-way discussions about their concerns and curiosity and listen to advice and guidance. These clients will be easier to work with. Listen carefully to all that is shared.

Inspection and monitoring: Conduct a thorough inspection. Record locations of monitors if used and retrieve after 48–72 hours. Any samples that were provided by the client or collected by the technicians should be packed and catalogued for future examination at the business or for submission to an entomologist or parasitologist. Never identify specimens unless certain. Misidentification is not uncommon [84], particularly when pressured by highly verbal intense clients with conviction. Clearly state that a report will be provided by a particular date to reduce repeated client calls to the business.

Reporting: If a pest is found which is covered by structural pest management, suggest an appropriate solution. If results are negative for parasites depending on the demeanor of the client, suggest that they contact a physician for help. Expect a negative reaction. With a frustrated angry client, maintain a courteous and professional demeanor. This usually reduces tensions and provides less ammunition for use against the technician(s) and/or business.

Treatment: Never "spray and pray!" Instead, "investigate and hesitate." Treating a space when there is no evidence of an infestation [84], even though there may be extreme pressure to do so by a client, is unethical. Using inert materials such as water may have a temporary placebo effect, and symptoms of the client may subside for a while, but inevitably they return resulting in follow-up calls. This may lock the PMP into a revolving cycle of treatment for nonexistent parasites and do more harm than good through erroneous confirmation.

Detachment: If a client continues to have a problem over an extended period without resolution and they refuse medical or other help, it may be necessary to politely end the relationship [55–60].

Case Example

A patient had called a PMP company complaining of being bitten and asked for a treatment at her home. She was recovering from a series of four joint replacement surgeries, and a close friend had recently died. She was on antibiotics and pain medications. The receptionist who answered the call said that she had "bird mites."

No bird mites were found.

20 Treatment

Treatment of patients with delusional parasitosis is notoriously difficult [61].

Heller et al. affirmed that rapport and engagement are important. They wrote:

"The most important step toward successful treatment of DOP (*delusions of parasitosis*) is to establish strong therapeutic rapport. … The practice gap is that many dermatologists are unaware or do not accept that establishing therapeutic rapport is of foundational importance in treating patients with DOP. Without adequate rapport, simply giving a prescription for an antipsychotic would be useless, because patients with DOP are unlikely to comply." [28, 62]

In most cases, a psychological belief system and a somatic condition in DI patients are concurrent. They need to be addressed simultaneously. Be aware that presumptive treatment of nonexistent contingent disorder distorts the picture. Treat dermatological or other causative complications, and if agreeable with patients, couple with low doses of antipsychotics to alleviate stress and anxiety. Understand patient's baseline, and make recommendations they can hear.

Partnering with patients will increase treatment acceptance, particularly if they are emotionally invested. When possible, allow patients to be actively involved in their care. Empower patients when moving toward actions by asking them questions such as:

"What are you willing to do?"

"Are you willing to do X, Y, Z?"

"We are working together on this; I am willing to do X, Y, Z; are you willing to do X, Y, Z?"

The word "willing" introduces choice and allows patients to feel empowered by being engaged in their own care. It inserts active responsibility compared to passively allowing the physician to lead. Use other phrases such as:

"Work with me, and we will try some treatments that I have found successful."

"You are part of a team working to resolve a mystery."

These shift patients' thinking from "defensive to cooperative."

Use of euphemisms and analogies: Reid and Lio [61] and Shmidt and Levitt [62] gained consent when a selected antipsychotic with antipruritic effects such as pimozide was presented as a treatment for dermal discomfort and itching while downplaying the primary function of the medication [63]. Freudenmann and Lepping [7] used a similar euphemistic strategy by referring to "antipsychotic" medications as "neuroleptics." Söderfeldt and Groß [63] suggest that when persuading a self-diagnosed "Morgellons" patient to consent to care, a less direct approach may be helpful. They wrote:

> "Explain that there is, as of yet, no evidence-based treatment for Morgellons disease that targets somatic causes, whereas there are examples where psychiatric therapy has been successful. Therefore, they should ask the patient if they are prepared to try the treatment options that the physician suggests and see if there is improvement."

Addictive psychiatric aspect of DI and Alcoholics Anonymous: Consider the addictive psychiatric aspect of DI. Many patients have become emotionally addicted to their beliefs and in some cases self-inflicted pain. Self-harm is often ubiquitous, and so techniques of peer support and redirection used by organizations such as Alcoholics Anonymous (AA), a worldwide fellowship of men and women who address alcoholism and addiction, may be helpful. Groups such as AA reprogram thinking in a supportive culture and restore normal healthy thinking, behavior, and habits.

Alcoholics Anonymous addresses the fundamental issues when addiction impacts personal quality of life. Addiction originates as a positive reward response from a good place. Scratching an itch can feel good. Ogden Nash once wrote:

> "Happiness is having a scratch for every itch."

But with DI, it is different. There is an itch, it is scratched and feels good, then there is another, and it too is scratched followed by a reward of feeling good; this continues to slowly evolve and build into a habit, which eventually in some people becomes a craving for self-inflicted pain which they easily tolerate. This increases until the quality of life is impacted. The patient's reward system is hijacked by evolved addictive behavior.

Consider that there may be issues in the brain limbic system. It regulates behavior and survival responses. With DI patients, more than most patients, it begins to interpret everything as a threat. DI patients can become locked into a cycle of fight or flight. Consider dynamic neural retraining system (DNRS) training with patients. This is cognitive and behavior modification therapy to reduce stress and panic attacks.

The internet: As part of care, instruct patients to stay away from the internet and explain why. Treat their relationship to the internet as addictive, and approach it accordingly. Make sure that patients do not employ third parties to gain internet access and avoid a physician's instructions.

Once patient internet access is restored following care, suggest the use of search engines that do not keep a record of past searches and preserve privacy. As of publication, available search engines which protect against intrusive algorithmic steering are DuckDuckGo, Mojeek, MetaGer, Ecosia, and Peekier.

Occlusive dressings, gloves, nail cutting, and photography as behavior modifications: Use wraps to cover injuries (occlusive dressings) with topical cream treatment to speed healing. Insist patients wear soft gloves such as cotton gloves, particularly at night. This intercepts unconscious grooming and self-injury. Instruct patients to cut their nails. This reduces dermal harm from scratching. Take photographs of injuries before and after this treatment to show patients. In most cases there will be evidence of healing, if patients have followed instructions.

By witnessing healed skin, patients may start to consider alternative causes for their discomfort. Behavior modification using topical anesthetics and anti-inflammatories with OTC drugs, especially acetylcysteine, may be helpful with compulsions.

Rest and vision: Make sure patients are rested. Many are sleep deprived. The downstream effects on the body of sleep deprivation are deleterious mood changes, difficulty thinking, and depression. Other deleterious effects can be loss of concentration and judgment, anxiety, a weakened immune system, skin aging, hypertension, poor balance, clumsiness, vulnerability to diabetes, heart disease, weight gain or loss, reduced sex drive, alcoholism, and premature death.

Encourage patients to "look up." Sky gazing is recognized as beneficial in leveling emotions and reducing stress. DI patients are often deeply engaged in the minutia of their disorder searching for their tiny persecutors by "looking down." Encouraging patients to look up is calming and stress reducing. Studies have shown that vision plays a role in threat response and those with high anxiety experience increased visual scanning and stress [64].

Relationship with food: The relationship between diet and mental health is bidirectional. Bremner et al., in their review of diet, stress, and mental health, wrote:

> "… changes in diet may influence psychiatric disorders through direct effects on mood, while the development of psychiatric disorders can lead to changes in eating habits [65, 66]."

Address diet with patients. When necessary, refer them to qualified dieticians for additional help.

In any treatment plan, encourage healthy eating. Monitor this during treatment and follow-up.

Exercise: Inactivity can lead to low stamina, poor health, physical weakness, and suppressed feelings of well-being. Encourage patients to exercise to the best of their ability. The benefits of exercise can complement and support other aspects of their treatment.

Psychotropic medications:

> "We can treat successfully! Antipsychotics work." Lepping et al. [23].

A study in 2008 by Freudenmann and Lepping suggests that low-level doses of second-generation antipsychotics (SGAs) can provide favorable outcomes for DI patients along with extended periods of follow-up [67]. These follow-ups may be up to 5 years post-treatment.

Dermatologists and primary care physicians may consider SGA use because psychiatric referral is usually not acceptable to patients. Explain the use of antipsychotics in small doses as analogous to taking aspirin or other mild painkillers. Earlier in 1993, Lynch suggested the use of psychotropic medications not to treat the disorder but to lessen "symptoms produced by the perceived parasites" [68].

There may be advantages of dopaminergic drugs such as amisulpride over broad receptor spectrum drugs because of reduced risks of side effects [69]. Second-generation antipsychotic medications that include amisulpride, olanzapine [72], and risperidone in low doses (lower doses than in schizophrenia) over a period of several weeks are very effective in secondary DI cases because these patients lack true psychiatric illness. However, there was a report of olanzapine having long-term success with a 77-year-old woman who had suffered primary DI. She had believed that her skin was infested with "small black animals" [70].

First-generation antipsychotic medications such as sulpiride, haloperidol, and perphenazine are no longer first choice with DI patients, because of poor side effect profiles. Switching drugs can also work in cases of nonresponse, but also consider patient noncompliance in home settings.

A study by Wong and Bewley [71] provided cautionary information regarding relapse of DI in patients following cessation of therapy. Fifty-nine out of the 73 patients in the study received treatment. Fifteen of these patients took pharmacological treatments. Sixty-eight percent of patients reported "improvement or resolution of symptoms," while 4/15 [27%] of the drug-supported patients experienced recurrence of DI ideations. They too suggest that it is important that long-term follow-up be incorporated into DI treatment. This should be from 2 to 5 years with regular calls made during the first 9 months following cessations of supportive drugs, the period of highest risk for relapse [71]. Table 5 lists medications that can assist patients in recovery as well as side effects. In rare cases, self-correction of the condition does occur and resolution to full recovery can be achieved.

Dealing with patient requests: It is common for patients to demand medical tests and/or treatments against physician intellectual judgment. Compared to the late 1990s and early 2000s, patients have become more assertive in governing their personal medical care. Internet and social media influencers encourage patients to feel empowered, irrespective of their knowledge of medicine. Drug corporations sponsor direct-to-customer advertising of drugs encouraging unenlightened patients to self-diagnose, which can cause conflict with physicians. These patients misinterpret self-acquired medical information. At the behest of advertisers who often use phrases such as "go tell your doctor" or "speak to your doctor" about the benefits of a particular drug, these patients demand what they think is appropriate care. This form of medical information is harmful since patients make medical decisions irrespective of their physician's input. Unenlightened media publicity about new research findings is also problematic.

Physicians who face patient requests for non-beneficial treatments and/or services while preserving respect for patient autonomy have three options:

1. Invoke practice guidelines,
2. Provide suitable alternatives and/or,
3. Justify their positions.

In doing so, physicians may have an educational opportunity to forge a better working partnership with their patients. Antithetical to this is modern medical practice in the United States that is driven by for-profit corporate insurance entities,

Table 5 List of medications in alphabetical order which can assist DI patient recovery [11, 39, 72–78]. The USA and UK identifiers indicate the two countries where drugs have been approved.

Antipsychotic drug	Benefits for DI patients	Two common side effects
Acetylcysteine (USA)	For compulsive behavior	Cold-like symptoms, drowsiness
Alprazolam (USA)	Anxiety	Dizziness, drowsiness
Amisulpride (USA)	Schizophrenia, antipsychotic	Insomnia, elevated anxiety
Aripiprazole (USA) (SGA[a])	Psychosis	Drowsiness, weight gain
Blonanserin (SGA)	Antipsychotic	
Buspirone (USA and UK)	Anxiety	Nausea and nervous issues
Chlorpromazine (FGA[a])	Early antipsychotic	
Citalopram (USA)	Depression [79]	GI disturbance
Clozapine (SGA)	Early antipsychotics	
Duloxetine (UK)	Antidepressant	GI disturbance
Escitalopram (UK)	Antidepressant	GI disturbance
Fluoxetine (USA)	Depression, anxiety	Anxiety, nausea, nervousness
Gabapentin	Anticonvulsant	
Haloperidol (USA) (FGA[a])	Schizophrenia, schizoaffective disorders	Weakness, dizziness
Hydroxyzine (UK)	Antihistamine	Headaches, drowsiness, etc.
Mirtazapine	Antidepressant (serotonin)	
Olanzapine (USA) (SGA)	Schizophrenia, mental disturbances, dementia	Dizziness, sleep loss
Paliperidone (UK) (SGA) [80]	Serotonin-dopamine antagonist	
Paroxetine (USA and UK)	Depression, anxiety, OCD	
Pimozide[b] (USA) (FGA)	Psychosis	Weakness, dizziness
Pregabalin (UK)	Antiepileptic	GI disturbance, pruritus, dizziness
Propranolol (UK)	Antihypertensive	GI disturbance and dermal issues
Quetiapine (USA) (SGA)	Schizophrenia, antipsychotic	

(continued)

Table 5 (continued)

Antipsychotic drug	Benefits for DI patients	Two common side effects
Risperidone (USA) (SGA)	Bipolar and schizophrenia. Helps DI patients with hypoperfusion of temporal and parietal lobes [75, 81]	Weight gain, especially in women
Sertraline (USA)	Antidepressant (serotonin)	
Spiridon (USA)	Diuretic and antihypertensive	Drowsiness, dizziness
Sulpiride (USA) (UK) (FGA)	Schizophrenia. Disturbed unusual thinking	Increased motor agitation
Trifluoperazine (FGA)	Schizophrenia. Disturbed unusual thinking	
Venlafaxine XL (UK)	Antidepressant	GI disturbance
Ziprasidone (SGA)	Antipsychotic	

[a] See glossary

[b] Pimozide (first-generation antipsychotic (FGA)) generally has superior acceptance by DI patients over other supportive drugs in the United States. The US Food and Drug Administration (FDA) has no official indicator that it is an antipsychotic medication. It is labeled for use to treat Tourette's syndrome [76]. Though its chemistry is similar to antipsychotic medications, it is not considered in the same class. Thus, hesitant patients are more willing and likely to accept the drug. Pimozide has a relatively high success rate with DI patients, and it avoids the stigma of labeling patients as "crazy," which understandably many do not appreciate. Pimozide might be considered as an emotionally supportive bridge medication to help DI patients get through the early stages of recovery [74, 75, 77, 80–83]. Pimozide can cause prolonged QT intervals requiring periodic electrocardiographic monitoring [17, 84]

which punish physicians for spending time with their patients. Time is an integral part of medicine, and the current financial climate does not allow this [85].

Follow-up: Follow-up with DI patients is vitally important, particularly in high-risk cases. An email or call inserts the physician into the lives of patients outside the office or clinic and telegraphs a sense the physician cares. A follow-up call is suggested within 24 hours after a patient visit. Continue with subsequent follow-up calls as part of the care regimen.

This approach is a powerful emotional antidote against established habits and often negative experiences with previous providers prior to current care. DI patients easily fall back into established thinking and habits at home, and a call from the physician may divert this. It also reinforces compliance, particularly with prescribed medications. This form of "buddy system" keeps communications open, reduces patient flight, reinforces relationships, and builds trust. For patients who have been suffering, in some cases for decades, follow-ups even long term may not work. In these chronic cases, when treatments discontinue, it often results in relapse [86].

Pushy patients: If a patient pushes limits, use the "institution motif" to set limits. Cite "the institution has strict rules for X, Y, Z." It allows the physician to avoid a procedure, test, or action they are not comfortable with while not invoking negative criticism from the patient. Have a written policy available in anticipation of difficult patients. Be strict with do's and don'ts. This establishes up-front expectations for patients and avoids confusion and frustration later.

21   Resigning From Care

How to deal with patients who cannot be helped: There are times when it is permissible to resign from care. Patients that do not listen or repeatedly missinterpret information, do not follow instructions, or become overly aggressive may be cause for resignation from care. When a situation is deemed intractable, resign and close obligations. Do not allow patient requests to force actions outside expertise. Simply say, "I'm not your person" or "I'm sorry, I cannot help you." Expect a negative response. Following are two exemplar letters written by angry patients illustrating misinformation, anger, and frustration. Spelling and grammar are unedited:

Letter No. 1

"It's too bad your haughtiness is hurting so many people. Someday, you will go to hell for it, but until then let me give you a few facts: Many insects are nocturnal, many microscopic and others subcutaneous and others look like dust and just because idiot, know it all physicians don't see them or aren't there when they are is no reason to depress people further in their suffering which is what your smug malpractice is doing, not to mention promoting the continuance of their suffering and infestation. There is plenty of science in support of what you call DP, but who wants that when billions of dollars are available in mental health grants, right? How much money have you or your entity received? Pesticide poisonings with biological pesticides release microscopic and very small worms, gnats, mites, and other vectors into the environment that are emerging as human pathogens, along with their pathogenic cohorts like bacteria, fungi and baculviruses. Plus, they include pheremones which attract other insects and are genetically modified to with insect nuerotoxins from scorpians, spiders and bees, so when they bite there is no doubt they have bitten because it hurts like hell. Take Photorhabdus for example, an insect pathogen caused by agriculture sprays that mutates to bubonic, but by the time that happens the patients have quit going to physicians because of people like you. So, they suffer in absolute humiliating agony, and end up dying of strokes and heart attacks because they have been eaten alive and these sometimes-cystic parasites damage their nerves, muscles, organs, etc.

I have researched this extensively and am not even going to waste my time researching you and your grant funding because you are a classic ass hole. Since your so stupid, you probably don't know about biological skin films that trap insects underneath them and when these bugs wiggle or move around, they can't be seen. There is a way to remove the film and the best way is with the essential oil of clove to be rubbed on the skin which releases the insects. For deeper infestations the oil must be held for more time. Then kaboom, you see them. They also can cause soft tissue infections beneath them.

Why don't you take your ignorant self to the internet and research cases against Delusional Parasitosis. I hate stupid people like you."

Letter No. 2

"People like you sicken me & I hope you are infested with this very real physical condition called Morgellons which I have. Professors, doctors, shrinks all giving statement that it's DP and you're giving interviews saying the same & making statements saying there's an epidemic of DP? Ask yourself a question! Why would there be such increase in people presenting symptoms involving bites, stinging, crawling, and insects coming out of them, because I've got proof it's real but governments. media & medical professionals are lying hiding the truth it's not like you're all stupid you know it's real

I don't drink or smoke or do drugs never have and my body is being eaten alive but professor X at X (identity protected) and his twisted colleagues have stamped a DP diagnosis on me, so I'm left to be eaten alive, I had good skin not a blemish all my life now I'm scarred all over where this crap comes out. Yours as bad and I hope God punishes you all one day soon. Check out my Twitter feed (X) you'll see what's going on & here's an image or two all from my body the bug wrapped up in same fibre that comes out of my skin it's very real & there's thousands like me suffering it's a man made biologically engineered disease like hiv was and those poor people and insects too it's twisted beyond belief.

Yours sincerely,"

22   Physician Fallibilities

Two case histories illustrate misdiagnoses and subsequent treatment. The first was the identification of a tropical fluke, *Schistosoma mansoni*, from a photograph of intact human feces submitted to a well-known online identification service by a patient. The patient had not seen a physician and had never traveled outside the United States. The fluke is not endemic to the United States but is found across sub-Saharan Africa, the Arabian Peninsula, and some South American countries including Brazil, Venezuela, Suriname, and the Caribbean (CDC). The patient used the diagnosis as a trophy to prove that she had an unusual infestation and was lost to care.

The second case was a young adult male who had been diagnosed with scabies from a cursory visual examination. He was treated with permethrin cream and oral ivermectin. A follow-up visit with another physician had an entirely different result. Tests found that the man had developed an allergy to hypoallergenic clothes soap and had never been parasitized by scabies mites.

Medicine is populated by highly trained experts whose professions garner cultural reverence. "Doctoring" is one of the arts, with the same demands as fine art, literature, architecture, or music and is not immune to the quality of its practitioners. The focus in this section is to help physicians identify pitfalls when working with DI patients who are often highly articulate, functional [87], invested, and misguided. DI patients often lie outside the routine experience of many physicians and other medical practitioners. Due to the nature of DI, physician-held cognitive biases may become an unintended obstacle in doctor-patient relationships.

To quote an inscription under Dr. E. L. Trudeau's statue (Lower Saranac Lake, New York, USA [88]):

"Guerir quelquefois, soulanger souvent, consoler toujours

To cure sometimes, to relieve often, to comfort always ...."

.... can be hard for some physicians [89], not because many do not want to, but because of circumstances. The insertion of time-consuming computer technologies and "pressure to comply with third-party guidelines and computerized algorithms" contribute "to a silo mentality and a fragmentation of knowledge." Additionally, hi-tech tools coupled to Western medicine expectations against "not knowing" as a benign answer are particularly problematic with DI patients [90]. It can push physicians into a "problematic pathway to action," which can lead to errors [91]:

> "Making wise decisions in an uncertain environment requires balanced reasoning, critical thinking, compassion and common sense" [90].

During the interview and testing period with DI patients, information is gathered. At this point, it is important not to jump to conclusions to avoid Cognitive Disposition to Respond (CDR). The aim in the following section is not to lay blame, but to alert physicians of diagnostic pitfalls which DI patients might cause. Being aware of cognitive pitfalls while working with DI patients protects against errors and protracted differential diagnostics.

Cognitive Disposition to Respond (CDR's) and diagnostic errors: A typical DI patient may present with a baffling medical history and a puzzling physical examination. The patient's disorder is unlike anything seen before. Things do not quite add up, and the physician is imbued with unease and uncertainty. Routine laboratory tests are unrevealing or may lead diagnosis astray, yet the patient is clearly suffering. Amid this mystery, being aware of cognitive bias may avoid pitfalls.

The nature of the medical profession leads many physicians to an aversion of risk and ambiguity. Nevertheless, they feel compelled to respond, despite incomplete or unreliable results. This is where CDR's, also known as cognitive biases, may lead to mistakes, safety risks, and inadequate utilization of medical resources [79, 91–98]. The following are 36 CDR biases:

Cognitive dispositions to respond (CDR's) can result in diagnostic error [3, 79, 93, 94, 98–100].

Ambiguity tolerance: Internal process that influences the way a person structures information who is faced with an array of unfamiliar, complex, and incongruent cues.

Aggregate bias: When physicians believe that aggregated data such as those used to develop clinical practice guidelines do not apply to individual patients (especially their own), they are invoking the aggregate fallacy. The belief that their patients are atypical or somehow exceptional may lead to errors of commission, e.g., ordering X-rays or other tests when guidelines indicate that none are needed.

Anchoring: Staying locked on to an earlier diagnosis despite new evidence to the contrary. This CDR may be severely compounded by confirmation bias.

Ascertainment bias: Occurs when a physician's thinking is shaped by prior expectations. Stereotyping and gender bias are both good examples.

Availability: Jumping to a diagnosis that comes to mind too quickly because it is either common, serious, recently encountered, or otherwise noteworthy. Conversely, if a disease has not been seen for a long time (is less available), it may be underdiagnosed.

Base rate neglect: The tendency to ignore the true prevalence of a disease, either inflating or reducing its base rate, and distorting Bayesian reasoning. However, in some cases, clinicians may (consciously or otherwise) deliberately inflate the likelihood of disease, such as in the strategy of "rule out worst-case scenario," to avoid missing a rare but significant diagnosis.

Blind obedience: Showing undue deference to authority and technology.

Commission bias: Results from the obligation toward beneficence in that harm to a patient can only be prevented by active intervention. It is the tendency toward action rather than inaction. It is more likely in overconfident physicians. Commission bias is less common than omission bias.

Confirmation bias: Tendency to look for confirming evidence to support a diagnosis rather than look for disconfirming evidence to refute it, despite the latter often being more persuasive and definitive.

Context errors: A critical signal is distorted by the background against which it is perceived.

Diagnosis momentum: Once diagnostic labels are attached to patients, they tend to become stickier. Through intermediaries (patients, paramedics, nurses, physicians), what might have started as a possibility gathers increasing momentum until it becomes definite, and all other possibilities are excluded.

Feedback sanction: A form of ignorance trap and time-delay trap CDR. Making a diagnostic error may carry no immediate consequences, as considerable time may elapse before the error is discovered, if ever. Or, poor system feedback processes prevent important information on decisions getting back to the decision maker. The CDR that failed the patient persists because of these temporal and systemic sanctions.

Framing effect: Allowing how a patient's medical story is framed, whom the story came from, and where the patient was seen to influence a diagnostic decision.

Fundamental attribution error: The tendency to be judgmental and blame patients for their illnesses (dispositional causes) rather than examine the circumstances (situational factors) that might have been responsible. Psychiatric patients, minorities, and other marginalized groups tend to suffer from this CDR. Cultural differences exist in terms of the respective weights attributed to dispositional and situational causes.

Gambler's fallacy: Attributed to gamblers, this fallacy is the belief that if a coin is tossed ten times and is heads each time, the 11th toss has a greater chance of being tails (even though a fair coin has no memory). An example would be a physician who sees a series of patients with chest pain in a clinic or the emergency department, diagnoses all of them with an acute coronary syndrome, and assumes that the sequence will not continue. Thus, the pretest probability that a patient will have a particular diagnosis might be influenced by preceding but independent events.

Gender bias: The tendency to believe that gender is a determining factor in the probability of diagnosis of a particular disease when no such pathophysiological basis exists. Generally, it results in an overdiagnosis of the favored gender and underdiagnosis of the neglected gender.

Hindsight bias: Knowing the outcome may profoundly influence the perception of past events and prevent a realistic appraisal of what occurred. In the context of diagnostic error, it may compromise learning through either an underestimation (illusion of failure) or an overestimation (illusion of control) of the decision maker's abilities.

Information bias: Unconscious or conscious selective distortion when gathering information.

Multiple alternative bias: A multiplicity of options on a differential diagnosis may lead to significant conflict and uncertainty. The process may be simplified by reverting to a smaller subset with which the physician is familiar but may result in inadequate consideration of other possibilities. One such strategy is the three-diagnosis differential: "It is probably A, but it might be B, or I don't know C." Although this approach has some heuristic value, if the disease falls in the C category and is not pursued adequately, it will minimize the chances that some serious diagnoses can be made.

Omission bias: The tendency toward inaction and rooted in the principle of non-maleficence. In hindsight, events that have occurred through the natural progression of a disease are more acceptable than those that may be attributed directly to the action of the physician. The bias may be sustained by the reinforcement often associated with not doing anything, but it may prove disastrous. Omission biases typically outnumber commission biases.

Order effects: Information transfer is a U-function: we tend to remember the beginning part (primacy effect) or the end (recency effect). Primacy effect may be augmented by anchoring. In transitions of care, in which information transferred from patients, nurses, or other physicians is being evaluated, care should be taken to give due consideration to all information, regardless of the order in which it was presented.

Outcome bias: The tendency to opt for diagnostic decisions that will lead to good outcomes, rather than those associated with bad outcomes, thereby avoiding chagrin associated with the latter. It is a form of value bias in that physicians may express a stronger likelihood in their decision-making for what they hope will happen rather than for what they really believe might happen. This may result in serious diagnoses being minimized.

Overconfidence bias: A universal tendency to believe we know more than we do. Overconfidence reflects a tendency to act on incomplete information, intuitions, or hunches. Too much faith is placed in opinion instead of carefully gathered evidence. The bias may be augmented by both anchoring and availability, and catastrophic outcomes may result when there is a prevailing commission bias. Additionally, the belief that a particular test procedure is more accurate and effective than it is.

Playing the odds: (Also known as "frequency gambling") is the tendency in equivocal or ambiguous presentations to opt for a benign diagnosis on the basis that it is significantly more likely than a serious one. It may be compounded by the fact that the signs and symptoms of many common and benign diseases are mimicked by more serious and rare ones. The strategy may be unwitting or deliberate and is diametrically opposed to rule out worst-case scenario strategy (see base rate neglect).

Posterior probability error: Occurs when a physician's estimate for the likelihood of disease is unduly influenced by what has gone on before for a particular patient. It is the opposite of the gambler's fallacy in that the physician is gambling on the sequence continuing; for example, if a patient presents to the office five times with a headache that is correctly diagnosed as migraine on each visit, it is the tendency to diagnose migraine on the sixth visit. Common things for most patients continue to be common, and the potential for a non-benign headache being diagnosed is lowered through posterior probability.

Premature closure: A powerful CDR accounting for a high proportion of missed diagnoses. It is the tendency to apply premature closure to the decision-making process, accepting a diagnosis before it has been fully verified. The consequences of the bias are reflected in the maxim: "When the diagnosis is made, the thinking stops."

Psych-out error: Psychiatric patients appear to be particularly vulnerable to the CDRs described in this list and to other errors in their management, some of which may exacerbate their condition. They appear especially vulnerable to fundamental attribution error. Comorbid medical conditions may be overlooked or minimized. A variant of psych-out error occurs when serious medical conditions, such as hypoxia, delirium, metabolic abnormalities, CNS infections, or head injury, are misdiagnosed as psychiatric conditions.

Representativeness restraint: The representativeness heuristic drives the diagnostician toward looking for prototypical manifestations of disease: "If it looks like a duck, walks like a duck, and quacks like a duck, then it is a duck." Yet restraining decision-making along these pattern recognition lines leads to atypical variants being missed.

Search satisfaction: This CDR reflects the universal tendency to call off a search once something is found. Comorbidities, second foreign bodies, other fractures, and co-ingestants in poisoning may all be missed. Also, if the search yields nothing, diagnosticians satisfy themselves that they have been looking in the right place which may not be the case.

Sutton's slip: Takes its name from the apocryphal story of the Brooklyn bank robber Willie Sutton who, when asked by a judge why he robbed banks, is alleged to have replied, "Because that's where the money is!" The diagnostic strategy of going for the obvious is referred to as Sutton's law. The slip occurs when possibilities other than the obvious are not given sufficient consideration.

Sunk cost fallacy: The more clinicians invest in a particular diagnosis, the less likely they may be to release it and consider alternatives. This is an entrapment form of CDR often associated with investment and financial considerations. However, for the diagnostician, the investment is time, in created testing, and mental energy. For some, ego may be an investment. Confirmation bias may be a manifestation of an unwillingness to let go of a failing diagnosis.

Triage cueing: The triage process occurs throughout the healthcare system, from the self-triage by patients to the selection of a specialist by the referring physician. In the emergency department, triage is a formal process that results in patients being sent in particular directions, which cues their subsequent management. Many CDRs are initiated at triage, leading to the maxim: "Geography is destiny."

Unpacking principle: Failure to elicit all relevant information (unpacking) in establishing a differential diagnosis may result in significant possibilities being missed. The more specific a description of an illness that is received, the more likely the event is judged to exist. If patients are allowed to limit their history-giving, or physicians otherwise limit their history-taking, unspecified possibilities may be discounted.

Vertical line failure: Routine, repetitive tasks often lead to thinking in silos—predictable, orthodox styles that emphasize economy, efficacy, and utility. Though often rewarded, this approach carries the inherent penalty of inflexibility. In contrast, lateral thinking styles create opportunities for diagnosing the unexpected, rare, or esoteric. An effective lateral thinking strategy is simply to pose the question: "What else might this be?"

Visceral bias: Personal feelings toward a patient influencing a diagnosis.

Yin-yang out: When patients have been subjected to exhaustive and unavailing diagnostic investigations, they are said to have been "worked up the yin-yang." The yin-yang out is the tendency to believe that nothing further can be done to throw light on the dark place where, and if, any definitive diagnosis resides for the patient, i.e., the physician is let out of further diagnostic effort. This may prove ultimately to be true but to adopt the strategy at the outset is fraught with the chance of a variety of errors.

The terms used to describe the CDRs above are those by which they are commonly known in the psychology and general medical literature, as well as colloquially. Some, e.g., feedback sanction and hindsight bias, are indirect, reflecting more on processes that interfere with physician calibration.

There is considerable overlap among CDRs, some being known by other synonyms. These, together with further detail and citations for the original work, are described in Croskerry P. [79, 93] and earlier work by Einhorn and Hogarth [96] and Hogarth [100].

As highlighted above, both physicians and patients may possess shared biases which include sunk cost fallacy and confirmation bias. Both are susceptible to important cognitive dispositions which may collide. If these interact during a consultation, then medical errors may be more likely to occur. Ogdie et al. [98] interviewed a population of 41 internal medicine physicians at the University of Pennsylvania (USA) and explored cognitive and contextual components that might cause diagnostic error. In the study, anchoring, locking into an early diagnosis, and not changing irrespective of new contradictory evidence were shown to be a serious issue in 87.8% of the respondents. Additionally, based on Saposnik et al.'s review of doctor cognitive biases in 14 cited papers, the authors identified seven CDRs as predominate in physician decision-making, e.g., confirmation bias, availability, anchoring, framing effect, risk aversion, feedback bias, and omission [94]. Other common biases among physicians were information bias, blind obedience, ambiguity tolerance, and premature closure.

DI patients challenge most of these biases, often simultaneously. The potential burden of a physician's personality, experience, level of fatigue, and biases makes diagnosis of DI patients difficult, especially when patients possess strong attachments to emotional components and claim that other physicians confirmed diagnoses of infestation. Although patients often continue to seek help, many will accept it only if it conforms to their worldview and reject recommendations if it contradicts their beliefs.

Modeling on the work by Ely et al. [3], we have provided a "checklist" specific to typical DI patient presentations (Box 3). It is important to maintain an executive decision-making process and a clear head during the diagnostic period. Stepping away from the physical presence of a DI patient who may be particularly challenging into another space or room (the doorway effect) to rest is encouraged. Ely et al. [3] suggest, "Pause to reflect-take a diagnostic time out."

Iatrogenic disorders: The word "Iatrogenic" is derived from two Greek language roots. The first *iatros* means "the healer" and the second *genic* "to generate, produce, pertain to." Iatrogenic symptoms are disease or symptoms induced in patients by physician attitudes, written or spoken comments, examinations, treatments, therapies, and/or procedures [101]. One of the top causes for iatrogenic symptoms is adverse drug reaction (ADR) [27, 102, 103]. Krishnan and Kasthuri wrote:

> "The harm that a physician can do is not limited to the imprudent use of medicine or procedure but may include unjustified remarks and misinterpretation of investigational data. The physician should be aware of the properties of drugs that he prescribes and their potential dangers. Ignorance of the possibility of a reaction is clear evidence of negligence. The physician should warn the patient of the likely side effects" [102].

DI patients are particularly vulnerable to iatrogenic error because they often deal with both psychological frailty and one or more undiagnosed underlying medical conditions. Physicians can do more harm than good. The following account written by Bury and Bostwick [27] describes how a physician planted an idea in the mind of a patient who was struggling to find answers resulting in a cascade of suffering.

> "Mrs. B. had first presented with folliculitis for which she received intravenous antibiotics locally. After treatment, still scratching, she sought help at "a doc-in-the-box." The doctor informed her that she had lice, even extracting a specimen from her hair to prove it, and prescribed Lindane. (She later confessed that, to her, the specimen resembled nothing more than "a flake of skin.") When Lindane did not relieve her itching, another urgent care doctor told her that the first doctor had been mistaken; what she had were mites. He recommended permethrin. On her own, she escalated her use to six bottles a month, treatment that did not help and may have worsened her pruritus. Although Lindane and permethrin were ineffective, they did reinforce her belief that she was infested. Like many DP (DI) patients, she began seeking an explanation fitting an infestation narrative. Having heard news reports about hotel lice in a city she had recently visited, she wondered if she could have acquired them there. Alternately, she worried that her hairdresser or a grandchild with school-acquired lice could have infected her. She isolated herself for fear of transmitting lice to others. When shopping, she wore a face-obscuring hat to hide blemishes from her gouging at itching skin. As doctors had instructed, she repeatedly washed her linens and cleaned her home in hopes of eliminating the lice. After 2 months, she finally saw a dermatologist who discerned no evidence of lice despite her insistence that she had them, proposed a DP (DI) diagnosis, and referred her for a second opinion. Mrs. B. arrived at our institution with a positive "matchbox sign," material in a sandwich bag that she insisted was insect related. Her examination revealed excoriations: some healed, and some were hemorrhagic on her face, arms, and lower leg, but with no evidence of either lice or mites. When a punch biopsy showed only necrotizing folliculitis and fungal and microorganism cultures were negative, the dermatologist concurred with the DP (DI) diagnosis. Unlike typical DP patients, Mrs. B readily agreed to see a psychiatrist who reaffirmed the diagnosis and prescribed pimozide. Later, while reviewing outside records, the dermatologist noticed previous elevated serum calcium levels of 11.1 and 11.2. When a repeat calcium level was 11.6, he consulted an endocrinologist who quickly diagnosed primary hyperparathyroidism with hypercalcemia-induced itching. A month after parathyroidectomy, Mrs. B was "doing better, skin wise." Several months later, off pimozide and topical agents, she endorsed neither itching nor delusions. She eventually recalled that the urgent care doctors, rather than she, first thought that she had bugs. "They set it in my mind that it might be a parasite," she said. "If a doctor says it, it must be true."

Seven case histories illustrating iatrogenic care: Following is a series of case histories illustrating variability of DI and physician responses to these cases. Names and identities have been changed to preserve confidentiality.

Case 1. Jennifer and a Drug-Induced Psychosis.

Jennifer was a new mother. She had just given birth and was on 80 mg of steroids for a bone marrow disorder. She had been released from hospital and was highly anxious. While in hospital, a medical student had accidentally doubled her steroid dose to 160 mg resulting in hallucinations. She experienced auditory and visual imagery. She saw lint on the ceiling that morphed into insects. Realizing their mistake, the hospital reinstated the 80 mg dose. The woman had suffered a drug-induced psychosis which had continued following discharge from the hospital. Following her conversation with an entomologist, she readmitted herself to hospital for follow-up care.

Cases 2 and 3. Parkinson's Disease and Drug-Induced DI.
An 83-year-old male patient during treatment of Parkinson's disease developed drug-induced DI. He had been given amantadine and carbidopa/levodopa three times a day [104]. The second case was a German 61-year-old woman also suffering from Parkinson's disease. She was given 200 mg/day add-on piribedil. Four days into treatment, she developed an unshakable belief she and her belongings were infested with "creepy crawlies, vermin, live" or "black dots." A diagnosis of piribedil-induced psychotic disorder presented as DI was made, and the medication was stopped. Symptoms persisted for 12 days following cessation of the medication, and then slowly abated [105].

Case 4. The Expectant Father.
A healthy man in his 20s developed a dermal pruritic rash in late December. An initial diagnosis of scabies was made by a nurse practitioner at a walk-in clinic. Subsequently, four physicians and one dermatologist confirmed the diagnosis. No mites were ever collected and identified. Most diagnoses were visual examinations, one from the opposite side of an examination room. In this case, the physician would not come near him from a fear of becoming infested. Several prescribed ivermectin and 5% permethrin cream. The man asked one of the physicians if he personally would take the ivermectin he had prescribed and the physicians response was:

> "On a personal level, if it was me, I would not take the medicine, but I am prescribing it to protect my ass."

In March the following year, an emergency room physician suggested his pregnant fiancé was the person who was infested. The physician said that she was re-infesting him but would not show signs of scabies because she was pregnant. Since his diagnosis, he had had no physical contact with his fiancé for 3 months because he was concerned he would give her scabies. The emergency physician prescribed her ivermectin. The couple refused. An entomologist whom he then found said that he had no scabies. Later, he discovered that he had developed an allergy to prolonged use of a proton pump inhibitor (PPI) medication, omeprazole (Fig. 7). His primary care physician changed the medication, and within 24 hours, the pruritic rash was gone. The man had been sleeping on his dining room floor for several months prior to seeing the entomologist and described being "at the end of his rope." The couple went on to be married and, the following June, had a beautiful baby daughter.

Case 5. Lead Poisoning And Treatment In Absence Of Parasites.
A similar case as this one was recently resolved. The patient, a nurse, was suffering from a similar form of dermatitis. She asked, "Is this scabies?" The physician's

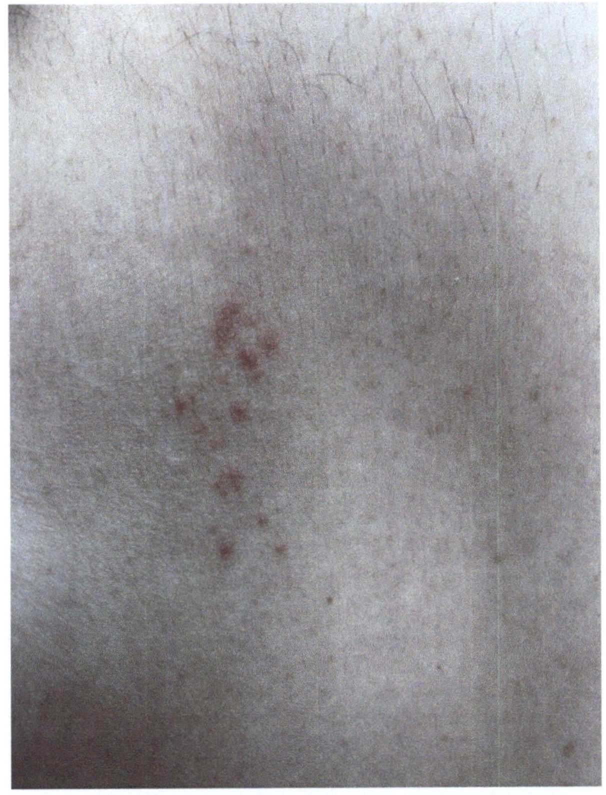

Fig. 7 A drug reaction to a proton pump inhibitor medication diagnosed as scabies

answer was, "No, but I'm going to treat it like it is," and he prescribed multiple courses of ivermectin and permethrin cream. She finally found out she was suffering from lead poisoning after drinking water from old lead pipes in her building.

Case 6. Daughter's Email About Her Struggling Mother and How a Doctor Got It Wrong.

"Thank you for that. Hopefully the call and setting up the appointment will help give her hope and will keep her out of the hospital. I hope you can help guide us and find some relief and happiness for my mom. It's hard to see someone suffer. I did get her list of meds, but she might get very angry if I share. I will ask her permission. Also, she speaks with a psychiatrist who is the one who first told her of Morgellons. I think he is a very odd person. Have a nice night."

Case 7. The Nurse.

Following is a letter written by a nurse describing her experience with DI ideation and her recovery. She pieced together several causes that set off the DI and was able to step back, rewind, and recover. This took courage and insight:

"I guess I will first explain the pictures below. The first is from last November 11th, 2020, when red rash appeared on my left cheek. The second and third are from this year. The recent car accident was Friday October 29th, 2021 (photo was taken the next day). Last picture is taken 2 weeks later, November 12th, when blisters surfaced on my right cheek. The red rash had appeared the day before, but unfortunately, I didn't take a picture. Unbelievably, it was exactly 1 year later. More unbelievably, my dad passed in January 2020, and his birthday is November 12th.

We first spoke last November 2020. I called you saying I believed I had delusional parasitosis. Before the rash on my left cheek even surfaced, I had sensations that bugs were crawling on my face and also under my skin. When the rash appeared, I was convinced I had scabies or mites. At that time, I was overwintering 8 or 9 caged squirrels (inside) and my building got infested with mice. The squirrels being unable to stop scratching (some with severe hair loss), really didn't help my cause. My Mom is also a wildlife Rehabber, and had a scraping confirmed diagnosis of scabies before in the past. This definitely fueled my paranoia. I covered the rash with hydrocortisone, triple antibiotic and kept a dressing on it.

This year, my PTSD was completely reactivated by the MVA (*motor vehicle accident*). I definitely saw the accident coming and I guess I screamed and threw my arms up before the impact. I just remember seeing the accident coming. I guess I disassociated for a bit after the accident. I was originally diagnosed with PTSD after a road rage accident that occurred in August of 2017. This accident left me with multiple sequelae of conditions that are treatable, but lifelong and not curable. Unfortunately, this second accident activated most of them, and my stress level and anxiety were unbelievably high.

When I started feeling that same feeling as last year of bugs crawling on and under my skin, my anxiety went through the roof. We weren't overwintering any squirrels, so it just couldn't be happening. I didn't speak to anyone about it. It had always really bothered me that I didn't know why it had happened the year before. I told myself it was the stress of my dad's upcoming birthday and my year-round allergies. I have a rosacea spot on my right cheek that gets dry and burns around this time every year. Then I realized it wasn't just that spot, but my entire right cheek felt it was on fire and was too painful to even touch. The rash appeared a couple days later. Once again, I applied hydrocortisone, triple antibiotic and covered with a dressing. The next day I removed the dressing and the blisters had surfaced.

I have no doubt (at least in my case) that shingles was induced by high stress/trauma. The car accident was unbelievably stressful and definitely created unbelievable anxiety. I also don't think it's a coincidence that I had shingles (both times) on my dad's birthday.

Probably more information than you wanted, but I hope it's useful to you and your colleagues. I give permission for you to share everything included in this email. My hope is that it can in some way help others. Feel free to contact me if you have any questions or need anything else. Thanks so much again for the help you gave me and for the work you and your colleagues are doing."

23 Support for Physicians, Caregivers, Medical and Nonmedical Professionals

DI is a complex interdisciplinary disorder [106, 107]. In the United States, many patients wander from one physician to the next [2]. Once they find a physician who accepts them into their practice, patients may present as proof of their infestation verbose accounts of their delusions accompanied by burdensome samples. Patients may also demand unreasonable tests such as biopsies. Others may threaten out of frustration. For those physicians who are unprepared, it may be overwhelming.

DI patients can trigger attending physicians emotionally. One technique to avoid this is to delay time between the "stimulus" (the fringe-like demands of a patient) and "response" (the physician's uncertainty when faced with suspect information and symptoms that lie outside the norms of medicine). The space created between stimulus and response provides time for consideration and for the physician to "take-a-beat." To quote Steven Covey/Victor E. Frankl (source uncertain),

> "Between stimulus and response there is a space. In that space is our power to choose our response. In our response lies our growth and our freedom."

To deescalate, allude to the patient's frustrations, and if necessary, find an excuse to leave the examination room. Use an excuse such as a prearranged phone call, checking the status of a time-sensitive test, but leave the room.

Temporary disengagement and the doorway effect: Stressful patients can trigger the amygdala and hijack it. Thus, a physician might make emotional decisions rather than executive ones. By isolating from a difficult patient, it allows the restoration of executive function. Take a moment, pause, walk away from a situation, breathe, and restore composure. It is a moment to clear thoughts and restore upper level cognitive function. It is an opportunity to disengage, rest, reconsider, and recoup. Read chapter "Treating the difficult patient: How to approach a patient with Delusional Infestation," by Michelle Magid, Nell Frackowiak, and Jason S. Reichenberg. They provide excellent advice on how to work with difficult DI patients.

Use the physical doorway to your advantage. The act of walking through a doorway and into another space can "reset" short-term memory. The mind routinely purges information in favor of new material, prioritizing what is happening in the new space over the previous one. It is a technique to clear mental clutter. This seperation also allows patients a few moments to reconsider their thoughts as well.

If this is a particularly difficult patient, it may help to view them as a drowning person. When a person is drowning, they instinctually grab anything, including the person who is trying to save them. The drowning person will literally climb over their rescuer to get above water, pushing them both down. Back away, reposition, and then return from a secure position.

With attention to the details, patience, and care, both patients and physicians can be successful finding a cure. It takes time, sometimes years, but in the end, it is worth the effort especially when patients can rejoin their families, return to their careers, and find their personal lives restored.

References

1. Lepping P, Freudenmann RW. Delusional parasitosis: a new pathway for diagnosis and treatment. Clin Exp Dermatol. 2007;33:113–7.
2. Sansone RA, Sansone LA. Doctor shopping: a phenomenon of many themes. Innov Clin Neurosci. 2012;9(11–12):42–6.
3. Ely JW, Graber ML, Croskerry P. Checklist to reduce diagnostic errors. Acad Med. 2011;86:307–13.

4. Bahmer JA, Petermann F, Kuhl J. [Psychosocial factors in psoriasis. A pilot study]. Hautarzt. 2007;58(11):959–965. doi:https://doi.org/10.1007/s00105-007-1371-4. PMID: 17701145.
5. Estrada E, Persaud-Sharma D, Corridor RG. Organic etiology of delusional parasitosis in the Hispanic population: a case report. Arch Clin Psychiatry. 2019;46(2):51.
6. Fitzpatrick T. The validity and practicality of sun-reactive skin types I through VI. Arch Dermatol. 1988;124:869–71.
7. Freudenmann RW, Lepping P. Delusional infestation. Clin Microbiol Rev. 2009;22:690–732.
8. Huber M, Kirchler E, Karner M, Pycha R. Delusional parasitosis and the dopamine transporter. A new insight of etiology? Med Hypotheses. 2007;68:1351–8.
9. APA dictionary of psychology, 2nd ed. 2015; pp. 1204. isbn:978-1-4338-1944-5.
10. Gordon-Elliott JS, Muskin PR. Managing the patient with psychiatric issues in dermatologic practice. Clin Dermatol. 2013;31:3–10.
11. Tey HL, Wallengren J, Yosipovitch G. Psychosomatic factors in pruritus. Clin Dermatol. 2013;31(1):31–40. https://doi.org/10.1016/j.clindermatol.2011.11.004.
12. Gawande A. The itch. Its mysterious power may be a clue to a new theory about brain and bodies. The New Yorker Magazine. 2008; pp. 58–65.
13. Oaklander AL. Shingles, postherpetic neuralgia and postherpetic itch. Int J Pain Med Palliat Care. 2002;2(1):1–10.
14. Miller G. Grasping for clues to the biology of itch. Chronic itch afflicts millions of people, but little is known about the underlying mechanics. Sci Mag. 2007;318:188–9.
15. Brewer KL, Lee JW, Downs H, Oaklander AL, Yezierski RP. Dermatomal scratching after intramedullary quisqualate injection: correlation with cutaneous denervation. J Pain. 2008;9(11):999–1005.
16. Kimsey LS. Delusional infestation and chronic pruritus: a review. Acta Derm Venerol. 2016;96:298–302.
17. Lepping P, Baker C, Freudenmann RW. Delusional infestation in dermatology in the UK prevalence, treatment strategies, and feasibility of a randomized controlled trial. Clin Exp Dermatol. 2009;35(8):841–4.
18. Hadhazy A. Think twice: how the gut's "second brain" influences mood and well-being. Scientific American Opinion; 2010.
19. Gershon MD. The second brain: a groundbreaking new understanding of nervous disorders of the stomach and intestine. New York: HarperCollins Publishers; 1999. p. 336.
20. Yadav KV, Balaji S, Suresh PS, Liu XS, Lu X, Li Z, Guo XE, Mann JJ, Balapure AK, Gershon MD, Medhamurthy R, Vidal M, Karsenty G, Ducy P. Pharmacological inhibition of gut-derived serotonin synthesis is a potential bone anabolic treatment for osteoporosis. Nat Med. 2010;16:308–12.
21. Cook EH, Leventhal BL. The serotonin system and autism. Curr Opin Pediatr. 1996;8:348–54.
22. Gabriele S, Sacco R, Persico AM. Blood serotonin levels in autism spectrum disorder: a systematic review and meta-analysis. Eur Neuropsychopharmacol. 2014;24(6):919–29.
23. Lepping P, Russell I, Freudenmann RW. Antipsychotic treatment of primary delusional parasitosis: systematic review. Br J Psychiatry. 2007;191:198–205. https://doi.org/10.1192/bjp.bp.106.029660.
24. Lyell A. Delusions of parasitosis (The Michelson lecture). Br J Dermatol. 1983;108:485–99.
25. Freudenmann RW, Lepping P, Huber M, Dieckmann S, Bauer-Dubau K, Ignatius R, Misery L, Schollhammer M, Harth W, Taylor RE, Bewley AP. Delusional infestation and the specimen sign: a European multicenter study in 148 consecutive cases. Br J Dermatol. 2012;167(2):247–51.
26. Freudenmann RW, Kölle M, Schönfeldt-Lecuona C, Dieckmann S, Harth W, Lepping P. Delusional parasitosis and the matchbox sign revisited: the international perspective. Acta Derm Venereol. 2010;90:517–9. https://doi.org/10.2340/00015555-0909.
27. Bury JE, Bostwisk JM. Iatrogenic delusional parasitosis: a case of physician-patient folie à duex. Gen Hosp Psychiatry. 2010;32:210–2.
28. Heller MM, Wong JW, Lee ES. Delusional infestations: clinical presentation, diagnosis and treatment (review). Int J Dermatol. 2013;52:775–83.

29. Foster AA, Hylwa SA, Bury JE, Davis MDP, Pittelkow MR, Bostwick JM. Delusional infestations: clinical presentations in 147 patients seen at Mayo Clinic. J Am Acad Dermatol. 2012;67(4):1–10.
30. Lee CS. Delusions of parasitosis. Dermatol Ther. 2008;21(1):2–7.
31. Marshall CL, Williams V, Ellis C, Taylor RE, Bewley AP. Delusional infestation may be caused by recreational drug usage in some patients, but they may not disclose their habit. Clin Exp Dermatol. 2016;42:41–5.
32. Lepping P, Huber M, Freudenmann RW. How to approach delusional infestation. BMJ. 2015;350:h1328. https://doi.org/10.1136/bmj.h1328.
33. Lepping P, Noorthoorn EO, Kemperman PMJH, Harth W, Reichenberg JS, Squire SB, Shinhmar S, Freudenmann RW, Bewley A. An international study of the prevalence of substance use in patients with delusional infestation. J Am Acad Dermatol. 2017;77(4):778–9.
34. Marschall MA, Dolezal RF, Cohen M, Marschall SF. Chronic wounds and delusions of parasitosis in the drug abuser. Plast Reconstr Surg. 1991;88(2):328–30.
35. Riggan MAA, Perreault G, Wen A, Raco V, Vassallo S, Gerona R, Hoffman RS. Analytically confirmed severe albendazole overdose presenting with alopecia and pancytopenia. Am J Trop Med Hyg. 2020;102:177–9.
36. Fuentes AV, Pineda MD, Nagulapalli Venkata KC. Comprehension of top 200 prescribed drugs in the US as a resource for pharmacy teaching, training, and practice. Pharmacy (Basel). 2018;6(2):43. https://doi.org/10.3390/pharmacy6020043.
37. Clincalc. 2019. https://clincalc.com/DrugStats/.
38. Drugreport. 2020. https://wdr.unodc.org/wdr2020/en/index2020.html.
39. Sambhi R, Lepping P. Psychiatric treatment in dermatology: an update (review article). Clin Exp Dermatol. 2010;35(2):120–5.
40. Ahmed AI, Alhajjar R, Bewley AP, Taylor R. Treatments for primary delusional infestation (Review). Cochrane Database Syst Rev. 2019;Rev. 12:CD011326. https://doi.org/10.1002/14651858.CD011326.pub1.
41. Merriam-Webster's dictionary. 11th ed. G. and C. Merriam Co. 2003; pp. 1664.
42. Kupfer TR, Fessler DM. Ectoparasite defense in humans: relationship to pathogen avoidance and clinical implications. Philos Trans R Soc Lond B Biol Sci. 2018;373:20170207.
43. Fowler E, Maderal A, Yosipovitch G. Treatment-induced delusions of infestation associated with increased brain dopamine levels. Acta Derm Venereol. 2019;99(3):327–8.
44. Lai J, Xu Z, Xu Y, Hu S. Reframing delusional infestation: perspectives on unresolved puzzles. Psychol Res Behav Manag. 2018;11:425–32.
45. Kim DI, Garrison RC, Thompson G. A near fatal case of pathological skin picking. Am J Case Rep. 2013;14:294–87.
46. Oaklander AL, Cohen SP, Raju SVY. Intractable postherpetic itch and cutaneous differentiation after facial shingles. Pain. 2002;96:9–12.
47. Lyman L. The remarkable life of the skin: an intimate journey across our largest organ. Atlantic Monthly Press. 2020; pp. 304.
48. Ryan ET, Wilson ME, Kain KC. Illness after international travel (review article). N Engl J Med. 2002;347:505–16.
49. Moryś JM, Jeżewska M, Koreniewski K. Neuropsychiatric manifestations of some tropical diseases. Int Marit Health. 2015;66(1):30–5.
50. Wang T, Chen JQ. Tinea capitis due to Microsporum canis. N Engl J Med. 2021;385:2077.
51. Schwartz E, Witztum E, Mumcuoglu KY. Travel as a trigger for shared delusional parasitosis. J Travel Med. 2001;8:26–8.
52. Maher S, Hallahan B, Flaherty G. Itching for a diagnosis—a travel medicine perspective on delusional infestation. Travel Med Infect Dis. 2017;18:70–2.
53. Rishniw M, Lepping P, Freudenmann RW. Delusional infestation by proxy—what should veterinarians do? Can Vet J. 2014;55(9):887–91.
54. Schwarzburg JB, Monso S, Huber L. How dogs perceive humans and how humans should treat their pet dogs: linking cognition with ethics. Front Psychol. 2020;11:1–17. https://doi.org/10.3389/fpsyg.2020.584037.

55. Holden W. Delusory parasitosis case review: what is the PMP's role? PCT Mag. 2021.
56. Nepper W. Delusion confusion: much ado about nothing. PMP Pest Man Prof Mag. 2015.
57. Ridge GE. Delusory parasitosis. The belief of being lived on by arthropods or other organisms. Guide for health departments, medical communities, and pest management professionals. Connecticut Agricultural Experiment Station. 2013. pp. 1–16.
58. Hinkle NC. Delusory parasitosis. Am Entomol. 2000;46(1):17–25.
59. Hinkle NC. Ekbom syndrome: the challenge of "invisible bug" infestations. Annu Rev Entomol. 2010;55:77–94.
60. Donabedian H. Delusions of parasitosis. Clin Infect Dis. 2007;45(11):e131–4. https://doi.org/10.1086/523004.
61. Reid EE, Lio PA. Successful treatment of Morgellons disease with pimozide therapy. Arch Dermatol. 2010;146(10):1191–3.
62. Shmidt E, Levitt J. Dermatologic infestations. Int J Dermatol. 2012;51(2):131–41.
63. Söderfeldt Y, Groß D. Information, consent and treatment of patients with Morgellons disease: an ethical perspective. Am J Clin Dermatol. 2014;15(2):71–6.
64. Balban MY, Cafaro E, Saue-Fletcher L, Washington MJ, Bijanzadeh M, Lee AM, Chang EF, Huberman AD. Human responses to visually evoked threats. Curr Biol. 2021;31(3):601–12. https://doi.org/10.1016/j.cub.2020.11.035.
65. Bremner JD, et al. Diet, stress, and mental health. Nutrients. 2020;12(8):2428. https://doi.org/10.3390/nu12082428.
66. Freeman MP, Rapaport MH. Omega-3 fatty acids and depression: from cellular mechanisms to clinical care. J Clin Psychiatry. 2011;72:258–9.
67. Freudenmann RW, Lepping P. Second-generation antipsychotics in primary and secondary delusional parasitosis: outcome and efficacy. J Clin Psychopharmacol. 2008;28(5):500–8.
68. Lynch PJ. Delusions of parasitosis. Semin Dermatol. 1993;12(1):39–45. PubMed PMID: 8476732.
69. Lepping P, Gil-Candon R, Freudenmann RW. Delusional parasitosis treated with amisulpride. Prog Neuro Psychcol. 2005;9:12–6.
70. Freudenmann RW, Schönfeldt-Lecuona C, Lepping P. Primary delusional parasitosis treated with olanzapine. Int Psychogeriatrics. 2007;19:1161–8. https://doi.org/10.1017/S1041610207004814.
71. Wong S, Bewley A. Patients with delusional infestation (delusional parasitosis) often require prolonged treatment as recurrence of symptoms after cessation of treatment is common: an observational study. Br J Dermatol. 2011;165(4):893–6.
72. Meehan WJ, Badreshia S, Mackley CL. Successful treatment of delusions of parasitosis with olanzapine. Arch Dermatol. 2006;142(3):352–5. PubMed PMID:16549712.
73. Delacerda A, Reichenberg JS, Magid M. Successful treatment of patients previously labeled as having "delusions of parasitosis" with antidepressant therapy. J Drug Dermatol. 2012;11(12):1506–7.
74. Narumoto J, Ueda H, Tsuchida H, et al. Regional cerebral blood flow changes in a patient with delusional parasitosis before and after successful treatment with risperidone: a case report. Prog Neuropsychopharmacol Biol Psychiatry. 2006;30:737–40.
75. Akahane T, Hayashi H, Suzuki H, et al. Extremely grotesque somatic delusions in a patient of delusional disorder and its response to risperidone treatment. Gen Hosp Psychiatry. 2009;31:185–6.
76. Seideman MF, Seideman TA. A review of the current treatment of Tourette syndrome. J Pediatr Pharmacol Ther. 2020;25(5):401–12.
77. Brownstone ND, Koo J. Recent developments in psychodermatology and psychopharmacology for delusional patients. Cutis. 2021;107(1):5–6. https://doi.org/10.12788/cutis.0144.
78. Nguyen CM, Danesh M, Beroukhim K, Sorenson E, Leon A, Koo J. Psychodermatology: a review. Pract Dermatol. 2015:49–54.
79. Croskerry P. Achieving quality in clinical decision making: cognitive strategies and detection of bias. Acad Emerg Med. 2002;9(11):1184–204.

80. Freudenmann RW, Kühnlein P, Lepping P, Schönfeldt-Lecuona C. Secondary delusional parasitosis treated with paliperidone. Clin Exp Dermatol. 2008;34(3):375–7.
81. Koblenzer CS. Pimozide at least as safe and perhaps more effective than olanzapine for treatment of Morgellons disease. Arch Dermatol. 2006;142(10):1364.
82. Driscoll MS, Rothe MJ, Grant-Kels JM, Hale MS. Delusional parasitosis: a dermatologic, psychiatric, and pharmacologic approach. J Am Acad Dermatol. 1993;29(6):1023–33. Review. PubMed PMID: 7902366.
83. Wilson FC, Uslan DZ. Delusional parasitosis. Mayo Clin Proc. 2004;79(11):1470.
84. Geary MJ, Russell RC, Moerkerken L, Hassan A, Doggett SL. 30 years of samples submitted to an Australian medical entomology department. Aust Entomol. 2020;60(1):172–97. https://doi.org/10.1111/aen.12480.
85. Brett AS, McCullough LB. Addressing requests by patients for non-beneficial interventions. JAMA. 2012;307(2):149–50. https://doi.org/10.1001/jama.2011.1999.
86. Martins AC, Mendes CP, Nico MM. Delusional infestation: a case series from a university dermatology center in São Paulo, Brazil. Int J Dermatol. 2016;55(8):864–8. https://doi.org/10.1111/ijd.13004. Epub 2015 Oct 16. PubMed PMID: 26475644.
87. Hinkle NC. Ekbom syndrome: a delusional condition of "bugs in the skin". Curr Psychiatry Rep. 2011;13(3):178–86. https://doi.org/10.1007/s11920-011-0188-0. PMID: 21344286.
88. Payne LM. Guérir quelquefois, soulanger souvent, consoler toujours. Br Med J. 1967;4:47–8.
89. Russell IJ. Consoler Toujours-to comfort always. J Musculoskeletal Pain. 2000;8(3):1–5.
90. Bransfield RC, Friedman KJ. Differentiating psychosomatic, somatopsychic, multisystem illness and medical uncertainty (review). Healthcare. 2019;7(114):1–28.
91. Maitra A, Verghese A. Diagnosis and the illness experience. Ways of knowing. JAMA. 2021;326:1907. https://doi.org/10.1001/jama.2021.19496.
92. Brennan TA, Leape LL, Laird NM, Hebert L, Localio AR, Lawthers AG, Newhouse JP, Weiler PC, Hiatt HH. Incidence of adverse events and negligence in hospitalized patients: results of the Harvard medical practice study I. N Engl J Med. 1991;324(6):370–6.
93. Croskerry P. The importance of cognitive errors in diagnosis and strategies to minimize them. Acad Med. 2003;78(8):775–80.
94. Saposnik G, Redelmeier D, Ruff CC, Tobler PN. Cognitive biases associated with medical decisions: a systematic review. BMC Med Inform Decis Mak. 2016;16:138. https://doi.org/10.1186/s12911-016-0377-1.
95. Thomas EJ, Studdert DM, Burstin HR, et al. Incidence and types of adverse events and negligent care in Utah and Colorado. Med Care. 2000;38:261–2.
96. Einhorn HJ, Hogarth RM. Behavioral decision theory: processes of judgment and choice. Annu Rev Psychol. 1981;32:53–88.
97. Wilson RM, Runciman WB, Gibberd RW, et al. The quality in Australian health care study. Med J Aust. 1995;163:458–71.
98. Ogdie AR, Reilly JB, Pang WG, Keddem S, Barg FK, Von Feldt JM, Myers JS. Seen through their eyes: residents' reflections on the cognitive and contextual components of diagnostic errors in medicine. Acad Med. 2012;87(10):1361–7.
99. Geller G, Tambor ES, Chase GA, Holtzman NA. Measuring physicians' tolerance for ambiguity and its relationship to their reported practices regarding genetic testing. Med Care. 1993;31:989–1001.
100. Hogarth RM. Judgement and choice: the psychology of decision. Hoboken, Wiley; 1987. p. 250.
101. Fauci SAS, Braunwald E, Kasper DL, Hauser SL, editors. The practice of medicine: iatrogenic disorders. Harrison's principles of internal medicine. 15th ed. New York: McGraw Hill. 2001.
102. Krishnan NR, Kasthuri AS. Iatrogenic disorders. Med J Armed Forces India. 2005;61(1):2.
103. Sarangi MP, Maini A, Sharma GK. Drug, and product liability. Ind J Clin Pract. 1995;5(9):94–6.
104. Swick BL, Walling HW. Drug-induced delusions of parasitosis during treatment of Parkinson's disease. J Am Acad Dermatol. 2005;53(6):1086–7.

105. Kölle M, Lepping P, Kassubek J, Schoenfeldt-Lecuona C, Freudenmann RW. Delusional infestation induced by piribedil add-on in Parkinson's disease. Pharmacopsychiatry. 2010;43(06):240–2. https://doi.org/10.1055/s-0030-1261881.
106. Mitchell J, Vierkant AD. Delusions and hallucinations of cocaine abusers and paranoid schizophrenics: a comparative study. J Psychol. 1991;125(3):301–10. https://doi.org/10.1080/00223980.1991.10543294.
107. Torales J, Almirón-Santacruz J, Barrios I, O'Higgins M, Melgarejo O, Navarro R, González I, Jafferany M, Castaldelli-Maia JM, Ventriglio A. "Cocaine bugs": implications for primary care providers. Prim Care Compan CNS Disord. 2022;24(2):39789.

Part III
Specific Issues

1.1 Introduction

Following is a series of chapters by the authors, experts in their own disciplines. The aim is to provide "pearls of wisdom" and insight supported by a considerable knowledge of DI gathered over years of experience, collected into one place. Our purpose is to impart knowledge and experience for the one purpose of providing information for physicians so that they may become empowered to better help their patients and also themselves. Chapters are in alphabetical order by author or lead author.

Practical Management of Delusional Infestation (DI)

Nicholas Brownstone, Sahithi Talasila, and John Y. M. Koo

1   Introduction

Delusional infestation (DI), also known as delusions of parasitosis (DOP) and sometimes by the eponym Morgellons disease or Ekbom syndrome, is a condition where a patient has a fixed false belief that he or she is infested with parasites [1]. Patients may also complain about foreign material, such as fibers, extruding from their skin. Clinical encounters with patients suffering from DI can cause unease, and even resentment, for dermatologists whose expertise lies in identifying objective evidence of skin disease. Dermatologists can become uncomfortable when placed in a situation where they must conduct their office visit mostly focused on the patient's thinking and behavior. This issue is exacerbated by the fact that few dermatology residency training programs offer formal training in psychodermatology, no official psychodermatology fellowship exists, and there is a paucity of resources for the practicing dermatologist such as specialized clinics for these patients [2]. This creates a situation where a dermatologist consciously or subconsciously may feel aversion for having to deal with patients who insist that their specific, and sometimes bizarre, ideation is a reality.

Furthermore, providers in the United States are now under the system of "value-based reimbursement," meaning that physicians' overall reimbursements are

N. Brownstone (✉)
Department of Dermatology, Temple University Hospital, Philadelphia, PA, USA

S. Talasila
Sidney Kimmel Medical College, Thomas Jefferson University Hospital,
Philadelphia, PA, USA
e-mail: sahithi.talasila@students.jefferson.edu

J. Y. M. Koo
Department of Dermatology, University of California, San Francisco,
San Francisco, CA, USA

G. E. Ridge (ed.), *The Physician's Guide to Delusional Infestation*,
https://doi.org/10.1007/978-3-031-47032-5_3

determined in large part by patient satisfaction scores. In this system, physicians may be penalized financially if their satisfaction score is below the 50th percentile compared to their peers. The total patient satisfaction score is based on the input from patients who are motivated enough to spend their time and energy rating their dermatology providers. It is well known that DI patients who are angry at their dermatology providers will be motivated to make sure that their dissatisfaction is well known, potentially causing patient satisfaction scores to decrease for dermatology providers. As such, even if a particular dermatology provider is not specifically interested in psychodermatology, it is prudent to learn a few tips regarding how to handle these patients. This, at a minimum, could help dermatologists from being globally penalized financially from these encounters. DI patients are known to "doctor shop" among many dermatologists in a certain area, hoping to find a provider who will take their concerns seriously. Thus, if one provider chooses to dismiss this patient, they will likely end up in your colleague's office. If no provider tries to treat these patients, then they will simply be passed down the line and an attempt to help or treat these patients will never come to fruition. Fortunately, dermatologists can, with training, diagnose DI, build rapport, and therapeutically treat DI patients.

2 Diagnosing DI

The evaluation of patients with suspected DI begins with a careful skin examination to exclude a true infestation or inflammatory disease as the cause for the reported symptoms. Every patient that presents with DI or is referred for this condition should be assumed to have a bona fide skin condition until proven otherwise. Patients should undergo a full body skin exam along with, when indicated, bacterial/fungal or viral culture and KOH preparations of any suspected areas of infestation. At the very least, these procedures help the patient feel cared for and mitigate negative feelings the patient may have towards the provider. Thus, a diagnosis of DI can only be considered after true dermatoses are ruled out after a thorough history and physical examination. These true dermatoses can include eczema, Grover's disease, and scabies.

Patients determined to have DI are categorized according to whether they have a spontaneous disorder, i.e., primary DI, or secondary DI consequent to conditions that include illicit substance abuse or withdrawal, organic brain syndrome, schizophrenia, major depression, hyperthyroidism, and vitamin B12 deficiency, among others [3–5] (Part 1, Table 1). In case of primary and secondary DI, the age range varies and does not show a predominance in female patients. Primary DI should be treated by the psychiatrist, while secondary cases are best co-managed by a psychiatrist and a dermatologist. It is important to realize that both primary and secondary DI patients generally end up in the dermatologist's office because they believe they have a skin problem and simultaneously can be aversive to any inkling of mental health or psychiatry [6]. The following laboratory tests may be considered in aiding

Table 1 The Koo-Brownstone staging system for delusional infestation patients

Koo-Brownstone stage	Name	Description	Patient quote example	Management pearls
1	Formication only (patient has insight)	• Patients only complain about strange skin sensations and do not believe that they have any parasites • Patients do not usually bring in specimens	"I feel crawling, biting and stinging sensations on my skin but it's probably not a parasite because I haven't seen any on my skin"	• Requires general pruritus workup • Treat for pruritus, if appropriate • Patients still may respond to pimozide despite having no ideation regarding parasites or foreign bodies • Can talk to these patients normally
2	Overvalued ideation of parasitosis (OIP), (patient has mild decline in insight)	• Patients experience more pronounced formication and mental preoccupation • Patients have insight into their disease and are open to etiologies other than parasites • Often apologetic	"I know this sounds crazy and sorry for mentioning this, but I worry that my crawling and biting sensations are due to parasites"	• Can openly discuss several diagnostic possibilities • Can even warn the patient about the possibility of becoming delusional • Can introduce oral medication even on the first visit
3	Pre-delusional (patient has moderate decline in insight)	• May initially appear delusional but with good rapport, the patient proves to be open-minded • Patients are worried about parasites but not entirely fixated on parasitosis and frequently are willing to accept treatment	"I am very worried about parasites but ultimately I really don't care if I have parasites or not, as long as we can get rid of this problem"	• Requires good rapport before risking asking the critical question of "how important is it for you that this problem is caused by something alive" (helps to better categorize the patient) • If the patient proves not fully delusional, then the rest of the discussion regarding treatment can be a normal conversation

(continued)

Table 1 (continued)

Koo-Brownstone stage	Name	Description	Patient quote example	Management pearls
4	Delusional (patient has very severe decline in insight)	• Patients absolutely believe that they have a parasite • Strongly averse to any inkling of mental issues • Patient may still consider "trial and error" treatment after a strong rapport is built and especially if the patient is encouraged to realize how important it is to find a way out of their miserable condition	"I am infested with parasites, and I am so miserable, so I am willing to try this medication to try to get out of the present misery even though you tell me that nobody really knows how this medication works for my condition"	• Requires strong rapport to treat patient • Solely for practicality, recommend using eponym such as "Morgellons disease" or similar nomenclature not "delusional infestation" so as not to offend the patient • Offer oral medication as "trial and error approach" • Emphasize their misery and de-emphasize etiology to help encourage oral therapy • If the patient is still fixated on etiology and not ready to try oral medications, encourage the patients to see other specialists such as parasitologists, entomologists, and infectious disease experts, but still offer future assistance

Table 1 (continued)

Koo-Brownstone stage	Name	Description	Patient quote example	Management pearls
5	Terminally delusional (patient has absolutely no insight)	• The only thing these patients want is validation about their parasitic beliefs • As soon as they sense that the providers are not going to agree with them, the patient may storm out of the room • Very difficult to treat • Have seen multiple providers already	"The only thing I care about is that you confirm for me that my symptoms are caused by parasites"	• Requires short, frequent follow-up visits to show support (short visits also help the provider to avoid burnout) • Ok to offer patient referral to parasitologist, entomologist, etc • May not be possible to introduce oral medication, but a supportive/positive rapport may still be beneficial • Consider interdisciplinary care with a psychiatrist, but this may be very difficult

to elucidate a secondary cause of DI and to establish a patient's general health status, especially if the patient is starting a systemic medication: Complete blood count with differential, serum electrolytes, liver function tests, thyroid function tests, serum calcium, serum glucose, serum creatinine, blood urea nitrogen, vitamin B12, urinalysis, toxicology screen, HIV, hepatitis C, and PRP for syphilis. Still, the three most useful tests in elucidating etiology are thyroid function tests, vitamin B12 level, and toxicology screen. Although further data are needed, there is some evidence that DI symptomology may in part be caused by dysregulation of the dopamine transporter system (DAT) as age-related loss of striatal DAT is much more prominent in older women and primary cases of DI are also seen mostly in postmenopausal older women [7].

Finally, not all patients who present with a chief complaint of "parasites" are fully delusional. It is important to know which stage the patient is in because this will help guide management, treatment, and appropriate communication. This is especially important in the rapport-building stage. Patients who are terminally delusional in contrast to patients who present complaining only about formication should be handled in different ways. The "Koo-Brownstone" staging system for delusional infestation offers advice on how to communicate and manage DI patients based on the stage or severity of DI disease and aims to help dermatology providers connect effectively with them (Table 1) [8].

3   How to Build Rapport with a DI Patient

Building a strong rapport with these patients is one of the most important principles of effective management. There is a high likelihood that the patient has had several negative encounters with other physicians or other healthcare providers. These patients may already feel quite skeptical or even hostile, even before presenting to the office. As such, first impressions count. Before entering the room to meet a patient with a chief complaint of "parasites coming out of my skin," take a few moments outside of the door to make sure that you are in an open and positive mindset. If an adequate rapport is not established, then it does not matter if we can prescribe appropriate or effective treatment, because the patient is unlikely to accept any of our advice or fill their prescriptions.

It is also important to avoid the diagnosis of "delusional infestation" either as diagnostic billing codes or for the purposes of documentation. Instead, we most often use the ICD diagnosis of "formication" or "(cutaneous) dysesthesia." Moreover, we deliberately avoid using terms that are offensive to the patient, such as "psychosis," "crazy," "psychogenic," or "delusional" in our notes. Medical records can now be accessed on demand by patients in many states without having to put a request in with medical records or their providers. Using such terminology for documentation, coding, and diagnosis risks undermining therapeutic rapport, which takes much effort to build with DI patients. Regarding the need to communicate to other clinicians when the patient is truly delusional and may even have bizarre delusions, a good way to communicate effectively in the medical record without using offensive terms is to use verbatim quotes of the patient's ideation (especially the most bizarre aspects), which any reasonable practitioner can recognize as delusional. For example, you might write in the history of present illness, "patient reports he sees a purple and pink parasite with 100 legs and 50 eyes fly out of his skin every night before bed." Furthermore, we do not refer to our clinic as "psychodermatology clinic" or our patients as "psychodermatology patients." These patients are usually referred to as "VIP" patients. The office staff knows what a "VIP" patient refers to, and they also know to schedule these patients at the end of the clinic, to not disrupt a busy clinic day.

Patients may bring in a wide variety of samples such as bundles of dirty paper towels, plastic containers with floating material inside them, random trash, debris, or anything else that one can imagine. If this is the case, we do not dismiss the samples as this could damage rapport. However, we also do not try to put them on our microscope. Instead, the patient should be thanked for bringing in the specimens, but nicely told that each specimen needs to be prepared in a very specific way as it cannot adequately be analyzed without examining them under a microscope. Patients should be given a few slides (with a warning to not break them in their pocket) and told that the most critical step of slide preparation is to use "absolutely clear," transparent, tape to fix the specimen on the slide and not to use opaque tape, which does not allow detail to be seen. This method helps to reduce the amount of material that DI patients tend to bring to the office while helping to further build trust and rapport [9, 10].

Sometimes, patients may appear as if they are only interested in getting validation that their skin problem is caused by a parasite. The best way to handle a patient with this disposition is to shift their focus from etiology to therapy by helping them become more aware of and acknowledge how miserable this condition is and how much it is distressing them. Their problem can be referred to as a "mysterious condition" for which no one knows the etiology, which is technically true because an exact etiology for this disease has not been delineated. Rather than focusing on etiology, you can inform the patient that there is a medication which usually works very well for this condition if they are willing to be pragmatic and try it on a "trial and error" basis. At the end of the day, the patient will usually appreciate the fact that there is a chance to treat and possibly even cure their condition, and if therapy and recovery are emphasized, rather than the etiology, the patient will likely be willing to accept treatment. On the other hand, if they still insist on investigation, the patient can be referred to an infectious disease specialist, entomologist, or parasitologist with the reassurance that if these providers cannot help the patient, they are always welcome back for follow-up and the possibility of a "trial and error" approach to treatment [11]. Rapport is the key. If the rapport is not there, then it does not matter what medications are prescribed because the patient will not take them.

4 Practical Psychopharmacology of DI

As stressed above, it is well known that DI patients can be reluctant to accept anything that has a psychiatric indication or is linked to mental health in anyway. Luckily, there is one medication, namely pimozide (Orap®), which works very well for the treatment of DI [12, 13]. Not only is pimozide efficacious, but it is also uniquely acceptable to these patients due to the fact it has no official U.S. Food and Drug Administration (FDA) psychiatric indication: it is only indicated in the United States for Tourette's syndrome, which is a neurological disorder. Therefore, patients who refuse all psychiatric medications are much more willing to try pimozide for this reason. The dosing strategy resembles a trapezoid figure (Fig. 1). Pimozide is usually started with very low dose of 0.5–1 mg per day. Then the dose is increased by no more than 0.5 mg increments every 2–4 weeks until an efficacious dose is achieved.

This dose is continued for 3–4 months before any further titrations are made. There is no absolute maximum dose; however, most patients show clinical response by the time the dose of 3 mg per day is reached. Patient usually does not have recurrences if the medication is tapered slowly (Fig. 1), but if the patient experiences a recurrence during tapering of the dose, the same trapezoid dosing method can be used in reverse (i.e., increase the dose again) to treat these recurrent symptoms effectively. The same dosing strategy can be used for risperidone, an atypical

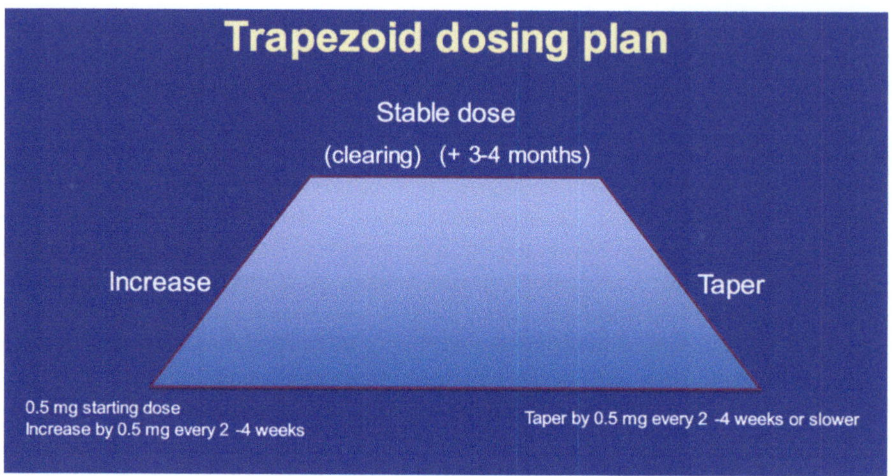

Fig. 1 Trapezoid dosing regimen for pimozide and risperidone in the treatment of DI

antipsychotic agent which can also be used to treat DI. Risperidone does have an FDA-approved indication for schizophrenia, a psychiatric disease.

Common side effects include extrapyramidal side effects such as minor stiffness or restlessness. If restlessness occurs, the medication should be given in the morning. Patients should also be instructed to carry diphenhydramine (which can be used up to four times per day) to help combat stiffness that is sometimes experienced on the medication (pseudoparkinsonism or extrapyramidal side effects).

One of the feared complications of antipsychotic therapy is tardive dyskinesia (TD), which is irreversible and involuntary movements that are usually concentrated in the face. Tardive dyskinesia is commonly seen as rhythmic movements of the lips. While cases of TD have practically never been reported with the dosing used in dermatology, there are now two FDA-approved medications approved to treat TD. Valbenazine and deutetrabenazine, VMAT-2 inhibitors, are thought to decrease dopamine-mediated neurotransmission in the brain by preventing dopamine in the presynaptic terminal from being packaged into vesicles [14]. While TD is usually treated by a neurologist or psychiatrist, it is good for dermatologists to know about these options to make them more comfortable in using antipsychotic medications in the treatment of DI.

5 Conclusion

Many dermatology providers may find it challenging when seeing a patient with DI. However, these patients are most often reachable when approached in a deliberate, sympathetic, and systematic manner as outlined here. Understanding the rapport-building techniques and with the guidance on the use of pharmacotherapy presented here, DI is often treatable and sometimes even curable. With preparation and successfully executing proper treatment, including the use of the trapezoid dosing strategy for pimozide or risperidone, these patients could become some of the most grateful seen in your practice.

References

1. Murase JE, Wu JJ, Koo J. Morgellons disease: a rapport-enhancing term for delusions of parasitosis. J Am Acad Dermatol. 2006;55(5):913–4.
2. Patel A, Jafferany M. Multidisciplinary and holistic models of care for patients with dermatologic disease and psychosocial comorbidity: a systematic review. JAMA Dermatol. 2020;156(6):686–94.
3. Alves CJM, Martelli ACC, Fogagnolo L, Nassif PW. Secondary Ekbom syndrome to organic disorder: report of three cases. An Bras Dermatol. 2010;85(4):541–4.
4. Durand C, Mary S, Brazo P, Dollfus S. [Psychiatric manifestations of vitamin B12 deficiency: a case report]. L'Encephale. 2003;29(6):560–565.
5. Nakamura M, Koo J. Drug-induced tactile hallucinations beyond recreational drugs. Am J Clin Dermatol. 2016;17(6):643–52.
6. Brownstone ND, Koo J. Special consideration for patients with Morgellons disease, even among psychodermatology patients? JAMA Dermatol. 2020;156(10):1142.
7. Huber M, Kirchler E, Karner M, Pycha R. Delusional parasitosis and the dopamine transporter. A new insight of etiology? Med Hypotheses. 2007;68(6):1351135–8.
8. Brownstone N, Koo J. The Koo-Brownstone staging system as a tool to assist in the management of patients with a possible diagnosis of dermatological delusions: an expert's suggestion. J Dermatol Treat. 2022;17:1–3.
9. Jeon C, Nakamura M, Koo J. Examining specimens brought in by a patient with delusional parasitosis. J Am Acad Dermatol. 2018;78(1):e1.
10. Koo J, Lee CS. Delusions of parasitosis. A dermatologist's guide to diagnosis and treatment. Am J Clin Dermatol. 2001;2:285–90.
11. Cheng C, Ridge G, Koo J, Brownstone N. Improving care for delusional infestation patients: what can dermatologists learn from an entomologist? Dermatol Online J. 2021;27(11).
12. Lorenzo CR, Koo J. Pimozide in dermatologic practice: a comprehensive review. Am J Clin Dermatol. 2004;5(5):339–49.
13. Hamann K, Avnstorp C. Delusions of infestation treated by pimozide: a double-blind crossover clinical study. Acta Derm Venereol. 1982;62(1):55–8.
14. Cheng C, Brownstone N, Koo J. Treatment of tardive dyskinesia: a review and update for dermatologists managing delusions of parasitosis. J Dermatol Treat. 2022;33(3):1319–23.

Delusional Infestation

Dirk Elston and Elliott Harrison Campbell

Delusional infestation (DI) commonly masquerades as true infestation and causes misery for the patients as well as those around them. It is classified as a monosymptomatic hypochondriacal disorder. Patients present with a fixed, but false belief that they are infested with a parasite. Patients are typically older adults, although the condition may appear in younger individuals as well. Patients are often highly intelligent, high-functioning individuals who are in control of every other aspect of their lives but suffer terribly from the delusion and present with excoriations and ulcers where they try to dig out what they perceive as organisms. They commonly present with a collection of what they believe to be "bugs" they have removed from their skin.

While many patients have no other psychiatric disorders, associated comorbid psychiatric conditions are not uncommon and patients deserve a full evaluation. The condition dominates all aspects of their lives, typically disrupting personal relationships out of proportion to professional relationships. They may describe an initial exposure that they believe to be the origin of the infestation. These patients suffer tremendously—it is like living in their own personal horror movie where they are consumed by a parasite, and no one will believe them. Acknowledgement of their stress and suffering is often the first step towards developing an effective doctor–patient relationship [1–5]. If the patients are willing to trust and try the medications that are prescribed, most will respond quite well and regain control of their lives.

D. Elston (✉)
Department of Dermatology and Dermatologic Surgery, Medical University of South Carolina, Charleston, SC, USA
e-mail: elstond@musc.edu

E. H. Campbell
Department of Dermatology, Mayo Clinic, Rochester, MN, USA
e-mail: campbell.elliott@mayo.edu

G. E. Ridge (ed.), *The Physician's Guide to Delusional Infestation*,
https://doi.org/10.1007/978-3-031-47032-5_4

Each patient deserves a thorough evaluation, including history and physical examination. A dermatologist has expertise in the diagnosis of skin disorders and is in the best position to evaluate for other causes of the patient's symptoms. Scabies is a common cause of genuine infestation, and the signs of infestation may be subtle, especially in older patients. Involvement of the genitalia, umbilicus, and web space is common, and application of a drop of India ink or gentian violet with subsequent removal by rubbing alcohol may make burrows easier to identify. Zoonotic or environmental mites can cause human disease, and the specimens that the patient provides should be examined by someone skilled in medical microscopy. Zoonotic mites range in size from 0.1 to 2 mm and may be difficult to see without magnification. A biopsy can be very helpful, as wedge-shaped perivascular polymorphous infiltrates with eosinophils and endothelial swelling are characteristic of genuine arthropod reactions. The dislodged hairs of caterpillars and moths can cause itch in those exposed. Tape stripping of the involved skin can be helpful to identify the urticating hairs or environmental fiberglass that is causing itching.

Illicit drug use is also an important cause of itching and crawling sensations. Formications are a particular cause of itch secondary to cocaine addiction, but many illicit drugs as well as prescription medications cause itching. Medical causes of itch should also be considered. Diseases such as hyperthyroidism and renal and liver disease may present with generalized itch or a crawling sensation in the skin [6, 7].

Some patients will require a complete blood count, comprehensive metabolic panel, thyroid-stimulating hormone, biopsy of lesioned skin, and perilesional skin for direct immunofluorescence that may help to rule out organic disease. Examination of the house and office for evidence of nesting birds or vermin, evaluation of the vacuum cleaner bag contents, and examination of pets by a qualified veterinarian can help to rule out authentic infestation when a biopsy suggests the possibility of true infestation. If mites are identified, they should be sent to an acarologist or entomologist for identification.

For optimal diagnosis, specimens should be cleared and mounted properly. Clearing is done with lactic acid or lactophenol, and then the specimen is washed with water and mounted in Hoyer's medium, which remains clear regardless of whether mites are prepared with water or alcohol. Hoyer's medium is typically prepared by the laboratory handling the specimen. A typical formula is to add 30 g of gum Arabic to 50 cc of distilled water, and then add 200 g of chloral hydrate followed by 20 g of glycerin. The mixture is then filtered (Fig. 1).

Fig. 1 Hoyer's oil preparation

Hoyer's oil medium is typically prepared by the laboratory handling the specimen. A typical formula is to add 30 gm of gum arabic to 50 cc of distilled water, then add 200 gm of chloral hydrate followed by 20 gm of glycerin.
The mixture is then filtered and bottled.

Table 1 Common antipsychotic medications and their side effects

Medication	Usual effective dose (mg)	Common side effects	Additional considerations
Pimozide	1–10	Sedation, weight gain, extrapyramidal effects	EKG recommended
Aripiprazole	2–10	Sedation	
Risperidone	0.5–4	Weight gain, breast swelling, sedation	
Olanzapine	5–10	Sedation, weight gain, extrapyramidal effects	

Once real infestation or other organic disease has been excluded, patients typically require medical management by a qualified physician. Patients often seek multiple physicians in search of someone who will believe them that the infestation is real. Some are angry and confrontational. Establishing rapport and treating them effectively are a true art.

Topical antipruritic agents containing pramoxine, or a mixture of camphor and menthol, may provide temporary relief of itching. Second-generation "atypical" antipsychotic agents are well tolerated by most patients and are often required for effective control. In the past, pimozide was used as the first-line agent, but it has a less favorable side effect profile compared with second-generation agents. Aripiprazole, risperidone, olanzapine, paliperidone, ziprasidone, and quetiapine have been used, but some have greater risk of side effects. Risperidone at doses titrated up to 4 mg daily is very effective, but some patients complain of weight gain, breast swelling, or actual lactation. Aripiprazole at doses titrated up to 10 mg daily avoids both the weight gain and lactation and is generally very well tolerated. Olanzapine has a somewhat greater risk of extrapyramidal side effects. All patients should be monitored for tolerance as well as common metabolic side effects including alterations in serum lipids. Commonly used medications and their side effects are listed in Table 1 [8–40].

Discontinuation of therapy after clinical response commonly results in relapse of symptoms, so patients should be monitored, and an effective relationship maintained. While serious side effects are very uncommon, extrapyramidal symptoms and weight gain can be particularly troubling for patients, and those who have other cardiac metabolic risk factors should be monitored and have their cardiac risks managed appropriately. Side effects are typically dose-dependent, but uncommon even at the maximum doses listed in the table, and those doses are commonly needed to control symptoms. Movement disorders occur more frequently in elderly patients, and they should be monitored to ensure full functionality and avoid falls. Psychiatrists are very adept at managing side effects and comorbidities and can be very helpful if the patient is willing to see a psychiatrist. Usually, it is best to establish a relationship with the patient prior to making such a suggestion.

References

1. Lynch PJ. Delusions of infestation. Semin Dermatol. 1993;12(1):39–45.
2. Gee SN, Zakhary L, Keuthen N, Kroshinsky D, Kimball AB. A survey assessment of the recognition and treatment of psychocutaneous disorders in the outpatient dermatology setting: how prepared are we? J Am Acad Dermatol. 2013;68(1):47–52. https://doi.org/10.1016/j.jaad.2012.04.007. Epub 2012 Sep 3.
3. Reichenberg JS, Magid M, Jesser CA, Hall CS. Patients labeled with delusions of infestation compose a heterogeneous group: a retrospective study from a referral center. J Am Acad Dermatol. 2013;68(1):41–6. https://doi.org/10.1016/j.jaad.2012.08.006. Epub 2012 Oct 8.
4. Foster AA, Hylwa SA, Bury JE, Davis MD, Pittelkow MR, Bostwick JM. Delusional infestation: clinical presentation in 147 patients seen at Mayo Clinic. J Am Acad Dermatol. 2012;67(4):673.e1–10. https://doi.org/10.1016/j.jaad.2011.12.012. Epub 2012 Jan 20.
5. Boggild AK, Nicks BA, Yen L, Van Voorhis W, McMullen R, Buckner FS, Liles WC. Delusional parasitosis: 6-year experience with 23 consecutive cases at an academic medical center. Int J Infect Dis. 2010;1(4):e317–21. https://doi.org/10.1016/j.ijid.2009.05.018. Epub 2009 Aug 15.
6. Brewer JD, Meves A, Bostwick JM, Hamacher KL, Pittelkow MR. Cocaine abuse: dermatologic manifestations and therapeutic approaches. J Am Acad Dermatol. 2008;59(3):483–7. https://doi.org/10.1016/j.jaad.2008.03.040. Epub 2008 May 7. Review.
7. Flann S, Shotbolt J, Kessel B, Vekaria D, Taylor R, Bewley A, Pembroke A. Three cases of delusional parasitosis caused by dopamine agonists. Clin Exp Dermatol. 2010;35(7):740–2. https://doi.org/10.1111/j.1365-2230.2010.03810.x.
8. Kenchaiah BK, Kumar S, Tharyan P. Atypical anti-psychotics in delusional parasitosis: a retrospective case series of 20 patients. Int J Dermatol. 2010;49(1):95–100. https://doi.org/10.1111/j.1365-4632.2009.04312.x.
9. Wong S, Bewley A. Patients with delusional infestation (delusional parasitosis) often require prolonged treatment as recurrence of symptoms after cessation of treatment is common: an observational study. Br J Dermatol. 2011;165(4):893–6. https://doi.org/10.1111/j.1365-2133.2011.10426.x.
10. Freudenmann RW, Lepping P. Second-generation antipsychotics in primary and secondary delusional parasitosis: outcome and efficacy. J Clin Psychopharmacol. 2008;28(5):500–8. https://doi.org/10.1097/JCP.0b013e318185e774.
11. Manschreck TC, Khan NL. Recent advances in the treatment of delusional disorder. Can J Psychiatry. 2006;5(2):114–9.
12. Driscoll MS, Rothe MJ, Grant-Kels JM, Hale MS. Delusional parasitosis: a dermatologic, psychiatric, and pharmacologic approach. J Am Acad Dermatol. 1993;29(6):1023–33.
13. Bak R, Tumu P, Hui C, Kay D, Peng D. A review of delusions of parasitosis, part 2: treatment options. Cutis. 2008;82(4):257–64.
14. Generali JA, Cada DJ. Pimozide: parasitosis (delusional). Hosp Pharm. 2014;49(2):134–5. https://doi.org/10.1310/hpj4902-134.
15. Hamann K, Avnstrop C. Delusions of infestation treated by pimozide: a double blind crossover clinical study. Acta Dermatol (Stockholm). 1982;62:55–8.
16. Mothi M, Sampson S. Pimozide for schizophrenia or related psychoses. Cochrane Database Syst Rev. 2013;5(11):CD001949. https://doi.org/10.1002/14651858.
17. Gowda BS, Heebbar S, Sathyanarayana MT. Delusional parasitosis responding to risperidone. Indian J Psychiatry. 2002;44(4):382–3. PubMed PMID: 21206607; PubMed Central PMCID: PMC2955314.
18. Laidler N. Delusions of parasitosis: a brief review of the literature and pathway for diagnosis and treatment. Dermatol Online J. 2018;24(1):13030/qt1fh739nx.
19. Elmer KB, George RM, Peterson K. Therapeutic update: use of risperidone for the treatment of monosymptomatic hypochondriacal psychosis. J Am Acad Dermatol. 2000;43(4):683–6.
20. Meehan WJ, Badreshia S, Mackley CL. Successful treatment of delusions of parasitosis with olanzapine. Arch Dermatol. 2006;142(3):352–5.

21. Martins AC, Mendes CP, Nico MM. Delusional infestation: a case series from a university dermatology center in São Paulo, Brazil. Int J Dermatol. 2016;55(8):864–8. https://doi.org/10.1111/ijd.13004. Epub 2015 Oct 16.
22. Ladizinski B, Busse KL, Bhutani T, Koo JY. Aripiprazole as a viable alternative for treating delusions of parasitosis. J Drugs Dermatol. 2010;9(12):1531–2.
23. Duarte C, Choi KM, Li CL. Delusional parasitosis associated with dialysis treated with aripiprazole. Acta Med Port. 2007;24(3):457–62. Epub 2011 Aug 12.
24. Kumbier E, Höppner J. The neuroleptic treatment of delusional parasitosis: first experiences with aripiprazole. Hautarzt. 2007;59(9):728–30.
25. Altınöz AE, Tosun Altınöz Ş, Küçükkarapınar M, Coşar B. Paliperidone: another treatment option for delusional parasitosis. Austral Psychiatry. 2014;22(6):576–8. https://doi.org/10.1177/1039856214546390. Epub 2014 Aug 21.
26. De Berardis D, Serroni N, Marini S, Rapini G, Valchera A, Fornaro M, Mazza M, Iasevoli F, Martinotti G, Di Giannantonio M. Successful ziprasidone monotherapy in a case of delusional parasitosis: a 1-year follow-up. Case Rep Psychiatry. 2013;2013:913248. https://doi.org/10.1155/2013/913248. Epub 2013; May 16. PubMed PMID:23762722; PubMed Central PMCID: PMC3670508.
27. Milia A, Mascia MG, Pilia G, Paribello A, Murgia D, Cocco E, Marrosu MG. Efficacy and safety of quetiapine treatment for delusional parasitosis: experience in an elderly patient. Clin Neuropharmacol. 2008;31(5):310–2. https://doi.org/10.1097/WNF.0b013e3181587ce4.
28. Szepietowski JC, Reszke R. Psychogenic itch management. Curr Probl Dermatol. 2016;50:124–32. https://doi.org/10.1159/000446055. Epub 2016 Aug 23.
29. Wykoff RF. Delusions of parasitosis: a review. Rev Infect Dis. 1987;9(3):433–7.
30. Risperdal (risperidone) [Prescribing information]. Titusville, NJ: Janssen Pharmaceuticals Inc.; 2017.
31. Orap (pimozide) [prescribing information]. Sellersville, PA: Teva Pharmaceuticals; 2014.
32. Aripiprazole orally disintegrating tablets [prescribing information]. Bridgewater, NJ: Alembic Pharmaceuticals; 2017.
33. Zyprexa (olanzapine) [prescribing information]. Indianapolis, IN: Lilly USA LLC; 2018.
34. Koo J, Lee CS. Delusions of parasitosis. A dermatologist's guide to diagnosis and treatment. Am J Clin Dermatol. 2001;2(5):285–90.
35. Lepping P, Russell I, Freudenmann RW. Antipsychotic treatment of primary delusional parasitosis: systematic review. Br J Psychiatry. 2007;191:198–205.
36. Möller HJ, Riedel M, Jäger M, Wickelmaier F, Maier W, Kühn KU, et al. Short-term treatment with risperidone or haloperidol in first-episode schizophrenia: 8-week results of a randomized controlled trial within the German Research Network on Schizophrenia. Int J Neuropsychopharmacol. 2008;11:985–97.
37. Thomson SR, Chogtu B, Bhattacharjee D, Agarwal S. Extrapyramidal symptoms probably related to risperidone treatment: a case series. Ann Neurosci. 2017;24(3):155–63. https://doi.org/10.1159/000477153.
38. Woerner MG, Alvir JMJ, Saltz BL, et al. Prospective study of tardive dyskinesia in the elderly: rates and risk factors. Am J Psychiatry. 1998;155:1521–8.
39. Saltz BL, Robinson DG, Woerner MG. Recognizing and managing antipsychotic drug treatment side effects in the elderly. Prim Care Companion J Clin Psychiatry. 2004;6(suppl 2):14–9.
40. Adler L, Angrist B, Peselow E, et al. Noradrenergic mechanisms in akathisia: treatment with propranolol and clonidine. Psychopharmacol Bull. 1987;23:21–5.

Delusions of Infestation in the Media and on the Internet

Erika Engelhaupt

In the summer of 2018, an Idaho woman we will call Lisa was treating her backyard chickens for poultry mites when she saw mites land on her own skin and clothing. Not long after, she started to feel an itch, a crawling sensation that would not go away. She remembered the mites and wondered if they could be the problem.

"I was desperate to feel better," Lisa said, so she got on the Internet to search for information about mites. "I feel like I got catfished," she said. "I started reading everything about mites and people with mites." She came across a book called "The Year of the Mite," by Jane Ishka [1], a San Francisco woman who contends that she was infested with bird mites and developed extensive cleaning protocols to rid herself and her home of them.

From there, Lisa (whose privacy I am protecting with a pseudonym) went down a rabbit hole that will be familiar to any physician or expert who has encountered people with delusional infestation (DI). Because she thought her problem was the mites, it never occurred to her to search for information about a psychiatric condition. In fact, when I spoke to her the following year as a journalist covering the condition, Lisa had never heard of the term "delusional infestation."

Instead, she found a wealth of misinformation online about mites and insects infesting the bodies of humans. She tried to follow the cleaning procedures she read about: vacuuming several times a day, throwing away piles of bedding and soft goods, and spraying her home with vinegar and essential oils. She called exterminators, and eventually moved out of her home.

Lisa could have easily followed the path of many who suffer from DI, sinking deeper into despair and becoming entrenched in the belief that she was hopelessly infested. Instead, she was one of the lucky few who found a doctor willing to think outside the box to help her and an expert—entomologist Gale Ridge, who is the

E. Engelhaupt (✉)
M.S. Environmental Science, Knoxville, TN, USA

© The Author(s), under exclusive license to Springer Nature Switzerland AG 2024
G. E. Ridge (ed.), *The Physician's Guide to Delusional Infestation*,
https://doi.org/10.1007/978-3-031-47032-5_5

editor of this book—who could kindly and patiently guide her in seeking the real cause of her suffering.

Lisa's case is unusual in that early intervention allowed her to recover from DI, but the misinformation she first encountered online was all too common. What is more, coverage in the news media and popular culture (such as film and television) has not been kind to people suffering from DI. When the condition is discussed at all, which is rare, people with the condition often appear as hostile and deranged.

It is no wonder that patients do not see themselves in such portrayals. Instead, many naturally turn to sympathetic online communities formed by those whose lives have been overtaken by their condition. These websites and forums appear to the sufferer to offer practical advice as well as validation while promoting a mistrust of doctors and experts who suggest that the infestations are not real.

The situation is bleak, but there is hope for improvement. Doctors, scientists, and other professionals who encounter patients with DI can play an important role in improving public understanding of the condition, both through communication with patients and by coordinating with other health care professionals. This chapter gives an overview of the landscape of information available to the public about DI, in media portrayals of DI (in both fictional and news coverage) as well as online resources. For professionals, it is important to understand the information landscape that patients are exposed to and to acquaint themselves with the information sources that patients are likely to have seen.

I first became interested in DI as a science journalist through my contacts with entomologists, who are sometimes approached by members of the public who believe that they have an insect infestation. I became fascinated with the subject and subsequently wrote a story on the subject for the National Geographic website in 2018. Later, I wrote about the condition in a section of my book "Gory Details: Adventures from the Dark Side of Science" [2].

For the book, I visited and interviewed Dr. Nancy Hinkle of the University of Georgia and Dr. Gale Ridge of The Connecticut Agricultural Experiment Station about their work on DI. Dr. Hinkle showed me the extensive collection of samples that have been sent to her by people who are looking for a species identification of insects that they believe are infesting them. I was also able to interview two people who were affected by DI, either through family members or directly.

In searching for mainstream media about DI, I quickly found that the coverage was sparse and that only a few journalists had followed the story in any depth. The most prominent article at the time had been published in 2017, when journalist Eric Boodman [3] profiled Dr. Ridge and described the problems faced by entomologists who are thrust into the role of "accidental therapists" for people with DI. Boodman won an award for that story and others from the Council for the Advancement of Science Writing later that year.

In 2018, DI got a little more attention in the news media upon the publication of a study in *JAMA Dermatology* [4] that estimated that about 27 out of every 100,000 people in the United States have DI. This was the first population-based estimate of the condition's prevalence and likely a very conservative number since only those who sought medical treatment would be counted.

In surveying the coverage of DI, it soon becomes clear that most is framed in one of a few ways. One typical form of framing is that of a medical mystery. People love mysteries, and readers are curious about whether a condition like DI is "real," meaning based in physical disease. For instance, Morgellons disease, in which people believe that there are fibers emerging from their skin, is often presented as a mysterious illness shrouded in questions of whether this is a physiological or psychological disorder. Examples of this framing include a 2006 Associated Press article titled "Bizarre Morgellons Condition: Is It Real?" [5] and a 2008 Newsweek article titled "Is Morgellons a Real Disease?" [6].

One prominent example of such medical-mystery framing is the documentary film "Skin Deep" [7]. The film focuses on people with Morgellons disease and centers around the question of whether the condition is "real" or not, i.e., whether people with the condition actually produce mysterious fibers or whether they are experiencing delusions. The film is interesting, and I suggest watching it to see the perspective of people who believe that this is a biologically caused disorder and the very real suffering from it. Dr. John Koo, a co-author of this volume, was interviewed in the documentary.

Another example in popular fiction appears in the television show "Chicago Med" [8]. In the episode "Things Meant to Be Bent Not Broken," a female patient presents in the emergency department with lesions and formication and hands over a "sample" that she claims contains the larvae of mites that have infested her body. The sample is empty, no mites are found, and the patient is swiftly diagnosed with DI and responds to antipsychotic drugs within 24 h. The speed with which the patient is diagnosed and treated is clearly unrealistic, but the episode does present some of the real symptoms and challenges to treatment of DI. For example, the psychiatrist treating the patient struggles with the ethics of prescribing an antipsychotic drug without informed consent.

Another common framing that the media uses for discussing DI focuses on mental illness. One example is seen in the play "Bug," which was turned into a movie of the same name [9] starring Ashley Judd. It features two people who sequester themselves in a motel room based on the shared belief that they have been infested with bugs as part of a government conspiracy. It frames the progression of the disorder as a "descent into madness" story, ending with the deaths of both characters in a fiery attempt to eradicate themselves of their infestations.

Some news coverage also focuses on the more bizarre aspects of the conditions. For example, a 2018 article in The Straits Times used the headline "She burned her skin to kill insects and dust-infested body" [10]. This framing has the unfortunate effect of painting sufferers of the conditions as bizarre or freakish, which distracts from the useful information in the text.

Apart from the mainstream media, much of what is publicly available about DI lives on the internet as a mix of information, misinformation (which is false or out of context), and even disinformation (which is purposely false or misleading). Informational websites range from the publicly sourced Wikipedia entry [11] to fact sheets on medical and university websites such as the Mayo Clinic [12]. Academic journals are also important resources for professionals but are largely inaccessible to the public.

Finally, publicly available health websites and academic sources tend to frame DI as a mental health issue. For example, two of the largest health-oriented websites, Healthline.com [13] and Health.com [14], have articles on DI that cite reputable sources such as the American Psychological Association. While these websites provide useful information, the framing around mental illness is likely to turn off people who are suffering from the condition. Indeed, it is hard to imagine that very many people with DI would even find such websites, since they are typically searching for information about infestations, not delusions.

In contrast, a great deal of misinformation and disinformation targets DI sufferers directly, including blogs and websites written by patients, sellers of sham solutions and pesticides, and support groups and organizations that perpetuate the idea that people's bodies can be infested by mites and other arthropods.

This is why it is important for more scientists and health care providers to engage in publishing accurate information about DI online. First, keep in mind that much of the best information about DI is relatively inaccessible to the public. To read an article in an academic journal, one must sign in and pay for a subscription or per-article fees. If you are an academic, doctor, or other expert planning to publish anything on this subject, think about your audience and how you are going to reach them. If you write a journal article intended for peers, for example, think about how that information could be presented on your own website or that of your institution in another way that is useful for the public.

One good option is to publish fact sheets. For instance, Dr. Gale Ridge has written an excellent fact sheet aimed at professionals who encounter DI for The Connecticut Agricultural Experiment Station [15]. She provides information about some of the underlying medical problems, skin conditions, and neurological disorders that can create the sensation of having insects on the skin. Her fact sheet directly addresses the symptoms people may be experiencing and, importantly, conveys that there are medical reasons for why one might have biting or crawling sensations. It is important to reassure people who are seeking information that what they are feeling is real and valid while steering them away from insect infestation as the likely explanation of their symptoms.

There are many different guides and websites on the internet that purport to show how to eradicate infestations. These can send susceptible people down a dangerous rabbit hole. Entomologists get very frustrated with some of these sites because they present inaccurate accounts of biology. It is useful for experts to be familiar with the websites that patients with DI encounter to counter them, including pest management websites aimed at eradicating mites.

I recommend doing what I call "The Google Experiment." Before you publish anything about DI, think about what information people are searching for. What search terms are they likely to use? If a person is starting to believe that they are infested, you would not expect them to Google search terms such as "delusional infestation." In their minds, they are not delusional but are feeling actual bugs biting or crawling on them. Imagine yourself in their shoes, and try searching online for answers as they would.

You might search something like "feels like bugs are biting me," "how to stop bug bites," or "mites on humans." Try it for yourself. See what your search engine serves up on the first page of results. Most likely, it is a grab bag of accurate information and misinformation.

So, here are a few recommendations from a journalist's point of view. First and foremost, foster empathy for those suffering from DI. If you are writing for or speaking to a person who might have DI, try to put yourself in their shoes and think about how they might react to your message. Consider their information landscape and how you can contribute to improving it.

More specifically, think carefully about how you use language and terminology depending on your target audience. Are you speaking to patients, or the professionals that work with them such as doctors, epidemiologists, health workers, or entomologists? For instance, using the phrase "delusional infestation" may be useful to scientists and doctors, but it will be off-putting for people who may have this disorder. They are not likely to click on or read something that suggests that their feelings are "all in their head." They are much more likely to be drawn to solution-oriented information.

In written communications, tailoring your message to your audience can start with steps as simple as using appropriate keywords and titles. By using search engine optimization (SEO) techniques, the information you want to provide can come up earlier in algorithmic searches and better compete in the vast ocean of the internet. If you do not feel comfortable doing this or do not know how to, talk to your organization's media specialists and those who manage the organization's website. You may have a community of web-savvy communicators at your organization who can help.

Finally, I will leave you with one last thought: Ignoring DI will not make it go away. It is understandable that many professionals who encounter DI patients feel that it is not their job to work with people who are delusional. But if you are willing to put yourself out there, take the time to talk to people with DI and try to provide good information for them; you really can make a difference. Not only are you helping the sufferers, but you are also helping their families, friends, colleagues, and the community they live in.

To conclude, this counsel from Boodman's STAT article sums it up: "You may not have signed up for the job of 'accidental therapist.' But if you are willing to take it on, there truly is an opportunity to help people and possibly even save lives."

References

1. Ishka J. The year of the mite. Bitingduck Press; 2016.
2. Engelhaupt E. Gory details: adventures from the dark side of science. Nat Geographic Books; 2021.
3. Boodman E. Accidental therapists: For insect detectives, the trickiest cases involve the bugs that aren't really there [Internet]. 2017. https://www.statnews.com/2017/03/22/insect-delusional-parasitosis-entomology/. Accessed 26 Oct 2022.

4. Kohorst JJ, Bailey CH, Andersen LK, Pittelkow MR, Davis MDP. Prevalence of delusional infestation—a population-based study. JAMA Dermatol. 2018;15(5):615–7. https://jamanetwork.com/journals/jamadermatology/fullarticle/2676941. https://doi.org/10.1001/jamadermatol.2018.0004.

5. The Associated Press. Bizarre Morgellons condition: Is it real? [Internet]. 2006. https://www.nbcnews.com/health/health-news/bizarre-morgellons-condition-it-real-flna1C9476993. Accessed 26 Oct 2022.

6. Newsweek. Is Morgellons a real disease? [Internet]. 2008. https://www.newsweek.com/morgellons-real-disease-93707. Accessed 26 Oct 2022.

7. Skin deep. [film] directed by: pi ware. USA: 2019. Watch options available from: https://www.imdb.com/title/tt10146070/?ref_=ttco_co_tt.

8. Things meant to be bent not broken [TV series episode]. Chicago Med Season 7, Episode 15. 9 March 2022. https://www.imdb.com/title/tt18260404/?ref_=tt_mv_close. Accessed 26 Oct 2022.

9. Bug. [Film] Directed by: William Friedkin. USA: Lions Gate Films; 2006.

10. Chian HP. Delusional parasitosis: she burned her skin to kill 'insects' that infested body [internet]. The straits Times 2018. https://www.straitstimes.com/singapore/health/she-burned-her-skin-to-kill-insects-that-infested-body. Accessed 26 Oct 2022.

11. Delusional parasitosis [Internet]. 2022. https://en.wikipedia.org/wiki/Delusional_parasitosis. Accessed 26 Oct 2022.

12. Morgellons disease: Managing an unexplained skin condition [Internet]. 2022. https://www.mayoclinic.org/morgellons-disease/ART-20044996. Accessed 26 Oct 2022.

13. What Is delusional parasitosis? [Internet]. 2022. https://www.healthline.com/health/mental-health/delusional-parasitosis. Accessed 26 Oct 2022.

14. Delusional Parasitosis Can Make People Think They Have Parasites—What to Know About the Psychiatric Condition [Internet]. 2022. https://www.health.com/condition/mental-health-conditions/delusional-parasitosis. Accessed 26 Oct 2022.

15. Ridge G. Delusory parasitosis: guide for health departments, medical communities, and pest management professionals [Internet]. 2022. https://portal.ct.gov/-/media/CAES/DOCUMENTS/Publications/Fact_Sheets/Entomology/DelusionsofParasitosis1pdf.pdf. Accessed 26 Oct 2022.

Animal Suffering and Euthanization Caused by Human Psychosis

Nancy C. Hinkle

1 Introduction

While the main focus of delusional infestation (DI) is on human suffering, involvement of pets and domesticated animals is gaining coverage [1]. In many cases, pet owners perceive (erroneously) that their animal is infested with the same organism they are convinced infests their own body, leading to treatments applied to the animal that can severely affect its health [2, 3].

Animals such as dogs and cats are maintained as pets by humans for their companionship and because we derive pleasure from them. Unfortunately, animals are at the mercy of their owners and often are subjected to cruelty by humans suffering from DI.

2 Deflecting "Infestation" onto Animals

A dairyman called his veterinarian and asked how he should treat worms in the eyes of one of his cows, describing tiny worms crawling across the animal's eyeball. When the veterinarian suggested that she come out and examine the animal, anticipating that she would find *Thelazia*, typical in North America, the dairyman reluctantly confessed that the worms were actually in his eyes. He figured that if he got the medication to treat eye worms in cattle, he could use it for his self-treatment. The veterinarian, realizing that *Thelazia* rarely infest humans, probed his story and found that he had already visited several physicians, none of whom were able to detect any evidence of worm presence, driving him to find his own treatment. The

N. C. Hinkle (✉)
Department of Entomology, University of Georgia, Athens, GA, USA
e-mail: nhinkle@uga.edu

G. E. Ridge (ed.), *The Physician's Guide to Delusional Infestation*,
https://doi.org/10.1007/978-3-031-47032-5_6

veterinarian was familiar with DI and characterized this situation as an example of DI.

When DI sufferers are projecting onto their pets, they are sucking their pets into the delusion, partially to reinforce their "evidence" and partially to gain additional sympathy. Likely what we were seeing is a variation on Munchausen's by proxy, where the human gains more concern by presenting a "suffering" animal. Pet owners do not plot this strategy, but they discover that having an "infested" animal gains them sympathy and emotionally gratifying attention:

> "I had this problem- everyone really started to believe I was crazy- if it was not for my dog who was reacting the same—I would have thought I was crazy—I went from dr to dr and very harsh medicines and ANYTHING else that I thought might ease this itching crawling."

Clearly, DI sufferers interpret visual input differently from how the rest of us perceive it, envisioning lint as arthropods and normal skin debris as parasites. If a dog scratches once a day, for the DI person it is evidence of the animal being infested. Typically, the owner is treating the animal with all sorts of materials, most of which are dermal irritants. One of the worst is diatomaceous earth (DE), a desiccant that soaks up skin oils, dries out the skin, and makes the animal feel even itchier. This creates a vicious cycle, as the owner keeps dusting on the DE and the animal scratches more.

People suffering from DI have their ideas of where the bugs come from. Remember, these are rationalizations of delusional individuals, but it is necessary to understand their view of reality to assist them. These are, obviously, attempts by delusional individuals to explain what they are experiencing. While the sufferer may attempt to make the explanation reasonable for the hearer (physician, entomologist, friend, family member), what seems possible to them generally is biologically impossible. This observation often leads to them assuring the scientist, "Well, if you identify the cause, you'll win a Nobel Prize!"

> "I've been having the same symptoms for about 8 weeks now, a feeling like bugs are crawling across my scalp, torso, ankles, and face. I have been to the Doctor twice without any solution. It is driving me nuts, it seems to start after I gave my dog a bath, I wonder if it is a parasite?"

Probably the most common source claimed is animals. It may be the neighbor's dog, a wild animal like ducks at the park, or a squirrel in the attic. Of 178 random DI callers, 28% said that they acquired their infestation from an animal, including chipmunks, a raccoon found in the attic, birds nesting in the eaves, mice, chickens, rats, pet birds, an opossum denning under the house, flying squirrels in the attic, and so on. Because of their close relationship in our lives, pets are the animals most likely to become enmeshed in their owners' delusions. Logically, DI sufferers conclude that to eliminate their infestation, they must simultaneously eradicate it from their pets, from which they assume they are continually reinfested:

> "My dog had another lime dip bath last week and he seems to be much better. I think once the dog is free, my cure will come fast."

"They are black triangle shaped bugs that bite just my dog and I. My doctor told me there are no bugs and for me to make an appointment with mental health. I'm tired of being shrugged off by the medical community. I already have health problems and am getting worse. I need help fast."

Seeking advice online:

"Help! We are experiencing pinpoint black bugs. They bite and get on us and our puppy. There are no trails; we just see them in groups. We found them on our mantel, our clothes, our hair, and our toilet. We were recommended to use bed bug spray, but they keep coming back."

"We don't know what they are. We thought maybe bird mites, but we have no birds, only city birds. What can we do to get rid of them? Please help. This is awful."

Pet owners request euthanization of animals they perceive to be infested. While these are perfectly healthy animals, the owner will often refuse rehoming the animals due to their misapprehension that the infestation might be perpetuated, and new owners infested [4]. They feel they have a moral obligation to destroy the animal to prevent their own reinfestation and to prevent the bugs moving to someone else. Typically, the loss of their pet and resultant depression worsens their situation.

As has been demonstrated, DI sufferers treat their animals with all sorts of concoctions, many highly destructive to their skin, including insecticides, kerosene, gasoline, turpentine, and other harsh chemicals; if the animal didn't have dermal issues to begin with, it certainly will after these treatments. In order to procure specimens to prove that their animal is infested, often they will scrape or dig into the animal's skin (Fig. 1).

Generally, we concur that adults have the right to make decisions (even bad decisions) about themselves; concern develops when others, especially those unable to protect themselves like children, the disabled, and pets, are affected by the delusional individual's choices. Animals, considered property of their owners, are particularly at risk because they are completely vulnerable to the whims of their owners' delusions.

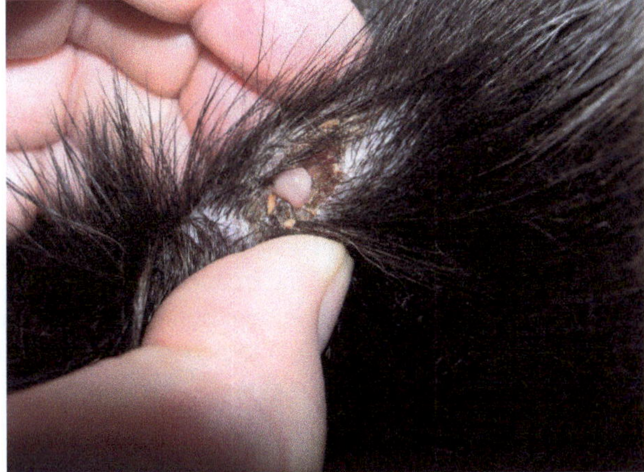

Fig. 1 Delusional pet owners will dig into their pet's skin just as they do into their own

3 It's Not Bird Mites

For over 30 years, I have worked on avian mites on both domestic poultry and wild birds. Most of this field work, in which my collaborators and I were handling mite-infested birds (mostly *Ornithonyssus* spp.), had us frequently in close and sustained contact with high numbers of mites. It was not at all uncommon for hundreds of mites not only to crawl on us but also to actively probe our skin, producing transient pruritus. However, despite regular prolonged exposure, none of this group (over two dozen people over a period of 30 years) had sustained a mite infestation.

Even more telling, mites are very common in laying hen facilities around the world. Workers in these operations are constantly exposed to northern fowl mites, *Ornithonyssus sylviarum* (Canestrini and Fanzago). Yet none of them acquires the infestation, nor do they carry mites home to members of their household. There are well over 320,000 workers on U.S. egg farms who daily are exposed to chicken mites, yet I have never had one of them call me to complain of a mite infestation on them or another human. Conversely over 1200 people from every other walk of life—teachers, engineers, bus drivers, construction workers, physicians, office workers, etc.—have contacted me about their "invisible mite" infestations. It is impossible to prove a negative, of course, but the overwhelming evidence indicates that avian mites are *not* capable of sustaining themselves on a mammal (including humans).

4 What is the Role of the Veterinarian in Delusional Infestation Cases?

Unfortunately practicing veterinarians encounter DI clients in their work. How do they respond when a client demands that an animal be treated for "parasites" when there are none present? And how does a veterinarian respond when the client demands the animal be euthanized to avoid reinfesting its owner?

One of the most revealing papers about veterinary DI was written by a practicing veterinarian sharing first person accounts of other veterinarians who have encountered DI in their clients [4]. In these situations, veterinarians are challenged to treat animals for non-existent infestations. Even more disturbing, often the client wants to have the animal euthanized, supposedly to relieve the animal's suffering, or to prevent the animal from harboring the parasite and serving as a source of reinfestation to the human. In the first instance, the owner is convinced that the animal is suffering and, since none of the treatments are working, the animal should be put down. Probably the more common rationale is that the animal is sustaining the parasite and allowing the infestation to persist, even when it is successfully eliminated from the human body.

When you tell the client it is a medical condition and they need to consult their physician, they reply that they have been to a dozen physicians and none of them were able to diagnose their problem, which is true. Overall, physicians unfamiliar with delusional infestation are reluctant to expend the time and effort to diagnose and treat these patients because

1. DI sufferers are not cooperative and will not work with the physician to try to get better.
2. DI sufferers require more time and effort than other patients.
3. Because DI sufferers typically are uninsured, the practice will never get paid.
4. DI sufferers are more likely to sue their practitioner than are other patients.

Sadly, because the animal's suffering is a direct result of the client's delusion, the animal has no chance for relief until the client's condition is successfully treated.

Delusional infestation should be considered when the condition has persisted for months or years (conditions not characteristic of valid arthropod infestations), and the pests are invisible or visible only to the sufferer.

Delusional infestation sufferers are not imagining or pretending, they are fully convinced that the bugs are real. Although itching is discussed as the main complaint of people suffering from DI, actually a biting feeling is a more common condition. Of the DI individuals we surveyed who called our office, 66% said they felt like they were being bitten, 46% described a crawling sensation (formication), 35% said they itched, 12% said it was stinging, and 5% described it as burning. Obviously many callers described a mixture of sensations, such as biting and itching (as one would expect from a blood-feeding insect like a mosquito).

If the animal is, indeed, scratching unduly, the underlying cause should be addressed, and it may be found to be imposed by the owner. As mentioned earlier, for instance, diatomaceous earth (DE) is a desiccant, absorbing moisture and skin oils and drying out the skin. It should never be applied to any animal, including humans. Use of DE on pets explains their increased scratching; their dry skin itches. When people treat their homes, and especially their bedding, with DE, the resultant skin dryness enhances itching.

Frequently the client will present the veterinarian with a sample of numerous arthropods taken from the home, but this is meaningless. Insects and other arthropods are commonly found in homes [5], with 100% of sampled homes having gall midges (Cecidomyiidae) and book lice (Liposcelididae) in them. Of course, neither of these infest humans or have any impact on human health. Likewise, ants, carpet beetles, and cobweb spiders are likely to be found in any home, but none of these can infest a living body, despite how many samples the DI sufferer brings you claiming infestation. Again, the more knowledge one has about actual arthropods, their biology, and behavior, the more confident one can be in determining what is *not* infesting a body.

While DI is frequently termed "rare," entomologists, pest control companies, and veterinarians find it quite common. For instance, samples submitted to a medical entomology laboratory for identification yielded no arthropod at all in 21% of the submissions [6]. Lint, scabs, soil particles, plant fragments, and other debris may look like insects or mites to DI sufferers, but adamant assertions that they are what is infesting the sufferer may be considered additional evidence of the delusion.

Fig. 2 A formerly pampered pet can become the detested perceived source of infestation

Typically, veterinarians seek legal and ethical means by which they can protect the animal. Simultaneously they are looking for ways to ensure the human receives the required treatment to recover from the delusion. I would argue that the two are inextricably linked. This is a human affliction that is inflicted on the animal, and until the human recovers, the animal remains defenseless (Fig. 2).

As has been pointed out elsewhere in this book, DI sufferers display doctor shopping, going from one physician to another, seeking one who will confirm their conviction of being infested. They also exhibit veterinary doctor shopping, looking for a veterinarian who can eliminate the putative infestation from their animal. Physicians who work with these patients have found that for a DI sufferer, it is more important that they be proven right than that they recover from their discomfort.

5   Identifying Delusional Infestation

Medical articles always caution that it is imperative to rule out the possibility of a valid infestation in patients complaining of invisible bug infestations. So we're going to exclude all (a) invisible bugs that (b) live in/on human skin, (c) can be seen by the sufferer but not by anyone else, (d) are able to persist for years, (e) that can feed on metals, wood, linoleum, and other synthetics, but simultaneously are capable of infesting and living in human bodies, (f) are not killed by parasiticides or insecticides/acaricides, and (g) have never been seen or described by any scientist. Again, no matter how impassioned the sufferer's pleas or how persuasive his story, the veterinarian must base diagnosis on science and facts.

Arthropods boast an extraordinary repertoire of behaviors, but they do not possess supernatural abilities. Knowledge of real insects and their capabilities allows discernment, to eliminate as possibilities bizarre claims of what "invisible bugs" can do.

An infection, of course, is caused by microbes living in host tissue. By comparison, an *infestation* involves a metazoan population living in (or on) the host, feeding, reproducing, and sustaining the population. Only two arthropods are capable of infesting the human body, scabies mites, *Sarcoptes scabiei* var. *hominis,* and lice (three species of human lice: the head louse, *Pediculus humanus capitis*, the body louse, *Pediculus humanus humanus*, and the pubic louse, *Pthirus pubis*). Being permanent ectoparasites, they depend on their host for both food and a home. By comparison, temporary parasites like ticks, mosquitoes, and other hematophagous arthropods approach the host, take a blood meal, then leave. They do not sustain a population on the host and do not use the host as habitation.

While it is perfectly logical for someone to say, "It feels like bugs moving in my skin," for someone to adamantly assert, "I have bugs in my skin," should signal alarm. Valid arthropod-induced skin irritations are transitory, seasonal, and self-resolving. The windborne shed skins of urticating caterpillars can cause skin irritation, but this is a seasonal, non-persistent phenomenon. Likewise, thrips may be found in clothing or bedding dried outside on a clothesline. In these cases, thrips will not survive more than a day or two, but clothing put on fresh from the clothesline may contain these small insects that can irritate skin:

One popular claim is that avian mites have evolved the ability to live on humans:

"One of the most important things to happen was the call I got from one of the doctors that I saw at the V.A. This was the day after we had our dogs put down. I was very upset and was feeling desperate. He confided to me:" … there is some type of mite that is doing this to people, but we don't know if it's coming from the birds or the squirrels."

Different theories exist: One is that a bird mite morphed to accept mammals, including humans, dogs, cats, and wildlife.

Accelerated evolution, with avian mites developing the ability to survive on and feed on mammals within a few decades, is biologically impossible. Additionally, bird mites are not active in the winter and do not persist for months. Bird mites abandon the bird nest when fledglings depart; without an avian host they can survive less than a month [7], so bird mites are not going to be found in homes after the spring fledging. Bird mites cannot survive on mammals (including humans), so do not infest dogs or cats.

6 Only Physicians Should Make Human Diagnoses

While it may be apparent to the veterinarian (and everyone else) that the individual is delusional, only a medical doctor can make the diagnosis, so it would be inappropriate to tell the individual they are mentally ill. And, practically, there is no benefit in telling a delusional individual that they are delusional; an integral aspect of the condition is that sufferers are incapable of recognizing reality:

> "Have you ever heard of Mirgellons disease? It is being caused by the GMOs in our food. The CDC dismisses it as DP because Monsanto owns the world's food supply. I too was diagnosed with DP in 2010 and I know that I was not crazy. I went to several doctors and dermatologists in an attempt to help myself. The microscopic bugs were keeping me up all night for months. The only way I was able to get rid of them was with food grade Diatomaceous Earth on my body, my cats, and my environment. I had to add ammonia, borax, and vinegar to my laundry. I was left to fend for myself because the medical community failed me. I lost my reputation with my friends and family who still think I am crazy to this day. I am right now going through something similar, but this time I have physical bites. The bugs are just as invisible as the first time this happened. They may be bird mites or cheylatiella. I am trying to collect samples to send them to an entomologist. I don't think you realize how much harm you are doing to an individual by telling them they are delusional when they are already living an unbearable nightmare. My newest nightmare is now going on 8 months. I sympathize with everyone that are being tormented by bugs they cannot see, and being diagnosed with DP. You can't know how devastating it is to your life until you experience this for yourself."

Itching and skin prickling as well as many other DI symptoms can indicate serious medical conditions, so this condition must not be taken lightly. Delusional infestation sufferers rarely are willing to consider any explanation other than "bugs" for the symptoms they are experiencing and may thereby miss early warning of life-threatening conditions. Underlying causes of pruritus and paresthesia should always be investigated to ensure significant health problems are not missed.

In summary, veterinarians should remember that (1) delusional infestation is a medical condition and can be treated effectively only with medications; we cannot talk someone out of a delusion (nor can they be persuaded with data, reasoning, or evidence). (2) We must recognize our limits. We cannot treat the root of the problem, a human medical condition. The most compassionate action we can take is to attempt to get DI sufferers to a physician. (3) The first step is diagnosis; if we cannot find an arthropod, then we have gone as far as we can go. (And I caution against trying to identify an arthropod from a client's description!).

Professionals who deal with DI sufferers are advised to demonstrate sympathy and concern, to let the sufferer know that their situation is being taken seriously. However, it may be difficult to demonstrate that their complaint is being taken seriously without reinforcing their delusion. Of course, the veterinarian should not treat for non-existent parasites, agree that the animal is infested, or take any other action that would validate the sufferer's claim of infestation.

Obviously, the human's health takes priority, but a responsible society should also be concerned about unnecessary animal suffering and seek to prevent it by

Fig. 3 Animals are dependent on their owners and vulnerable to their decisions

finding ways to treat the source—the human illness. Until the human's mental illness is treated and resolved, the animal is vulnerable and will likely bear the brunt of the delusion, being unfairly blamed and penalized (Fig. 3).

7 Summary

Delusional infestation is a medical condition. Anything that affects the body constitutes a medical problem and should be addressed by a physician. Remember, a delusion is an *unshakable* belief, and no one can talk them out of it.

Only human lice (three species) and human scabies mites can infest human bodies. Both of these are strictly human medical issues, and there is no role for veterinarians to play in their management. They do not originate from non-human animals or the environment, but are acquired only from other humans. Ectoparasites such as *Cheyletiella* mites that move from animals to humans are transitory pests that may produce pruritus or rash by test bites but cannot survive on humans. Once the host animal's infestation is treated, the human's symptoms will self-resolve. Despite sufferers' persuasive arguments, stick with science; there are no winged insects that infest human bodies and there are no facultative parasites (arthropods that can live off the host, feeding on inorganic materials such as furniture and appliances, then switch to parasitizing humans).

Delusional infestation is an under-researched medical condition that deserves more attention. People experiencing this illness require treatment that will restore their mental health and enable them to regain their lives. It is only by curing the human sufferer that hurting animals will avoid needless euthanasia.

References

1. Nel M, Schoeman JP, Lobetti RG. Delusions of parasitosis in clients presenting pets for veterinary care. J S Afr Vet Assoc. 2001;72:167–9. https://doi.org/10.4102/jsava.v72i3.643.
2. Rishniw M, Lepping P, Freudenmann RW. Delusional infestation by proxy—what should veterinarians do? Can Vet J. 2014;55(9):887–91.
3. Lepping P, Rishniw M, Freudenmann RW. Frequency of delusional infestation by proxy and double delusional infestation in veterinary practice: observational study. Br J Psychiatry. 2015;206:160–3. https://doi.org/10.1192/bjp.bp.114.144469.
4. Nicholson F. Delusory parasitosis: a veterinary perspective. In: Proceedings 2012 National Conference Urban Entomology. 2012;171–174. https://ncue.tamu.edu/wp-content/uploads/sites/9/2017/03/2012NCUEProceedings.pdf.
5. Bertone MA, Leong M, Bayless KM, Malow TLF, Dunn RR, Trautwein MD. Arthropods of the great indoors: characterizing diversity inside urban and suburban homes. Peer J. 2017;4:e1582. https://doi.org/10.7717/peerj.1582.
6. Geary MJ, Russell RC, Moerkerken L, Hassan A, Doggett SL. 30 years of samples submitted to an Australian Medical Entomology Department. Austral Entomol. 2020;60:172–97. https://doi.org/10.1111/aen.12480.
7. Chen BL, Mullens BA. Temperature and humidity effects on off-host survival of the northern fowl mite (Acari: Macronyssidae) and the chicken body louse (Phthiraptera: Menoponidae). J Econ Entomol. 2008;101:637–46. https://doi.org/10.1603/0022-0493(2008)101[637:taheoo]2.0.co;2.

Delusional Infestation: A View from Europe

Peter Lepping

1 Definition

Delusional infestation (DI) is a relatively rare delusional disorder [1]. People with DI characteristically have a fixed belief that they or their immediate environment is infested with insects, parasites, inanimate objects or small living creatures. This is in the absence of medical evidence for such an infestation. Patients often believe that the infestation is on or underneath their skin, in their whole body or in their immediate environment. This could entail anything from bedding or clothes and the entire house to their surroundings. The characteristic hallmark of a delusion is that the belief of being infested is held with a high degree of certainty. This is called delusional intensity in psychiatry, and refers to delusional beliefs: it can only be shifted temporarily, if at all. The evidence put forward by a patient with DI is usually not logical or in keeping with real infestations. In its long history, DI has been known under numerous terms. The most famous name is Ekbom syndrome, named after the Swedish psychiatrist who coined the German phrase Dermatozoenwahn back in the 1920s. Amongst numerous different terms that have been used in the last three centuries to describe this illness, the most common one until recently was delusional parasitosis. Since 2010, the term delusional infestation has been used and is now the most widespread term because it leaves room for any future type of alleged pathogen.

P. Lepping (✉)
Bangor University and Mysore Medical College and Research Institute, Mysore, India

Wrexham Maelor Hospital Psychiatric Liaison Team, Betsi Cadwaladr University Health Board, Heddfan Psychiatric Unit, Wrexham, UK
e-mail: PETER.LEPPING@wales.nhs.uk

© The Author(s), under exclusive license to Springer Nature Switzerland AG 2024
G. E. Ridge (ed.), *The Physician's Guide to Delusional Infestation*,
https://doi.org/10.1007/978-3-031-47032-5_7

2   The Evolution of the Terminology

In the twenty-first century, most people started to adopt the name delusional infestation which was introduced by Freudenmann and Lepping in 2009 [2]. The reason for this terminology was that the term infestation emphasises the constantly changing nature of the alleged pathogens and covers all present and future variations of the theme that are bound to arise [3]. When we look at the history of DI, the descriptors of the illness have changed over time. The alleged pathogens in the nineteenth century were scabies, typhus and pests, while in the twentieth century, patients chose to identify their pathogens as insects and bacteria, followed later by parasites and viruses as likely candidates. In the 21st century, fibres and other non-living pathogens have been increasingly implicated by patients [2]. Historically, there have been approximately 40 terms used to describe DI (Part 1, Table 1.2). All were dependent on their historic period, influenced by information at the time and social conventions.

Bacteria were a common suspect in the early twentieth century, while non-living pathogens have become increasingly common in the twenty-first century. In a 2010 paper, 77% of alleged pathogens were organic with insects seen as the leading cause followed by mites, parasites, lice and worms [4]; 23% of patients suspected non-organic pathogens like fibres and threads. A subtype of this is an alleged infestation with threads that some authors have called Morgellons disease (MD), a terminology not supported by most experts. This descriptor is prevalent in English- and German-speaking countries. Its popularity in English- and German-speaking countries is because the two languages are the most commonly used internet languages. Because of this, MD has been described as "an internet-transmitted disease" [2]. In particularly, in this sub-type of DI, patients believe that they suffer from non-organic pathogens, e.g. man-made threads (see Part 1, "Understanding the Morgellons phenomenon, a sub-type of delusional infestation"). Morgellons does not exist as a true illness outside the context of DI, despite attempts by a small group of authors to argue for this. Several large studies have confirmed Morgellons to be a sub-type of DI, most notably a CDC study published in 2012 [5]. Results of the study showed no infections, positive Borelli or infestations in patients with self-declared Morgellons disease [5].

3   Non-clinical Professionals Who See DI Cases

Outside the medical professions, there are other professional groups who find themselves dealing with DI cases, most notably, pest control officers/pest management professionals (PCOs/PMPs). When a DI patient becomes convinced that they are infested with alleged pathogens, they will inevitably call a PCO/PMP for help. As a non-clinician, it is very important not to go along with the clients' delusional beliefs. Do an inspection, and if nothing is found, be frank and say, "No parasites or infestation were found". Immediately, a DI-suspected client is likely to become upset and angry with the findings. It is not helpful to give in to the client's beliefs, but equally, long arguments or discussions should be avoided. It may be most useful to agree to disagree on this matter and leave. *Never* treat to placate customer demands. Seen as "the expert" in the situation, if a PCO/PMP treats for non-existent pests to please the customer, there may be a placebo effect for a few days or weeks, but the underlying issues are never addressed. Inevitably, the customer will call back ordering more treatments. This sets in motion a vicious cycle of feeding customer beliefs with repeated treatments, but as time goes by, they may become increasingly disappointed and angry because the problem was not corrected. In the event a PCO/PMP technician finds nothing, it is *most appropriate to advise the client to seek medical help*. This applies to all non-medical professionals. Therefore, if no alleged pathogens are found, do not placate client beliefs; advise them to seek medical help.

4 Clinical Presentation

Clinical presentation depends on the alleged pathogen and where it allegedly occurs. Typically, patients present with itching and other skin symptoms such as wounds (often superinfections), which are described as "bites". Patients may also present with generalized symptoms, such as crawling sensations on or underneath their skin accompanied by general malaise. Patients usually present to primary care physicians or dermatologists, but rarely to psychiatrists, because they do not believe that they have a psychiatric illness. The prevalence of DI therefore remains difficult to establish.

Patients often bring alleged evidence of their infestations with them to the physician's office. They often point to skin debris collected from different parts of their bodies as evidence of a systemic infestation. These may be pieces of skin they picked off, household debris, digital photographs or videos. The volume of sampling is usually high and is described as "specimen sign". Specimen sign was once called matchbox sign as matchboxes were used as containers for specimens, but this is rarely the case now. The number of people who come with specimen is in the region of 50–75% of patients seen. Digital photography on mobile phones has become increasingly common.

Excessive cleaning, scratching and use of pesticides are common. Secondary itching and superinfections are not unusual and need treatment in their own right. Patients often think that they are contagious and will self-isolate in an act of conscientiousness. They do not want to infest others who are important to them. Additionally, it is not uncommon for family members and significant others to end personal contact with DI patients for fear of becoming infested themselves. Reduced social contact and avoidance of one's own accommodation cause a much-reduced quality of life. Occasionally, patients embark in dangerous attempts to get rid of their alleged pests by trying to cut the pathogen out of their skin or "kill them" using pesticides, bleach or other harmful chemicals. If this is confirmed, a patient should receive immediate medical care.

5 Types of Presentation

DI can generally present as primary or secondary disorders. 40% are primary disorders, which means that the illness arises de novo, whereas in secondary delusional infestation, the illness arises in the context of another illness. With primary delusions making up 40% of DI, the majority of DI arises as part of other conditions. Particularly common are neurological conditions and other medical illnesses, especially brain disorders or illnesses that can cause generalised itching such as diabetes, uraemia, jaundice or certain cancers. DI also occurs in the context of substance misuse, particularly stimulants such as cocaine or amphetamines, but also in association with certain antibiotics, steroids or non-steroidal anti-inflammatories. Psychiatric illnesses can also present with DI as a secondary problem, particularly dementia, schizophrenia and major depression. In addition to this, there are a number of different forms in which DI can present. Up to 10% of presentations are a shared delusional disorder. Here, one or more people, usually relatives, share the delusional belief with a primary DI patient who is described in psychiatric terminology as "the inducer". DI can also present by proxy when a person believes that someone else is infested. If this other person cannot confirm whether they share this belief or not because they are either small children or pets, this is called "double-delusional infestation" [6]. In DI by proxy, it is normally children who patients believe are infested. If patients believe that their own children are infested, this can have serious consequences for the children if the parent uses bleach or other aggressive substances to get rid of the alleged pathogens. It may necessitate child protection procedures.

6 New Developments

A 2017 paper showed that drug use is about twice as common in DI patients as in the normal population [7]. Not surprisingly, it is the most common in younger men which is in keeping with the normal population. It is unclear whether this patient group is particularly difficult to treat. A 2022 study showed that patients working in healthcare themselves have a particularly bad prognosis because they tend not to engage with therapy [8]. A 2015 study has shown that DI by proxy with pets is not uncommon for veterinarians to see [6]. Like public health clinicians, they are often unsure how best to help.

Several MRI studies support the notion that DI is caused by a dysfunctional neuronal network mediating symptom of DI. There appears to be a distinct pattern of MRI findings that separate DI from other mono-delusional disorders and schizophrenia [9]. The data supports the hypothesis that dysfunctional somatosensory pathways mediate delusions in patients with DI, in contrast to delusional disorders without somatic content. A 2020 study has shown that the duration of untreated psychosis changes the outcomes in DI. The results suggest that longer duration of untreated psychosis in patients with DI is associated with significantly less favourable clinical outcomes. However, improvements can still be achieved even after many years of untreated illness. The same study showed that older age at presentation is associated with less favourable outcomes than younger age at presentation. A 2018 study analysing 75 patients assessed at the School of Topical Medicine in Liverpool, England, suggested that in this specialist clinic setting, almost 70% of patients were given the diagnosis of DI, suggesting a reasonably good screening process. About 60% of DI patients engaged with treatment, most of whom had significant improvements with antipsychotic medication. Health anxiety was the commonest diagnosis seen in those patients who did not have a diagnosis of DI. In new ICD-11 WHO terminology, this is best described as "bodily distress disorder". It suggests that a diagnosis of DI needs to be made by a medical professional as a number of other differential diagnoses are possible.

7 Treatment

Antipsychotic medication is particularly efficacious. Most people respond to antipsychotic medication. Full response rather than partial response is more likely with first-generation antipsychotics based on very limited data. Most clinicians would try second-generation antipsychotics first to minimise side effects; for example amisulpride, olanzapine or risperidone. However, sulpiride, haloperidol and perphenazine are also reasonable options. Pimozide was used in one very small Hungarian randomised controlled trial in the 1980s with 11 patients and has thus long been promoted [2]. It has the advantage that it does not have a licence as an antipsychotic in the United States, but its side effect profile is relatively unfavourable compared to more modern medications.

The biggest problem in the treatment of DI is building up trust between patient and physicians. To persuade patients to try medications is entirely dependent on a strong trusting relationship between the two. Patients rarely respond positively to the suggestion of a psychiatry referral because they are adamant that they do not have a psychiatric illness but a real infestation. Getting the patient to see a physician who is already familiar with DI is therefore the best way forward. This physician should ideally be someone the patient already knows and trusts. In the general medical profession, we find that many primary care physicians, dermatologists or clinicians whom patients see are reluctant to start antipsychotics because they lack familiarity with the drugs. However, the doses needed for the treatment of DI patients are about one-third that of the normal maximum doses for the treatment of schizophrenia [2]. This gives us the opportunity to say to DI patients that we understand that they do not suffer from schizophrenia, but we have found that small doses of the same medications used to treat schizophrenia are helpful in alleviating distress. The medications have been found to greatly reduce DI symptoms. At the end of the day, the main and most important aspect of treatment of any delusional disorder is to try and initiate antipsychotic treatment.

8 Health Systems and DI

Clearly, health systems that are largely private are very expensive, but they provide patients a maximum of choice. There are more opportunities for patients to pick and choose their care. If patients can fund a series of consultations until they find a physician of their liking, it has been found that physicians in these private systems are not well placed to treat DI patients. This is because patients can move from one clinician to another in the hope of getting a diagnosis of being infested. These systems allow patients to think that they can dictate their own care when they are not equipped and often lack the capacity to make major medical decisions. Boundaries need to be set for care to successfully treat patients with DI, and doctor-hopping needs to be avoided [2]. By and large, more restrictive health systems such as health systems found in most European countries make it more difficult for patients to "doctor hop". Patients have a primary care physician who coordinates care with other specialists as needed. Patients remain with their primary care doctor and do not travel from physician to physician. Additionally, by staying with one physician, treatment can be coordinated, and medical records kept in one place.

In Europe, there are specialist clinics and specialists who work with DI patients, but they are scarce. There are various good articles and podcasts about DI on the internet such as a BMJ (*British Medical Journal*) article that summarises the approach to DI, also available as a podcast [10]. The first ever national guidelines were published in 2022 by the British Association of Dermatologists [11].

References

1. Lepping P. Eine seltene Störung mit hohem Leidensdruck. CliniCum derma. 2017:3/17.
2. Freudenmann RW, Lepping P. Delusional infestation. Clin Micro Rev. 2009;22(4):690–732.
3. Lepping P, Walker SL, Bewley AP. Why delusional infestation should remain delusional infestation. Letter to the Editor. Creative Commons Attribution. 2020; (CC-BY).
4. Freudenmann RW, Kölle M, Schönfeldt-Lecuona C, Dieckmann S, Harth W, Lepping P. Delusional parasitosis and the matchbox sign revisited: the international perspective. J Compil Acta Derm-Venereol (Letters to edit). 2010;90:517–9. https://doi.org/10.234 0/00015555-0909.

 5. Pearson ML, Selby JV, Katz KA, et al. Clinical, epidemiologic, histopathologic, and molecular features of an unexplained dermopathy. PLoS One. 2012;7(1):e29908.
 6. Lepping P, Rishniw M, Freudenmann RW. Frequency of delusional infestation by proxy and double delusional infestation in veterinary practice: observational study. Br J Psychiatry. 2015;206:160–3.
 7. Lepping P, Noorthoorn EO, Kemperman PMJH, Harth W, Reichenberg JS, Squire SB, Shinhmar S, Freudenmann RW, Bewley A. An international study of the prevalence of substance use in patients with delusional infestation. J Am Acad Dermatol. 2017;77(4):778–9.
 8. Frewen J, Lepping P, Goulding JMR, Walker S, Bewley A. Delusional infestation in healthcare professionals: outcomes from a multi-centre case series. Skin Health Dis. 2022;2:e122. https://doi.org/10.1002/ski2.122.
 9. Huber MR, Wolf C, Lepping P, Kirchler E, Karner M, Sambataro F, Herrnberger B, Corlett PR, Freudenmann RW. Regional gray matter volume and structural network strength in somatic vs. non-somatic delusional disorders. Prog Neuropsychopharmacol Biol Psychiatry. 2018;82:115–22.
10. Lepping P, Huber M, Freudenmann RW. How to approach delusional infestation. BMJ. 2015;350:h1328. https://doi.org/10.1136/bmj.h1328.
11. Ahmed A, Affleck A, Angus J, Assalman I, Baron S, Bewley A, Goulding J, Jerrom R, Lepping P, Mortimer H, Shah R, Taylor RE, Mustapa FM, Manounah L. British Association of Dermatologists guidelines for the management of people with delusional infestation 2022. Br J Dermatol. 2022;187:472–80. https://doi.org/10.1111/bjd.21668.

Treating the Difficult Patient: How to Approach a Patient with Delusional Infestation

Michelle Magid, Nell Frackowiak, and Jason S. Reichenberg

1 Introduction

It is a common experience for a doctor to come across patients that are "difficult to treat."

Difficult patients are often sorted into three different categories. The first category is the patient that is challenging to treat due to the complexity of his/her illness. They often stump multiple providers, not because they are interpersonally difficult, but rather because of the severity or rarity of their presenting problem. Once the diagnostic dilemma is solved, these patients tend to be delightful and thankful.

The second category is the patient that is difficult to treat due to socioeconomic hardship or legal status. These patients tend to face obstacles in getting to appointments or affording medications. The illness often takes a backseat to more pressing issues, such as paying for electricity so as to live and finding transportation.

The last category of patients, and possibly the most problematic, are those that disrupt the doctor-patient relationship. These difficult patients are often demanding, are tough to satisfy, and lack the ability to see the world through others' eyes [1]. They also tend to see the world as black and white. As such, the doctor is either placed on a pedestal, which ultimately crumbles, or devalued after a perceived slight.

M. Magid (✉)
Department of Psychiatry, Dell Medical School, University of Texas at Austin, Austin, TX, USA

Austin PsychCare, PA, Austin, TX, USA

N. Frackowiak
Austin PsychCare, PA, Austin, TX, USA

J. S. Reichenberg
Department of Internal Medicine (Dermatology), Dell Medical School, University of Texas at Austin, Austin, TX, USA
e-mail: jreichenberg@ascension.org

G. E. Ridge (ed.), *The Physician's Guide to Delusional Infestation*, https://doi.org/10.1007/978-3-031-47032-5_8

When the doctor and patient are getting along well and seem to be on the same page regarding the treatment plan, it is known as a "good therapeutic alliance" [2]. With the third category of "difficult patient," it is near impossible to maintain a good therapeutic alliance; the patient is often fighting with the doctor, rather than working together to solve a common problem. The doctor becomes frustrated at his/her inability to help the patient, threatening the provider's self-worth, self-control, and professional identity [1]. In this chapter, we will be focusing on the third category.

Patients suffering from DI are frequently placed in the third category of treating the difficult patient [3]. They may be experienced as unreasonable and impossible to please. They may be insistent of their condition, despite the lack of medical evidence, and may request medications harmful to humans to kill the bugs that they believe are crawling on their skin. When their expectations are not met, they often go to a different doctor and continue the cycle. These patients can be very frustrating as they often go over the set appointment time and might leave bad online reviews about the provider, despite the provider feeling like he/she went above and beyond care for these patients [4].

A litmus test for the doctor of whether or not a patient fits this profile is if the doctor becomes distraught when the patient's name shows up on his/her schedule, or if the doctor frequently fantasizes about how to end care without causing patient backlash. In other words, a patient is difficult because of how he/she makes the doctor feel before, during, and after the interaction. However, thinking this way may place undo blame on the patient and absolve the provider of the responsibility of maintaining a good relationship. Even with the most challenging of patients, the provider can take certain steps to optimize the relationship and improve the chances of a good treatment outcome. This chapter presents the ten strategies that the authors have found to be effective in their practices (Table 1).

Table 1 Summary of clinical pearls for treating the difficult patient

10 Clinical pearls for treating the difficult patient	
1. Set expectations at the beginning	6. Delay time between stimulus and response
2. Shift into automatic pilot mode	7. Know when to resign
3. Ally around a common enemy	8. Know the rules of your state when terminating care
4. Blame a policy	9. Know when to do nothing
5. Identify the conflict and choose the lesser of two evils	10. Never worry alone

2   Set Expectations at the Beginning

One issue with difficult patients is their (usually unintentional) disregard for boundaries [5]. They become caught up in the visit and go past their set appointment time. This then leads to the doctor being late for a following appointment and feeling irritated by the end of the appointment.

The first suggestion for facilitating relationships between doctors and difficult patients is to set clear and firm boundaries at the beginning of the appointment. This consists of informing the patient of the allotted time for the visit and making it known that as the provider of care, you will not make exceptions to these boundaries. Ask the patient if you can interrupt him/her when he/she begins to exhibit signs of getting off topic. This can be said in the following manner:

> "We have a lot to cover today and in order for me to get the information I need to help with your care, I may have to interrupt you."

Firm yet kind reminders of the time boundary may be helpful in aiding a patient to prioritize. At the end of the appointment, reassure the patient that whatever concerns were not mentioned during this visit, they can be addressed during a following appointment, but make it very clear that once the appointment is over, it is over. We advise against going over time, as the patient will learn that this is an acceptable practice, and this can be a difficult learned behavior to unlearn. Establishing the rules of the visit, what we call "the therapeutic frame" in psychiatry [2], can actually improve patient satisfaction in the long term, as boundary setting creates a safe space [5, 6].

Become an Automatic Pilot

Another way to prevent the overstepping of boundaries is to shift into "automatic pilot mode." In this mode, the doctor refrains from giving the patient any sort of special treatment. This could be extending the appointment time, as mentioned above, scheduling an appointment after hours, giving the patient a personal cell phone number, etc. While these exceptions to the rules may be justified in certain situations, we strongly advise against doing this when dealing with the difficult patient. The best way to handle unusual requests is to simply adhere to a set of strict policies. The provider needs to be very systematic in his/her way of approaching a patient. If the provider breaks his/her own boundaries, it reinforces that the patient can do the same, and the patient will begin to overstep.

Ally Against a Common Enemy

When difficult patients complain about some aspect of the treatment, i.e., the discomfort they are experiencing, the cost, and the struggle they face with insurance companies, it can be best for the provider to validate their point. It is suggested that the doctor does not engage in the conflict or make it a personal issue, but rather ally with the patient against a common enemy [4]. If the patient is upset that their procedure is not covered by insurance, it is okay to agree and to "blame" the insurance system. For example,

> "I agree with you that this should be covered. Our healthcare system is quite broken."

If the patient attacks you for the suffering that their condition is causing, isolate the condition as the real enemy. The patient's anger may feel personal, but after further reflection, patients are usually angry at a situation. The key is to get the patient to direct this anger away from you and towards whatever the patient is really angry with.

One effective strategy for this can be sitting down next to the patient and looking at a brochure, or a similar document with information, about the given situation. By doing so, you are on the same side (physically and mentally) as the patient and both are looking at the problem together. We use a tennis analogy to explain the shift in focus:

> "The patient is no longer playing a singles match, where the doctor is the opponent, and the ball is the illness being hit back and forth. Rather, the patient and the doctor are now playing doubles and are on the same team. The ball, and whatever is on the other side of the court, can be confronted as a team."

Blame a Policy

Another third party you can easily blame when dealing with difficult patients is an office or governmental policy. If a patient asks you to give them special treatment or do them a favor you do not feel comfortable with, you should cite an office policy as to why you cannot honor the patient's request. Blaming a policy depersonalizes the situation. It is not about you or about the patient, but rather about adhering to a well-thought-out regulation that was backed by multiple parties. It portrays that this is an organizational issue, and to make an exception, it is going to require the approval of a whole system. For example,

> "I'm sorry that I cannot come to your home for a visit. It is our policy that all visits must be done via telemedicine or in the office."

Another example:

> "I will not call an exterminator to your house; it is our policy not to do so."

And a third example:

> "I do not prescribe that medication for your condition, it is not in line with the Food and Drug administration (FDA) policies and guidelines, and we adhere to FDA guidelines."

Blaming a policy tends to trigger patients less than citing a personal preference for a decision. For example,

> "I'm sorry that I cannot come to your home for a visit as I don't feel safe doing so."

This may open up a discussion as to why you would not feel safe and what that says about you, the patient, and the therapeutic alliance.

For those in a solo practice, the patient may counterargue that "you can change the policy at any time." Consider blaming "community standard of care" or "an insurance/malpractice regulation" instead. The bottom line is to keep all decisions universal and based on a rule, versus personal and based on your opinion.

Identify the Conflict and Choose the Lesser of Two Evils

Difficult patients may put doctors in no-win situations. No matter what the doctor decides to do, someone will walk away unhappy. For example, the DI patient may threaten to write a bad review, if a specific treatment that the patient read about on the internet is not given. The doctor then must decide whether to give this patient a non-approved treatment option or to take the risk of a bad review. In another scenario, a 70-year-old female with DI may refuse treatment while her adult child is insisting that she take the treatment. Does the doctor listen to the patient or to the patient's kin? If the doctor provides the treatment, the patient may not follow through correctly on the treatment plan. On the other hand, if the doctor does not provide the treatment, the child would be angry, given that she is the one paying for the care. She might file a complaint against the doctor or not pay the medical bill.

One way to facilitate these decisions is to resort to an old-fashioned, yet effective, list of pros and cons. This list can help clarify the conflict and consider the many factors in acting one way or another. It may even be helpful to explain out loud these pros and cons to the patient, although the final decision is often made by the doctor. In this situation, it is not about making the right decision, because there is none, but rather about making the better of two not-so-great decisions [4].

Delay Time Between Stimulus and Response

Difficult patients have the potential to trigger a doctor's emotional part of the brain, also known as the amygdala [7]. When your amygdala is triggered, you begin to make emotional decisions rather than logical ones. It can be very challenging to respond to the patient in a calm way when the amygdala is hijacked. The only way to settle down the amygdala once it is activated is to give yourself some time.

With time, the executive decision-making part of the brain, also known as the frontal lobes, will start to gain control again [7]. If the doctor feels his/her heart racing or blood pressure rising or frustration emerging, it is suggested that the doctor remove himself/herself from the situation before escalating the conflict. We suggest stepping out of the room to delay the time between the stimulus (the frustrating interaction) and the response (your reaction). Tell the patient that you have received a page or an important phone call and will be back as soon as possible to give them your full attention. Leave the space, attend to other patients or work, or even just take a few minutes to regroup. When you feel you are ready to handle the patient in the best way, come back to the interaction. Hopefully, at this point, you have reengaged the frontal lobes and can accordingly calmly blame a policy.

Know When to Resign from Your Job as the Patient's Provider

Resigning as a patient's doctor, or terminating care, may be the best option in some cases. When a conflict between a patient and provider becomes too serious to handle in a professional manner and the risk of continuing treatment is high for either party, it is best to end the doctor-patient relationship. For example, if a patient refuses to treat a severely infected wound but their family member threatens to sue you if the patient's wound worsens, resigning from care is the best resolution. Another example is a patient who continues to harass staff despite warnings that verbal or physical harassment will not be tolerated. In this situation, it is important for the doctor to keep his staff safe, and termination of care is prudent.

If a patient insists on a treatment that is harmful and unethical, the provider should not comply. If the patient continues to insist, then the provider simply cannot do the job that he/she was hired to do and should consider resigning from that job. The idea of resigning from care is essentially stepping down from your job as a patient's provider. "Resignation from the job," rather than "firing of the patient," is our preferred terminology. After all, the patient hired the provider to do a job, not the other way around. If the job turns out to be impossible, the provider must resign from the position as doctor, just as anyone else would resign from an unsustainable work situation.

Know the Rules of Your State When Terminating Care

If you have determined that the best course of action is to terminate care, i.e., resign from your job as provider, make sure that you do so in compliance with your state's (or country's medical board) rules for terminating doctor-patient relationships. For instance, the termination letter will often include an agreement to treat emergently for 30 days, a willingness to send records to the new physician, and the

Table 2 Recommended guidelines of what to include in a termination letter [8, 9]

Termination letter key points
Explain the reason for termination (not mandatory)—can be a broad statement such as "I feel it is necessary to end our doctor-patient relationship due to inability to maintain rapport"
Provide an effective date—often 30 days from the date on the letter. This provides sufficient time to find another provider
Offer interim care—provide *urgent* care and medication refills *up* to the effective date of termination. For emergencies, refer the patient to the emergency department or 911 (see international emergency numbers at the beginning of part 1 if not in the United States)
Provide a referral for seeking care elsewhere—can be the patient's insurance company or medical society website. Do not recommend another healthcare provider by name
Offer to provide a copy of records for the next provider—can include a "release of information" for medical records, which the patient can sign and return to the office
Emphasize patient responsibility—specify that continued care is now the patient's responsibility and failure to do so may result in harm

consequences of not receiving continued care from another provider (Table 2). Many providers send the letter by certified mail to document that the patient received it. Following state (or country's medical board) guidelines is a safe way to avoid being accused of "abandoning" your patient [8–10].

Know When To Do Nothing

Some patients may usually be pleasant to work with but come into your office displaying the qualities of a difficult patient. This is true for anyone including DI patients. In this situation, it is important to remember that the patient is human, and he/she might just be having a bad day. You may want to ask the patient if there is anything causing their stress and difficulty. You may decide to do nothing except respond with kindness and concern.

Never Worry Alone

The last tip for dealing with difficult patients is to share a difficult patient dilemma with a colleague. *Psychiatrist Ned Hallowell emphasized on "never worry alone" because a burden shared is a lighter burden* [11]. Firstly, sharing with a colleague can ease the provider's mind. Secondly, from a legal perspective, it further establishes the standard of care. The modern definition of standard of care is *"that which a minimally competent physician in the same field would do under similar circumstances"* [12]. When a colleague in the same field agrees with your assessment, the standard of care has been met. It can be helpful to document this interaction by writing "I discussed this case with Dr. John Smith who agrees with my treatment plan," if the colleague consents to being named in the medical chart.

3 Conclusion

Of the many patients we treat, some are more difficult than others and create rifts in the doctor-patient relationship. Delusional infestation patients, the difficult patient group of interest, tend to be categorized as some of the most difficult patients encountered. Many doctors are unable to establish good rapport despite multiple efforts. By setting clear boundaries, following office protocol, shifting into automatic pilot mode, and using techniques like delaying the time between stimulus and response, even challenging patient encounters can be potentially enjoyable. Regardless of the situation, if an encounter is keeping one up at night, it is wise to call a colleague and not worry alone. When all attempts by the doctor to resolve the conflict fail, resigning from care is a reasonable way to handle the situation. Hopefully, the patient and provider will not see this as a personal failure, but rather as choosing the lesser of two evils.

References

1. Groves JE. Taking care of the hateful patient. N Engl J Med. 1978;298(16):883–7.
2. Gray A. An introduction to the therapeutic frame. Routledge; 2014.
3. Reichenberg JS, Magid M, Jesser CA, Hall CS. Patients labeled with delusions of parasitosis compose a heterogeneous group: a retrospective study from a referral center. J Am Acad Dermatol. 2013;68(1):41–6.
4. Magid M, Reichenberg JS. Treating the difficult patient: ten pearls to reduce resentment and regain control of the doctor-patient visit. Skinmed. 2019;17(1):11–4.
5. Krupp BH. Professional boundaries: safeguarding the physician-patient relationship. Dermatoethics: contemporary ethics and professionalism in dermatology. Springer; 2021. p. 161–8.
6. Aravind VK, Krishnaram VD, Thasneem Z. Boundary crossings and violations in clinical settings. Indian J Psychol Med. 2012;34(1):21–4.
7. Etkin A, Buechel C, Gross JJ. The neural bases of emotion regulation. Nat Rev Neurosci. 2015;16(11):693–700.
8. Schleiter KE. Difficult patient-physician relationships and the risk of medical malpractice litigation. AMA J Ethics. 2009;11(3):242–6.
9. Company TDs. Terminating patient relationships. The Doctor's Company online. https://www.thedoctors.com/articles/terminating-patient-relationships/.
10. Jung S, McDowell RH. Abandonment. Treasure Island (FL): StatPearls Publishing; 2021.
11. Hallowell N. Driven to distraction at work: how to focus and be more productive. Harvard Business Review Press; 2014.
12. Moffett P, Moore G. The standard of care: legal history and definitions: the bad and good news. West J Emerg Med. 2011;12(1):109.

Concerns for Infestation (CI): Dermatologic Evaluation of Patients with Unwanted, Uncomfortable, and Unexplained Sensations

Scott A. Norton

1 Introduction

Many chapters in this comprehensive volume are written for health professionals working with patients who already have the diagnosis of delusional infestation (DI). Those chapters address the behavioral and pharmaceutical management of this disorder. This chapter takes a different perspective: here, we discuss the early (and thorough) evaluation of patients who are concerned about a possible infestation. These are patients who have unwanted cutaneous sensations (such as generalized pruritus), and they are concerned that an infestation may be the cause.

The patient's abnormal sensations are new and unexplained, but their cause has not been identified. These patients have expressed concern for possible infestation, but we do not yet know if they are truly infested or if there is another explanation.

This chapter helps clinicians evaluate patients who have unwanted, unexplained, and uncomfortable cutaneous sensations—and who express concern that they are infested with some sort of organism (Fig. 1). The patients may have already seen a variety of providers ... with a variety of outcomes, they found unsatisfying. Imagine an emergency room (ER) where an otherwise healthy-appearing patient presents with widespread pruritus and expresses concern for possible scabies. In a busy ER, the provider is able to perform only a perfunctory examination and notes widespread excoriations. Scabies is certainly on the differential. But insufficient time, training, and equipment make the provider unable to perform a skin scraping for microscopic evaluation, which might confirm that diagnosis. Instead, the provider makes a presumptive diagnosis of scabies to account for widespread pruritus. The patient is then sent home with a standard scabies treatment, typically permethrin or ivermectin, and instructed to follow up with the patient's regular provider.

S. A. Norton (✉)
Departments of Dermatology and Preventive Medicine, Uniformed Services University of the Health Sciences, Bethesda, MD, USA

G. E. Ridge (ed.), *The Physician's Guide to Delusional Infestation*,
https://doi.org/10.1007/978-3-031-47032-5_9

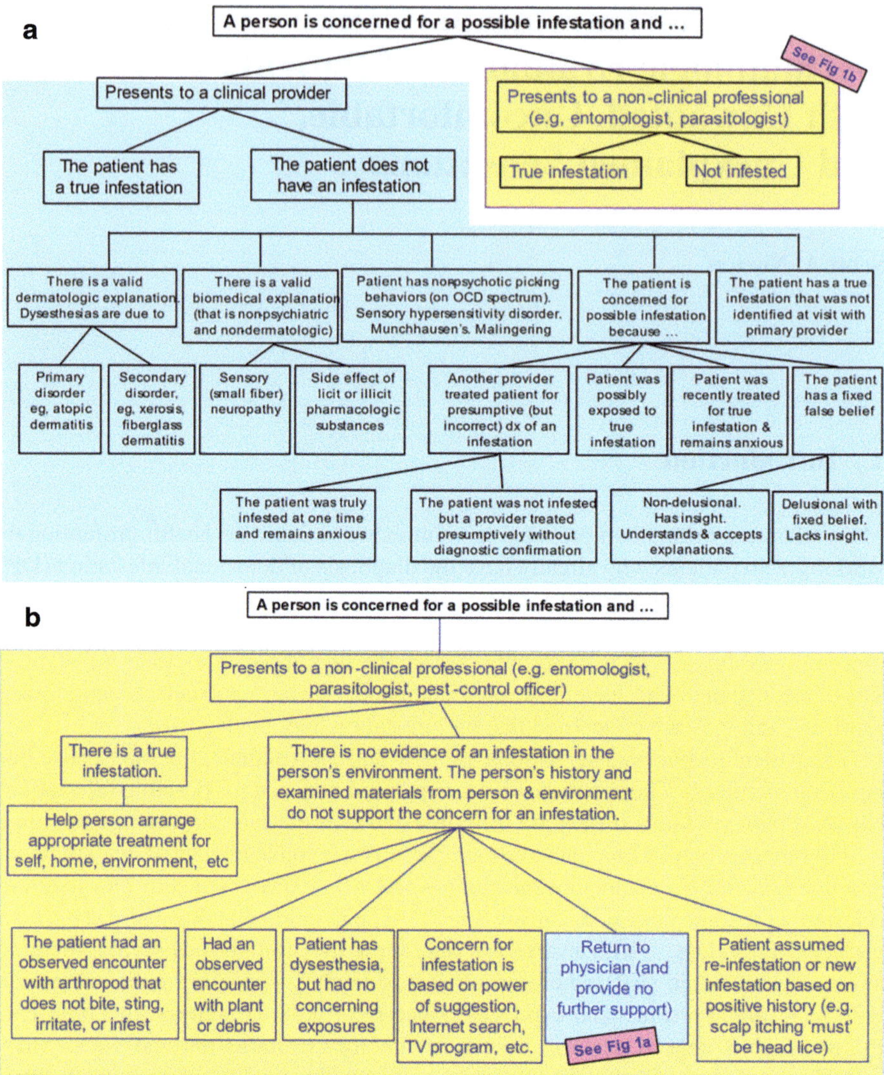

Fig. 1 Differential diagnosis flow charts when working with DI patients

If scabies is not the right diagnosis and the unwanted sensations persist, the patient may enter a cycle of healthcare visits with providers who, for one reason or another, have neither the time nor the patience to engage the patient for a full evaluation. Patients in this group often go without a thorough evaluation to determine the cause of the unwanted sensations—and still, no one has performed a skin scraping to detect scabies by microscopic examination. There are no diagnostic blood tests for scabies (not for most other ectoparasitic infestations). Hence, we still lack an objective determination if the patient is (or was) truly infested. In such situations, many patients are treated repeatedly for scabies (or other ectoparasitic infestations, such as head lice), despite the continued lack of a confirmed diagnosis. With each treatment, the abnormal symptoms may improve transiently but recur shortly afterward—or simply continue unabated.

Scenarios like this are exceedingly common and almost invariably lead to the patient believing three things: (a) the diagnosis of scabies (or similar infestation) must be correct because so many providers have made this diagnosis; (b) none of the treatments have resolved the unwanted sensations successfully; and (c) therefore, the patient's particular "strain of scabies" must be unusual. They wonder if their strain is resistant to standard therapies, necessitating higher and more frequent doses of mediation. They may also suspect that "their organisms" can somehow evade treatments by hiding elsewhere on or in the body.

This chapter takes the approach of the clinician, typically a dermatologist or an infectious disease physician, often based at an academic center, who accepts the task to conduct a thorough evaluation to identify the cause of the abnormal sensations that persist despite several rounds of anti-scabies medications. If the cause is indeed an infesting organism, this clinician reasons, then now is the time to find the organism, identify it, and provide the patient with definitive treatment. On the other hand, many skin diseases, many systemic diseases, many medications (and other pharmacologically active substances), and many psychological conditions can induce widespread dysesthesias.

Clinicians who thoroughly evaluate such patients are looking for evidence of an infestation as well as considering other possible causes of persistent dysesthesias (Fig. 1). These clinicians generally acknowledge that some patients in this situation have a fixed belief that they are infested. Despite the lack of evidence of an infestation, these patients continue to believe that an infestation—and only an infestation—can cause their symptoms. By definition, these patients have delusional infestation (DI). I believe that my evaluations are particularly thorough. I am a dermatologist with a background in medical entomology and medical parasitology. I have an interest in unusual infections, infestations, and environmental hazards worldwide that can affect the skin. If I am unable to determine the cause of the unusual sensations and the patient continues to insist that they are infested, I will consider DI as a possible diagnosis. There are occasions when I consider DI as a possible diagnosis much earlier in the evaluation.

When a patient with widespread pruritus presents for evaluation, the clinician must approach the diagnostic process without a priori bias regarding the sensations. The clinician understands that many itchy patients who believe that they have scabies are indeed infested with scabies. Of course, the unwanted sensations may be caused by something other than an infestation. For example, allergic contact dermatitis and adverse reactions to medications are well-recognized causes of pruritus and other dysesthesias. A patient with widespread itchy skin may be worried about a possible infestation, but something else unrelated to the infestation may be causing the itch. In these cases, a suspicion that the itch is due to an infestation may simply be incorrect, no matter how strongly patients may believe that it is the cause. In either case, a true infestation or an imagined one, the clinician's duty is to identify the correct (or likeliest) cause of the unwanted sensations and to treat the patient.

Identifying the cause of pruritus is often challenging and may require several medical visits before a diagnosis is reached. While evaluating a patient who is concerned that their unwanted sensations are caused by a possible infestation, I use the working diagnosis of *"concern* for infestation" and avoid terms such as *"delusional* infestation" or *"delusions* of parasitosis." I believe that it is premature, unfair, and illogical to conclude that a patient is delusional before the evaluation begins.

Another reason to use the phrase "concern for infestation" rather than "delusional infestation" is that this term is truthful and may be used openly and frankly with the patient. Many patients in this situation have encountered family members, co-workers, health providers, or public health professionals, who raised the possibility of delusions. Many patients have already heard, "It's all in your mind" or "Maybe you should see a psychiatrist," which they find offensive and intolerable. They start feeling rejected, scorned, disbelieved, and isolated—and will likely find the phrase "concern for infestation" refreshingly truthful and more acceptable. A clinician who uses the term "delusional" may create an irremediable impasse. The clinician who uses the term "concern for infestation" may create the climate for a more thorough understanding of the patient and the cause(s) of their distressing symptoms.

2 Terminology: Parasitism, Infestation, and Humpty-Dumpty

Throughout this chapter, I will use the abbreviations CI for "concern for infestation" and DI for "delusional infestation." The goal is to identify the cause(s) of a patient's dysesthesia, but there are circumstances when the diagnosis of DI is inescapable.

Trying to define DI reminds me of Supreme Court Justice Potter Stewart's famous attempt to define obscenity and pornography, "I know it when I see it." In this case, most clinicians agree that certain clinical presentations are "textbook cases" of DI, yet they do so in the absence of widely accepted diagnostic criteria. In this chapter, we will also discuss the clinical working definition of DI, propose the type of evaluation that CI patients should receive, and discuss when the diagnosis of DI is inescapable.

The editors and authors of this volume chose to use delusional ***infestation (DI)***, rather than delusions of ***parasitosis*** (DP). Both refer to the same group of patients, but some medical folks, pedants perhaps, point out that these patients report that "their organisms" reside in or on their skin. In medical parlance, such creatures (such as scabies and lice) are called ***ectoparasites*** and they ***infest*** the skin. These same authorities limit the use of the term ***parasite*** to organisms that reside internally, usually in intestinal, visceral, or parenchymal organs (such as hookworms and bancroftian filaria). Ectoparasites infest the skin; parasites (or endoparasites) infect internal organs. So say the medical folks.

The word *infestation* has several connotations [1]. It means different things to different people. Thus, *infestation* is a humpty-dumpty word. As the ol' egg told Alice in "Through the Looking Glass," by Lewis Carroll,

"When *I* use a word," Humpty-Dumpty said, in rather a scornful tone, "it means just what I choose it to mean—neither more nor less."

I use the term *infestation* to describe the condition where multicellular organisms, usually invertebrates, parasitize human skin. These organisms can also be called ectoparasites.

Of course, ecologists and evolutionary biologists use the term **parasitism** to describe a two-organism symbiosis in which one organism benefits from the relationship (the parasite) and the other organism is harmed by the relationship (the host) [1]. These biologists might argue (if they were aware of the semantic dispute) that the difference between DI and DP is biologically meaningless. Because these conditions are delusional, neither infestation nor parasitism is taking place.

Two organisms sharing space within a micro-ecosystem is called *symbiosis*. There are four types of symbioses: mutualism, commensalism, parasitism, and competition. In classic evolutionary biology, the term infestation is rarely used. For purposes of this book, however, we will adopt the term. In most cases, the infesting organism benefits by feeding on a human—and that human is harmed to some degree. This satisfies the biologist's definition of parasitism, even when the harm is limited to a few itchy bumps where the arthropod obtained its meal.

On the other hand, can dog or cat hookworms that cause cutaneous larva migrans benefit by infesting human skin? Absolutely not. A wayward hookworm larva that enters human skin is unable to continue maturing into a reproductively competent adult. Its life cycle ends within the purgatory of human skin, and the hookworm will leave no progeny to perpetuate the family name. So in this case, we have an unnamed fifth type of symbiosis: where both organisms are harmed.

My concept of infestation includes obligate and facultative ectoparasites that enter a symbiotic relationship with one or more humans, usually benefitting the ectoparasite and harming (to some degree) the human host. Scabies, head lice, body lice, trombiculid mites (chiggers), and many other non-scabetic mites; leeches; hematophagous dipterans; other dipterans that cause myiasis; jigger fleas (*Tunga penetrans*); and dog and cat hookworms (*Ancylostoma caninum* and *Ancylostoma braziliense*) can infect humans.

Examples of human-ectoparasite symbiosis include relationships where the infesting organism can:

- Live, feed, breed, and die obligately *in* the skin of a single human host (e.g., scabies, *Sarcoptes scabiei* var. *hominis*).
- Live, feed, breed, and die obligately *on* the skin of a single human host (e.g., head lice, *Pediculus humanus capitis*).
- Feed obligately on the skin of a single human host and live, breed, and die obligately within inches of that human host (e.g., body lice, *Pediculus humanus humanus*).
- Feed obligately on the skin of any number of human hosts and live, breed, and die indifferently to their proximity to a host (e.g., human feeding bed bugs, *Cimex lectularius*).
- Feed obligately on the skin of one or more human hosts but live, breed, and die elsewhere (e.g., the yellow fever mosquito, *Aedes aegypti*, as well as numerous other species).
- Feed on the skin of a human host opportunistically (facultatively) but live, breed, and die elsewhere (e.g., trombiculid chiggers).

- Enter human skin by mistake, where they reach a reproductive dead end and die within that human (dog and cat hookworms, *A. braziliense* and *A. caninum*).

A full list of known human ectoparasites is beyond the scope of this book, but some common human-infesting parasites are discussed in Chap. 13. This chapter focuses on patients who falsely believe that they are infested. The value of elaborating on these points is to help clinicians understand the complexity of humanity's symbiotic relationships and to assuage and reassure patients with CI that there is no evidence of an infestation. This background may also help a clinician state with confidence and certainty that the "infesting organism" a patient has described does not exist. Not only has there never been a report of a creature with such features and characteristics, but it is also biologically impossible for such a creature to live and thrive.

3 Synonyms? Morgellons, Delusional Parasitosis, and Delusional Infestation

The term Morgellons disease (or simply Morgellons) is a controversial one. Some individuals define Morgellons sensu stricto as a condition where patients report fibers (endogenous or foreign; animate or inanimate) in the skin (or deeper) that ultimately pass through the skin and emerge from the stratum corneum still in the form of fibers. Others use Morgellons in a very broad sense (sensu *lato*) to include delusions involving either fiber (Morgellons sensu stricto) or creatures of some sort that live in or on one's skin. Morgellons sensu stricto is a curiously distinct, highly stereotyped disorder. The reasons for this fairly narrow expression of Morgellons are not clear; it may be a highly conserved Dawkinsian meme that was transmitted largely unaltered over cyberspace and adopted by members of social networks, chat rooms, influencers, etc.

The National Library of Medicine, part of the National Institutes of Health, uses a hierarchically organized vocabulary called the "Medical Subject Headings (MeSH)" thesaurus to index and catalog medical literature (https://www.nlm.nih.gov/mesh/meshhome.html). They added the term Morgellons disease in 2009 (https://meshb.nlm.nih.gov/record/ui?ui=D055535), defining it as:

"An unexplained illness which is characterized by skin manifestations including non-healing lesions, itching, and the appearance of fibers."

It is classified as both a *skin disease* and as a *schizophrenia spectrum and other psychotic disorder*. Delusional parasitosis was adopted as a MeSH term in 2014 (https://meshb.nlm.nih.gov/record/ui?ui=D063726), defined as:

"A delusional disorder of belief in infestation by insects or other parasites. This formication is typically accompanied by dermatological manifestations such as pruritus that may lead to self-mutilation in order to remove the perceived parasites. It can be either primary or secondary to a somatic or psychiatric condition."

It is classified as a *schizophrenia spectrum and other psychotic disorder* but not as a synonym of Morgellons disease. The MeSH page with metadata for "delusional parasitosis" lists 50 synonyms for the condition, many of which use the word parasitosis or close variations. None use infestation. Most use the word delusional, or variations such as delusory or psychogenic. Two additional terms are dermatozoic and Ekbom [syndrome]. Ekbom syndrome is the eponymous name for this condition. Karl-Axel Ekbom, a Swedish neurologist, described the condition in 1937. Two of the best review articles on DI, DP, Morgellons, Ekbom syndrome, or whatever name you prefer were written by Nancy Hinkle, a contributor to this volume (Chap. 4). In her earlier publications, she used the term Ekbom syndrome [2, 3]. For her chapter in this book, she uses the term delusional infestation.

The CDC used the term *an unexplained dermopathy* in its sole publication on the matter [4]. A decade ago, around 2010, the CDC's website had an information page on Morgellons disease and the "unexplained dermopathy." The CDC's website no longer has any information on this spectrum of conditions. NIH's website for genetic and rare diseases has a placeholder page for Morgellons disease (https://rarediseases.info.nih.gov/diseases/9805/morgellons) among nearly 6000 other rare diseases. This link leads to the NIH GARD Morgellons entry, which currently has the notification, "This section is currently in development."

Many DI patients are convinced that "their creatures" are as real as anything else in their life. They say that the presence of creatures infesting their skin or bodies is as real and non-debatable as the presence of a mouth or nose. Invoking the possibility that their malady might be delusional is so repugnant, and so off-putting, that they are unlikely to continue care with a provider who uses the "D word."

4 Delusional Infestation: A Working Definition

Around the world, literally more than a billion people have ectoparasites or internal parasites, so is it far-fetched for a patient to believe that they, too, are participating in the global symbio-palooza? If the patient is leaning that way, when should the clinician start to consider the diagnosis of DI in a patient with CI, and *how* does a clinician arrive at the diagnosis? We will answer these questions in two general ways: an abstract, philosophical manner and a practical, clinical manner.

First, let us review the criteria to make the diagnosis of DI. The diagnostic criteria found in other chapters list just two criteria:

The patient must have:

1. Abnormal sensations in the skin.
2. A conviction that the sensations are caused by an infestation, despite a lack of supporting evidence.

That definition does not sit well with me, mainly because Criterion #2 is like a run-on sentence. Two ideas are being expressed, and they should not be merged into a single criterion. My working case definition lets each idea stand alone. Furthermore, I have added Criterion #4, which is the converse of Criterion #2.

The patient must have:

1. Abnormal sensations in the skin.
2. A conviction that the sensations are caused by an infestation.

3. A lack of evidence to support that conviction.
4. Categorical rejection of alternative explanations for the dysesthesia.

In this chapter, I use the term *dysesthesia* to mean an abnormal cutaneous sensation, usually one that is perceived as unpleasant. Dysesthesias include pruritus (synonymous with itchiness); formication (the feeling that ants or other insects are crawling on or in your skin; derived from the Latin *formicare*, for "crawling like an ant"; this is further derived from *formica*, Latin for "ant"); hyperesthesia (overly sensitive skin); hypesthesia (decreased sensation, such as varying degrees of numbness); and allodynia (pain induced by something touching the skin, even if the touch is delicate).

5 The Patient's Experience

Let us start with a patient who has started to notice strange sensations on their skin. The sensations are new; they have not felt them before. The sensations are widespread, and during a typical week, nearly every area of their skin experiences them. The sensations are particularly prominent on the scalp and dorsal forearms. During the daytime, when the person is engaged in (or distracted by) work and other activities, the sensations go unnoticed. In the late afternoon and evening when the patient settles into bed, the sensations become particularly annoying. The patient finds the sensations hard to describe; they are slightly itchy, but scratching does produce the pleasurable sense of relief that people feel when they scratch an itch.

At times, the patient feels as if something small and delicate is crawling on their skin. That sensation reminds them of ants crawling up one's leg when sitting on the ground during an outdoor picnic. The patient examines their skin closely several times a day, but has never observed an ant, any other insect, or even dust, lint, or human hair on their skin. It is baffling but the patient is somewhat relieved because the sensations are not terribly uncomfortable. As weeks go by, the sensations persist and are perhaps more disturbing. The patient is now feeling a mix of fatigue and frustration, fatigue from trouble falling asleep at night and frustration at not knowing what is causing this. A suspicion arises that the sensations are caused by insects or another type of minute creature that crawls on (or perhaps within) their skin. To get rid of the creatures, the patient decides to schedule a visit with their healthcare provider … or considers going to the local emergency room for a more immediate visit.

If the provider finds evidence of multiple arthropod bites or that the patient is truly infested with an ectoparasite, she will likely refer the patient to dermatology. The corollary is that if the provider cannot find evidence of an infestation, they will also refer the patient for a dermatology visit. Either way, a patient with unpleasant dysesthesias will likely be seen by a dermatologist, especially if there is CI.

Because of my worldwide interest in unusual infections, infestations, and environmental hazards, many patients with CI are referred to me. They also hear of me by word of mouth or find my name by searching the Internet. In any event, during most weeks, I see at least two patients who are concerned that they are infested by some sort of annoying creature.

Of course, on some occasions, a patient has an infestation. More often, the patient has a plausible medical explanation for the strange sensations. Sometimes, the patient is delusional. Among the remarkable features of DI patients are their similar symptoms, clinical findings, behaviors, and responses to attempted treatments. Moreover, descriptions of the "creatures" they believe that are infesting them are also strikingly (and fascinatingly) similar.

6 Typical Case Presentations

Let us start with the cases of two typical, but unrelated, patients, a 26-year-old graduate student (Fig. 2a) and a 67-year-old retired librarian (Fig. 2b). Both describe the sensation of creatures living in their skin and moving across the skin's surface.

Both patients believe that they are infested by minute creatures, and both acknowledge never having seen one of the purported creatures. Consequently, they believe that their creatures are too small to observe with the naked eye. Otherwise, both individuals were reasonably healthy and had seen several physicians, including dermatologists, over the previous 2 years. Each has tried a variety of treatments without relief of their overall symptoms. The treatments had included oral and topical products obtained in several ways; some products were prescribed by physicians; other products were purchased over the counter at local pharmacies; and others were bought online. So far, the treatments either had been unhelpful or had provided transient relief of symptoms for periods of several hours to several days. Both remain concerned that their infestations continue unabated; both were worried that their creatures were multiplying on or within their bodies and that the creatures may be resistant to the remedies they had tried. While obtaining their medical histories, I asked both to describe the sensations they felt. Neither one mentioned pruritus, stinging, burning, or any other dysesthesia. When I pressed them to describe the feeling, neither patient answered the question. They changed the topic of conversation or returned to a previous question. Over several subsequent visits, I was never able to learn what specifically bothered them about their sensations.

In truth, neither patient had a true infestation. Both however were unwavering in the belief that they were infested—and both devoted a great deal of time to examining their skin, seeing physicians, searching the Internet, and attempting to treat their purported infestations. Both were frustrated that no treatment yet worked—and they were even more frustrated that most physicians had, in their eyes, dismissed their concerns by concluding that the unwanted sensations were "all in their head."

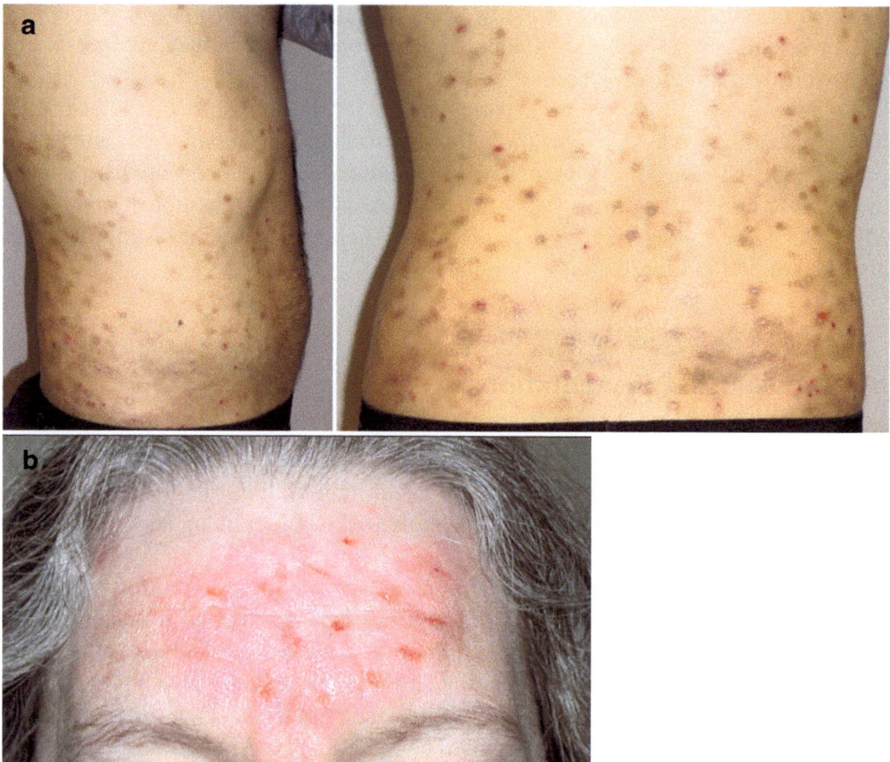

Fig. 2 (**a**) The lower back of a 26-year-old man. His lesions were distributed in a strikingly regular, yet unnatural, pattern. The lesions had uniform diameters, but they were of various ages (fresh, healing, scarred). He devoted time to hunt for the creatures, using various sharp household implements to inspect, probe, and excavate his skin. He had repeated the process several times a week for a period of 2–3 years. His goal was to "dig deep" in order to remove the parasites, which he acknowledged that he had never seen. The lesions taper off in the mid- to upper back, areas where he could not reach. (**b**) The forehead of a 67-year-old woman who used her fingernails to remove creatures that infested the skin of her forehead. She pursued them on a daily basis. She was self-conscious of the sores and retired from her position as a librarian because of it. She lives alone and has ceased contact with friends and family, fearing she might spread her condition to them

7 The First Visit Between the Patient and a New Provider

Conceptual Framework

At a patient's first visit, one of the dermatologist's initial diagnostic tasks is to address the patient's CI by searching for *evidence that might confirm* the patient's suspicions. In other words, when a patient expresses concern for a possible infestation, then one must include appropriate elements in the directed medical history and physical examination. At this stage, it is too early for the dermatologist to redirect that examination toward a diagnosis of a delusional disorder, even if one is alerted to the possibility of DI by the patient's primary physician or a doctor who saw the patient in the emergency room.

A Hypothetical Patient

We will start with a hypothetical patient who has abnormal, unpleasant, and unwanted sensations of the skin. The patient is concerned about a possible infestation and sees a medical provider. The clinician must evaluate the patient to detect any condition, systemic or purely dermatologic that might explain the abnormal sensations. A proper evaluation includes taking a medical history, performing a physical examination, and conducting any additional tests (e.g., usually laboratory tests or skin biopsies) that may help confirm (rule in) or negate (rule out) a diagnosis (see Part II).

The patient with CI will become frustrated if they sense that a new provider considers the matter a straightforward delusion. At this point, the unspoken communication changes dramatically. The patient feels that the provider is retreating from the expected role of being an advocate for the patient's health to an adversary who is trying to dismiss or disprove every remark the patient makes. For most CI patients, this is a familiar scenario, resembling several other medical visits that ended poorly. The patient now feels like they are in their own private *Groundhog Day* movie.

The situation starts to spiral downward; some physicians may feel that any additional time spent listening to the patient is a pointless waste of time. On the other hand, some providers will continue to listen and allow the patient to communicate their message. Most CI patients show up for medical visits with an agenda, written or mental, of topics they want to discuss. In previous visits, the providers may have terminated the discussion.

If you are the one who will be "the buck stops here" provider, it is important to let the patient talk! Let the patient tell you their history, let them tell you their theories, and let them tell you how they have tried to treat their condition. If the patient has DI, well, you have allowed the patient to tell their story. This greatly helps establish a potentially positive working relationship. Not allowing the patient to talk greatly reduces the chance of a good relationship. Many patient advocacy organizations encourage providers to listen to patients for at least 1 min without interruption. I recommend listening attentively to a new CI patient for at least 3 uninterrupted minutes. This may feel tedious, but it may help develop a working relationship that can lead to a therapeutic contract later in the patient's course of care.

The Dermatologic Evaluation

With the physical examination, I look for signs of possible infestation (Table 1). If none are found, I do not immediately conclude that the patient's condition is unrelated to an infestation. Instead, I consider several scenarios. Is it possible that:

1. The patient was infested (and perhaps still is) but my ability to confirm the diagnosis is inadequate for some reason, maybe because there are relatively few organisms?

Table 1 Directed physical examination for patients who are concerned for a possible infestation

Comet sign	Often seen in mite bites from straw itch mites, *Pyemotes* spp.
Cutting	Psychologic disorder expressed by self-mutilation with shallow cuts on skin
Dermatographism	People with dermatographism will develop linear, itchy, urticarial wheals at sites of forceful scratching. Firmly stroking the skin with a blunt object (such as a tongue depressor or spoon) will induce the phenomenon of "skin writing."
Evidence of arthropod bites (papular urticaria)	A typical reaction to the bite of a mosquito, bed bug, flea, etc, is known as papular urticaria. These are edematous round pink papules that often have a central punctum or crust representing the site of the insect's actual bite.
Fingernails	Evidence of picking at cuticles or biting the nails Blood under fingernails Excoriations and their patterns Distribution, spares mid-back? Streaks of three or four scratches of same length, caliber, and uniform inter-scratch distance, all of same age (Table 6) If the patient acknowledges scratching the skin, then examination of excoriations is usually unnecessary If the patient reports they are not scratching the skin, then it is appropriate to examine possible excoriations more closely Erosions and ulcer—and their patterns
Genital examination	Pubic lice; lichenified scrotum; nodular scabies Pruritus scroti, pruritus vulvae, pruritus ani
Neurological exam	Speech, affect, gait, tremor, fidgeting, picking, pupils, eye contact
Scabies evidence? (Table 7)	Examine volar wrists, male genitals Scabies do not produce discrete "mosquito bite"-type lesions (papular urticaria)
Scalp examination	Active lice, nits, excoriations along posterior hairline Other scalp disorders that may be itchy (e.g., seborrheic dermatitis, scalp psoriasis, tinea amiantacea, scarring alopecias, bacterial folliculitis) Contact dermatitis, e.g., allergic reaction to hair dyes; irritant reaction to caustic hair relaxers/straighteners Trichotillomania Hair casts
Skin exam	Full body (to include scalp hair, oral mucosa and teeth, nails, genitalia). May use dermatoscope
Thyroid	Hyper- or hypothyroidism
Trichotillomania	Forceful plucking of one's hair, usually from the scalp. This is usually a compulsion with no adaptive value. The result is partial alopecia with patches of missing or broken hairs. Regrowth leads to tufts of hair of varying lengths.
Xerosis cutis	Latinate for "extremely dry skin." In the elderly, dry ski is often disturbingly itchy. A frequent cause of xerosis is overbathing. Examples include taking two or more long, hot showers in a day; scrubbing with harsh soap; and soaking in a tub full of hot, soapy water.

2. Previous antiparasitic treatments reduced the ectoparasite burden, only to obscure the evidence, thereby making it more difficult to confirm the diagnosis?
3. The patient was infested and previous courses of antiparasitic medications have successfully treated the disorder, and however for some reason, the patient continues to perceive unpleasant sensations?

One must also consider that the patient is not infested (now or previously), but long-standing dysesthesias led the patient to consider infestation as the probable (although incorrect) diagnosis. In these cases, the dysesthesias have several possible explanations:

1. The patient has a *dermatologic condition* unrelated to infestations—and this skin disease can cause pruritus or other dysesthesias. Examples of such conditions are listed in Table 2.
2. The patient has a *systemic disorder* that induces dysesthesia via a conventional biomedical mechanism. Examples include diabetic neuropathy and other conditions listed in Table 3.
3. The patient is taking a pharmacologically active substance that can induce dysesthesia, very often formication. Both licit and illicit substances are known to produce pruritus or formication as a side effect. Examples include conventional prescription medication, such as the anticonvulsant topiramate, over-the-counter products, fraudulent or adulterated medications, and drugs of abuse. Examples are listed in Table 4.
4. The patient has a neuropsychiatric disorder, such as depression or anxiety, that can generate the perception of unwanted skin sensations (Table 3). Patients with delusions of infestation are part of this category and are classified in the DSM-10 as F22.0 (delusional disorder, somatic subtype).

Table 2 Other dermatologic disorders that can cause widespread pruritus

Check for dermatographism	
Existing skin disease	Eczema (atopic dermatitis), etc.
Hair dye	Allergic contact dermatitis, Para-phenylene diamine was once used to enhance or prolong the effect of the hair dye. It is not permitted in US-manufactured products but can be found in imported hair products.
Hyper- or hypothyroidism	Both underactive thyroid disease and overactive thyroid disease can induce generalized pruritus.
Perms or relaxers	Irritant contact dermatitis due to harsh or caustic chemicals in products intended to change the natural shape and curl of one's hair.
Scalp disorders that can itch	Bacterial folliculitis, eosinophilic folliculitis. Sometimes an itchy scalp, particularly of the occiput, is due to a compression neuropathy of the sensory nerves that emanate from the upper cervical spine. In this case, there is no skin disease; this is neuropathic pruritus.
Scalp disorders that produce flakes of skin	Seborrheic dermatitis (which is essentially an inflammatory variant of dandruff). It presents with a red, itchy, scaly scalp, often accompanied by a more subtle red, finely scaly rash of the mid-face.
Xerosis	Super-dry skin. Often caused by overbathing ("overly vigorous hygiene"). In the elderly, dry ski is often disturbingly itchy.

Do not be distracted by examining obvious excoriations. Check the middle of the back!

Table 3 Systemic disorders that may cause dysesthesias

Allergies	Food, environmental, etc.
Anemia	
Congestive heart failure	
Deficiencies	B12, folate, thiamine
Encephalitis	
Endocrinopathies	
HIV/AIDS	
Hyper- and hypothyroidism	
Hypertension	
Hypoglycemia	
Infections	
Liver disease	Hepatitis, cholestasis (especially due to autoimmune biliary disease), hemochromatosis, cholestasis of pregnancy
Malignancy	Leukemia
Meningitis	
Menopause	
Neuropathy	Small fiber polyneuropathy, diabetic neuropathy, notalgia paresthetica, meralgia paresthetica, diabetic neuropathy
Neuropsychiatric disorders	Huntington's disease, various dementias, Munchhausen's disease, malingering, obsessive-compulsive disorder (OCD), depression, phobias (especially entomophobia), conversion disorders, anxiety disorders, hyperawareness of normal sensations, schizophrenia and other psychoses, delusional infestation
Nutritional disorders	
Primary biliary cirrhosis	
Renal diseases/ uremia	
Rheumatoid arthritis	
Several autoimmune diseases	Hodgkin disease, lupus erythematosus, multiple sclerosis, etc.
Syphilis	
Toxicities	Heavy metal toxicity, fluoride poisoning, illegal drugs, carbon monoxide, niacin overdose

The Medical History: Is the Patient Completely Candid?

It is now time for the provider to enter the examination and meet the patient. The patient and the provider each have hopes, expectations, uncertainties, and fears for this visit. The patient hopes that this visit is the one that finds an explanation for

Table 4 Drugs (licit and illicit) that may cause formication, pruritus, or other dysesthesias. Many pharmacologic compounds have an unwanted side effect of generating widespread pruritus. In most cases, the itchiness is caused by the drug's effect on the central nervous system, not because of any skin condition

Amphetamine/methamphetamine
Cocaine
Cannabinoids (THC)
Opioids
Alcohol
Modafinil
Steroids, high dose
Topiramate
Monoamine oxidase inhibitors (MAOIs)
Dopamine agonists

Drug- or medication-induced pruritus with action on the skin, not the brain

their persistent, annoying sensations. More precisely, they hope that their new provider is the one who confirms their beliefs that the sensations are caused by an infesting creature of some sort. Correspondingly, a provider who knows that the patient has CI is there to help the patient and that the patient finds the visit positive and helpful. The provider may also be concerned that the CI patient may be a DI patient. These patients frequently provide lengthy, time-consuming, complicated histories full of disconnected, irrelevant, outlandish, and implausible details. Other chapters and Part III in this book offer useful recommendations on establishing a fruitful doctor-patient relationship with CI/DI patients.

The Clinical Evaluation

The clinical evaluation starts with the medical history. DI patients approach each new visit hoping that the provider will (a) believe their story and (b) successfully identify the organism that is causing the problems. However, confounding circumstances often hamper efforts to obtain a complete and accurate medical history, especially when it comes to previous diagnostic workups, coexisting mental health concerns, and prior use of antiparasitic medications. Patients reason that revealing negative results of previous evaluations (physical examinations, skin scrapings, hair analyses, blood tests, skin biopsies, etc.) will influence a new provider. Consequently, patients often choose to keep certain details close to the chest. But this guardedness also complicates the specialist's evaluation.

Patients also want to avoid unsettling encounters with providers who intimate that their CI stems from an underlying psychological cause or who declare plainly that the patient is delusional. For DI patients, a doubting physician usually generates feelings of helplessness, frustration, humiliation, or even anger. Each encounter with a provider who (a) cannot find evidence of an infestation, and (b) suggests that the patient is delusional, causes frustration and anger to mount.

Patients with a history of depression, substance abuse, psychosis, recent emotional crisis, or mental illness, in general, are often reluctant to disclose this information. They fear that a new provider who cannot confirm the presence of an infestation may attribute the dysesthesias to an underlying psychological cause. Although patients are usually aware that they have been diagnosed with a psychiatric condition in the past, they believe that mental health concerns are irrelevant and unrelated to their CI. They believe that the dysesthesias are truly caused by an infestation and not a manifestation of an unrelated mental health issue.

Finally, patients often conceal the list of medications they have taken to treat "their infestation." Such medications are obtained in two general ways: they are either prescribed properly by a licensed provider or obtained on the patient's initiative. The first group of medications is those obtained conventionally via a provider who examines the patient, decides to treat the patient for a possible infestation, and writes a prescription for an antiparasitic medication. The second group of medications (i.e., those obtained through the patient's initiative) includes an array of over-the-counter products; borrowed prescription medications; veterinary medicines; items bought from online pharmacies; therapies recommended by online forums; illegally imported treatments; and home or traditional remedies. In addition to pharmacologic therapies, many patients engage in physically destructive measures, such as digging at their skin to remove unwanted organisms (as seen in the two cases described at the beginning of this chapter) and self-treating using chemicals and pesticides.

Previous Treatments for Presumptively Diagnosed Infestation

In many cases, CI patients have already been treated presumptively for an infestation by a doctor who found their story plausible. For example, a patient may see their regular provider or go to an urgent care center saying:

> "Ever since I returned from vacation, I've been super-itchy all over. I think I picked something up when I stayed at [a particular location]."

The provider finds this as a plausible story, and noting multiple scratches may make a presumptive diagnosis of scabies. Primary care providers often lack the tools, time, or skills needed to confirm the diagnosis with a scabies preparation. Consequently, based on an empiric diagnosis of scabies, they may give the patient a prescription for a standard scabicide, such as topical permethrin or oral ivermectin.

In such cases, the provider may have been unsure or even doubted that there was an infestation. They may find the story implausible and suspect that the patient's CI is truly DI. Nevertheless, these visits often end with the provider prescribing antiparasitic medications or referring the patient to a dermatologist or both. Although providers may suspect that DI is the correct diagnosis, they also know that raising this possibility with a CI/DI patient may lead to either seemingly an interminable clinic visit or a bitter hostile impasse. The simplest way to expedite the visit is to prescribe one or more antiparasitic medications and recommend follow-up with another provider, such as a dermatologist.

Some patients obtain specific medications to treat "their infestation" in bulk quantities. Two common ways to do that are buying large amounts of equivalent veterinary formulations or buying products through online pharmacies.

Using online pharmacies is risky. Many online pharmacies that sell bulk medications claim to be based in Canada, but the Canadian Government reports that >75% of these pharmacies are neither based nor licensed in Canada. Those online distributors are engaged in the lucrative business of pharmaceutical fraud. The manufacture, sale, and distribution of fraudulent medications are hugely alarming and poorly known aspects of today's medical landscape [7]. People with unsubstantiated (and unsubstantiable) conditions, such as candida sensitivity syndrome, chronic Lyme disease, and DI/Morgellons, are often unwitting (or willing) prey for such enterprises. The medications they receive are often made under poor manufacturing practices, contain incorrect amounts or concentrations of the pharmaceutical agent, or contain contaminants and adulterants.

The Physical Examination

When performing an examination, CI patients generally try to help by pointing out all spots on their skin that they believe are part of their infestation. They typically point at large lesions or open sores or bloody areas. These certainly are the most dramatic in appearance, but those areas on the skin often contain very little useful information. Sores, ulcers, excoriations, scabs and crusts, scars, and thickened areas are what dermatologists call *secondary lesions*. These may be the most eye-catching places on the skin, but secondary lesions are greatly altered from their initial appearance, which is known as a primary lesion. Primary lesions reveal meaningful information, and secondary lesions do not. Let us learn more about what is meant by the terms *primary* and *secondary* lesions.

Primary lesions: The most informative skin lesions are *primary* lesions, those that are unaltered by scratching, rubbing, picking, or any other form of manipulation that changes the appearance of lesions. It helps to know what someone's skin lesions looked like at their onset, *before* scratching or rubbing. Skin lesions whose appearance has changed from the original form are *secondary* lesions, and these changes may arise naturally as primary progress toward healing or in response to treatment.

Secondary lesions: Secondary lesions usually arise from primary lesions. If you have an itchy insect bite, then scratch it. The surface of the bite can open up and ooze bloody fluid. Now we have a lesion with an excoriation, an erosion, and a hemorrhagic crust. Moreover, it can get secondarily infected. The original *primary* lesion is gone, and we do not know what it looked like.

This distinction is important because primary lesions usually reveal diagnostically meaningful information; secondary lesions often do not. With secondary lesions such as scabs, hemorrhagic crusts, and lichenified plaques, it is often difficult—or even impossible—to identify the cause because a great many disparate and unrelated conditions resemble each other as they pass through those secondary stages.

A consistent feature of DI is that primary lesions are often nonexistent. Worried patients may misconstrue a benign transient itch, the sort that everyone has dozens of times a day, as something caused by insects on or in their skin. Scratching and picking the skin can induce secondary lesions, even though there had never been a primary lesion. Patients with CI/DI often point to secondary lesions on their skin as evidence proving the existence of their infestation. Unfortunately, secondary lesions that arise de novo (and not from a preexisting primary lesion) cannot support the contention that an infesting organism was the cause.

Figure 3 is a montage of four photographs: one shows a primary skin lesion and three show secondary skin lesions. The secondary lesions have been altered in ways that make the predecessor primary skin lesions unrecognizable, thereby obscuring any underlying skin conditions. The only *primary lesions* in this montage are the insect bites in the upper left. Such bites, known to dermatologists as papular urticaria, are edematous, pink-to-red, slightly elevated papules, often with a central bite punctum, where the insect's mouthparts have disrupted the epidermal surface (Figs. 3 and 8).

The other pictures show secondary lesions that arise only after prolonged scratching, rubbing, or picking. Unfortunately, secondary lesions are much less useful for the dermatologist trying to figure out what is going on. I explain this to patients and ask them to help me find primary lesions. This may confuse patients who believe that the most noticeable lesions, say the longest excoriation or the thickest lichenified plaque, will prove the presence of an infestation. I hope that they will start to understand why it is that I want to see *unaltered* lesions.

Ironically, some DI patients present with hundreds of secondary lesions on their skin but completely lack primary lesions. Two explanations for this observation are as follows:

1. The patient may have already altered every primary lesion, for example by scratching them vigorously. The primary lesion might have been something innocuous and certainly not a sign of an infestation. I have seen patients with DI whose primary lesions were relatively minor, e.g., patches of dry skin, contact irritant dermatitis, insect bites, and dermatographic urticaria, but the patient perceived them as sites where creatures either entered or emerged from the skin. The patient's efforts to remove "their creatures" led to months of manipulating the skin with fingernails, kitchen implements, and miscellaneous items, such as unfolded wire cloth hangers. After a few weeks of these manipulations, no primary lesions could be found.

2. The second explanation is that many DI patients *never* had primary lesions. After all, many DI patients have dysesthesias that are unrelated to the skin. The psychological urge to rub, scratch, pick at, or dig into the skin can lead to many types of secondary lesions, even without an underlying skin condition (Fig. 4).

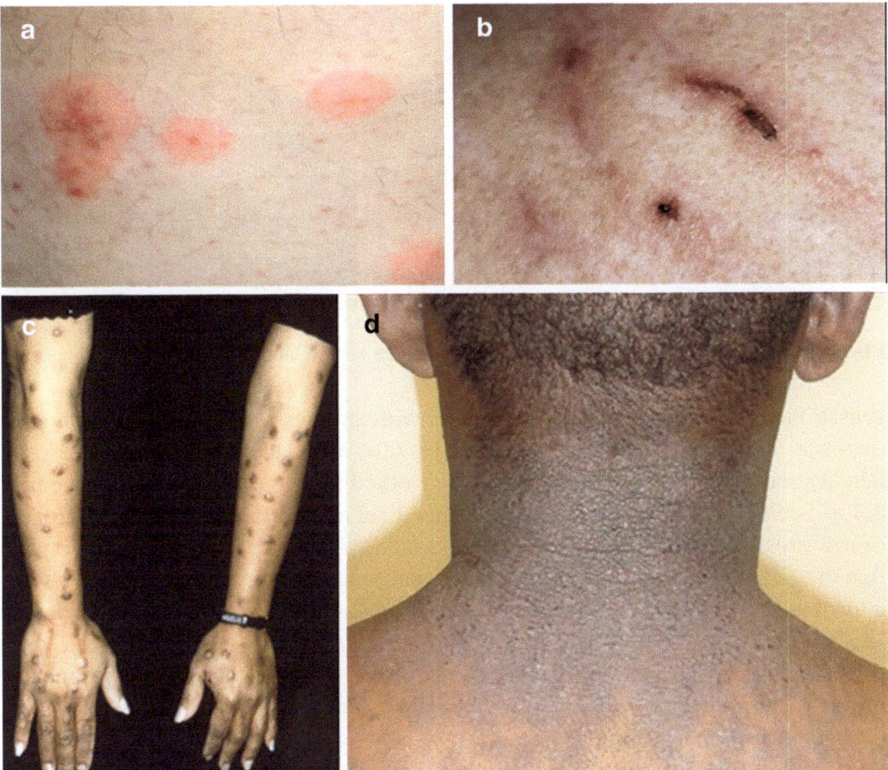

Fig. 3 Montage of primary and secondary skin lesions. (**a**) **Papular urticaria**: This is the dermatologist's term for the common reaction to insect bites, such as mosquito bites. Lesions are pink-to-red, edematous (subtly elevated but not firm), and itchy; some will have a central punctum (a minute papulovesicle or hemorrhagic crust). (**b**) **Excoriations**: Deep scratches that went completely through the epidermis and entered the superficial dermis, the most superficial level of the skin that has blood vessels (capillaries) and is capable of bleeding. (**c**) **Picker's nodules**, also known as prurigo nodularis (PN): Patients with PN typically have the habit or tendency to pick or scratch any sort of lesion or itchy spot on their skin. They pick at small, focal, itchy areas, which reinforces or perpetuates the itch-scratch cycle. PN lesions are focal, slightly elevated, and firm to touch and have a rough surface. Unlike lichenified lesions, PN is usually discrete and rarely coalesces to form plaques. (**d**) **Lichenification**: Lichenified skin feels thick, rough, and dry. The epidermis is markedly thickened with marked thickening of the epidermis. Lichenification is the end state for many itchy conditions that induce incessant rubbing or scratching. Lichenified plaques cover broader surfaces than do PN. In many cases, lichenified skin started as an eczematous eruption, but almost any type of inflamed skin can become lichenified if scratched or rubbed enough times. (**a**) is the only primary lesion in this montage. The other images show secondary lesions

The patient in Fig. 2a, discussed at the beginning of the chapter, had hundreds of secondary lesions, but the history he gave, and my physical examination, suggested that he never had a visible skin lesion, just the sense that something was moving inside his skin. He freely acknowledged that he never had primary skin lesions ("My parasites are under my skin, not on top of it") and that he created dozens of secondary lesions by digging for creatures that he believed were causing his strange

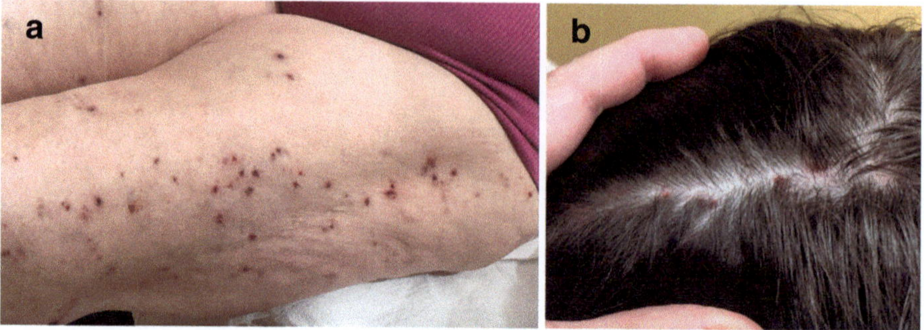

Fig. 4 (**a**, **b**) Marks made by picking at normal skin are not evidence of an infestation

sensations. Let us contrast this young man with another patient who was much the opposite. This patient was a gentleman in his 60s, who had undergraduate and graduate degrees from Ivy League schools. He worked for the US Government and held a Top-Secret security clearance. At each medical appointment, he revealed dozens of new erosions, excoriations, ulcers, crusts, and scars on his skin. He would explain that these had appeared after his previous visit roughly 2 months earlier. He always adds that the lesions were not itchy and that he did not scratch them. His account was that each lesion showed up suddenly on his skin—and they emerged de novo on pristine skin. He said that each lesion started out looking exactly as they appeared now: the scabs first appeared as scabs, the ulcers first appeared as ulcers, and so on. He emphasized that the spots were not itchy, and he adamantly denied scratching or picking them.

His denials of scratching and picking amazed me for two reasons. An ulcer with a thick hemorrhagic crust cannot arise overnight, yet he insisted that was exactly what happened. Also on many visits, there was dried blood under the nails of the middle three fingers on his dominant hand. These observations essentially negated his contentions that the lesions arose sui generis and that he did not scratch or manipulate the lesions.

Clues that Scratching is Taking Place

If a patient has abundant lesions that appear to be excoriations, yet denies ever scratching or manipulating their skin, then one should look for clues that scratching is indeed taking place. Excoriations usually appear on surfaces that a person can reach with their fingertips (or with simple tools that extend their reach). Conversely, excoriations are generally not found on surfaces that a person cannot reach easily. This produces a zone in the middle of one's back that shows no evidence of excoriations, specifically because patients cannot physically reach those areas to scratch (Fig. 5). The absence of excoriations from the mid-back and other areas that a patient cannot reach while there is an abundance of excoriations on accessible surfaces suggests that excoriations in the accessible areas are the result of scratching.

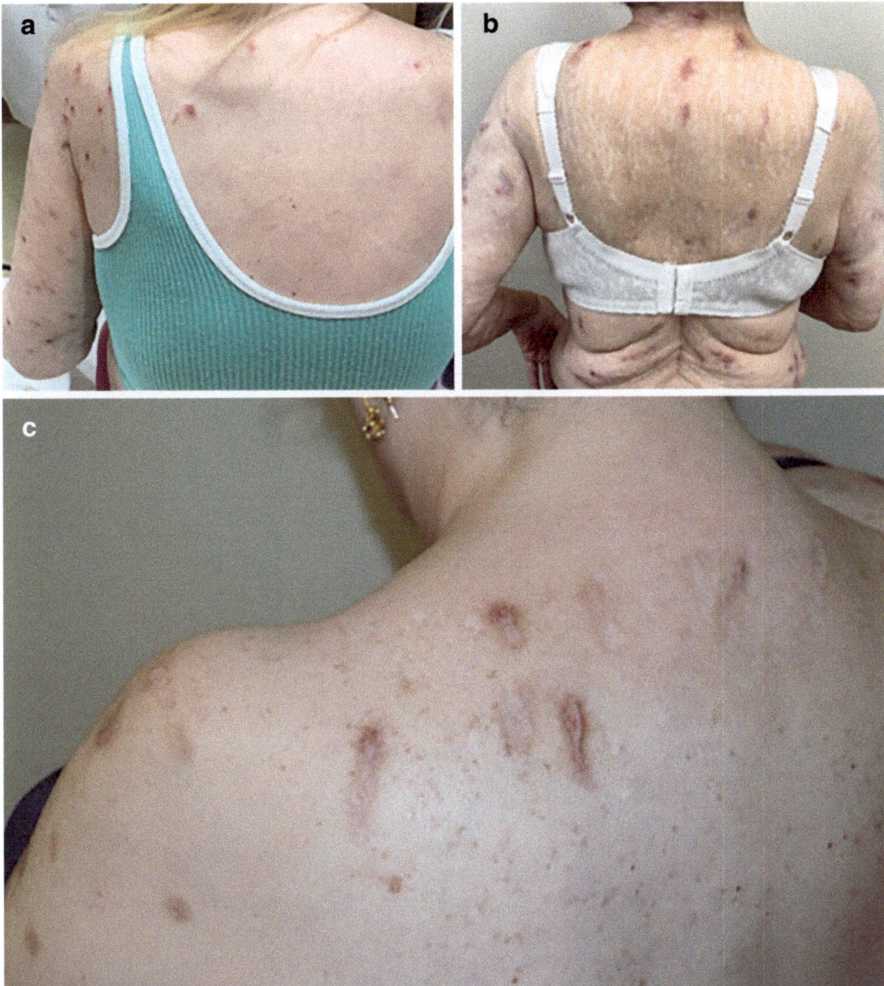

Fig. 5 These patients have scratched their skin for years with enough vigor and force to create linear ulcers that heal as scars. Patients with DI often deny that they are scratching their skin, postulating that their particular creatures are making these lesions. Note that the mid-back (between the shoulder blades) is largely free of excoriations or scars. If the patient truly had a widespread infestation, we would expect involvement of this surface. On the other hand, it is difficult to scratch the interscapular area with one's own fingers, thus leaving this area relatively untouched, unexcoriated, and unscarred. See the text for further details. Self-scratching produces few skin findings on the upper mid-back. These are examples of sparing of the upper mid-back by people who are vigorous scratchers. Most people have difficulty scratching between their shoulder blades unless they use an implement to reach the area. Consequently, that region is largely spared the effects of vigorous excoriations. This observation is important when CI patients claim that marks on their skin are caused by "their creatures" and insist that they do not pick at or scratch their skin. Organisms that infest the skin will not selectively spare the mid-back. If the mid-back lacks telltale lesions, one can generally infer that the lesions are made by the patient, not by an infestation. Therefore, a patient's contention that they do not pick at their skin can be viewed with skepticism

Both patients seen in Fig. 5a, b report picking at normal-appearing skin. Their lesions, hemorrhagic crusts in both cases, have no preceding lesions. These lesions start as secondary lesions, the result of scratching their own skin.

The 55-year-old woman in Fig. 5c felt as though ants were crawling on her skin (formication), even though she acknowledged never seeing an ant or other creature on her skin. She picked at normal-appearing places on her skin where she felt the odd sensations, hoping that she could dig out whatever creatures had entered her skin.

The 14-year-old girl in Fig. 5a had an anxiety disorder and had no thoughts about a possible infestation. She picked at her skin, particularly her scalp, as a self-soothing behavior. She said that the skin she picked appeared normal and was not itchy. This means the hemorrhagic crusts on her scalp had no predecessor lesions; they started out as secondary lesions; there were no primary lesions beforehand.

There are other clues (Table 5). The patient described in the previous paragraph often had dried blood under the fingernails of the three middle fingers on each hand. Another clue is the simple observation that a single vigorous scratch leaves a cluster of three lines that have roughly the same length and caliber (or width). They are parallel, and they start and end at roughly the same place. They appear to be in the same state of healing, suggesting that the initiating event for the three lines was simultaneous. For example, deep scratches that were newly made often show fresh serosanguinous exudate (Figs. 2b, 3b and 6). Older

Table 5 Evidence that scratching has occurred

Clustered excoriations	Grouped in clusters of three or four parallel lines
	    Lines have similar length and caliber
	    Areas of normal skin between scratches are equidistant
	Clusters spare mid-back (unless implements are used)
	Clusters are densest on dorsal forearms, deltoids, and suprascapular areas
	Lesions appear to be of same age
	If lesions appear fresh, look for blood under fingernails
	Elbow-angle sign: extending a line in the direction of parallel lines will lead to position of the elbow ipsilateral to the hand that made the scratch marks
	Nondominant side with more lesions as the result of scratching using the dominant hand

Evidence of other forms of cutaneous manipulation	– **Cutting:** Healed shallow lacerations on forearm of nondominant hand. Scars are short, usually <5 cm in length, and straight with evidence of suture marks. Several cutting lines are typically seen because individuals who cut usually do so on repeated occasions – **Fingernails:** Onychotillomania/onychodaxia, e.g., picking and biting of the fingernails – **Trichotillomania:** Pulling on one's hair to the point that the hair is ripped out – **Prurigo nodules:** Commonly known as "picker's nodules." these are often symmetrical on surfaces that can be reached with one's hands – **Areas of lichenification:** Lichen simplex chronicus or prurigo nodularis

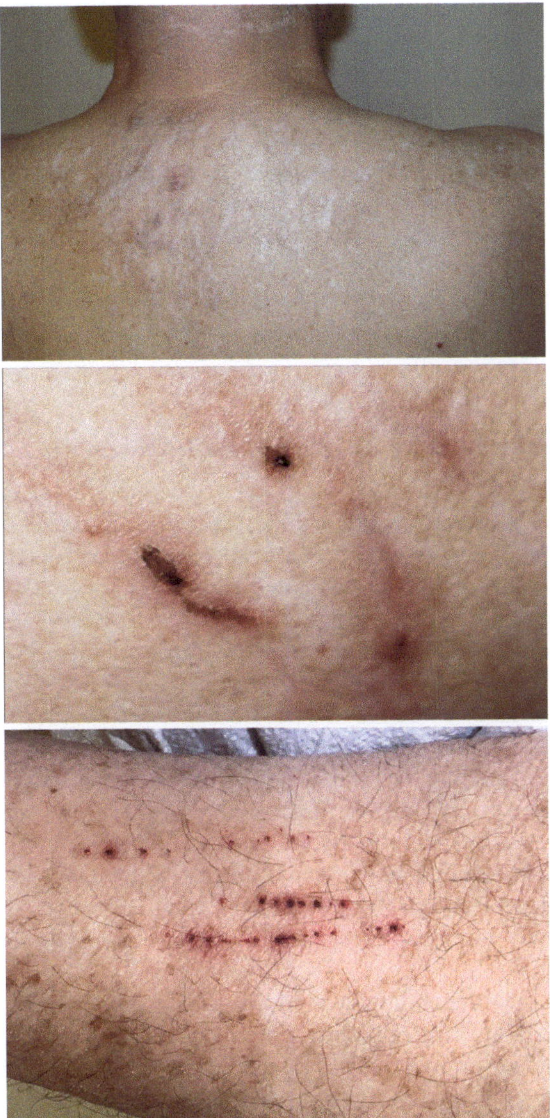

Fig. 6 Examining patients who deny scratching or picking their skin. Signs to look for when a patient says that skin lesions are caused by "their creatures" and deny that they are scratching their skin. (1) Lesions are parallel lines. (2) Lesions appear in clusters of three or four, i.e., the number of fingers used in scratching. People scratch mainly with the middle three fingers; some people also use their little fingers. The thumbnail is rarely part of the scratching motion. (3) Lines within a cluster are roughly the same length and caliber/width. (4) Uninvolved (normal-appearing) skin between adjacent lines is roughly the same width (1–2 cm) throughout. That width reflects the distance between fingernails when the hand is held in a scratching position. (5) The surface appearance of each line within a cluster is similar, indicating that the lesions occurred at the same time and with similar force, going to the same depth. Thus, lesions resemble each other. For example, they all have a dry, dusty look or are interrupted **streaks of dried blood**. (6) **If bloody lesions seem to be freshly made, check for** blood under the fingernails of the hand that would make the scratch. (7) No lesions in the mid-back (interscapular)

scratches often have a hypopigmented, scar-like appearance (Fig. 5b). The lines are parallel, and their interlinear distance remains constant from one end to the other. In addition, the interlinear distance matches the natural spacing of one's index, middle, and ring fingernails when the hand is held in a scratching posture. Occasionally, there are four parallel lines, not three; in this situation, the person used their fifth digit, i.e., the little finger, to help scratch. Good places to look for this sign are areas where it is easiest to scratch, e.g., the dorsal forearms, the lateral aspect of the upper arms, and the area just superior to the shoulder blades (Fig. 6).

To me, these features show convincingly that vigorous scratching is occurring. The reasons why patients deny scratching, … well, that is harder to pin down.

After the Physical Exam Finding Creatures

In general, the most common cause of widespread pruritus is excessively dry skin—and this topic will be discussed in more detail later in this chapter. Other itchy disorders might be common eczema also called atopic dermatitis and contact dermatitis. The scalp is often the site of persistently annoying itch, most often caused by common benign disorders such as seborrheic dermatitis (an inflamed dandruff-like disorder) and scalp folliculitis.

If the clinician finds evidence of a possible infestation, they should take steps to substantiate the evidence and confirm the diagnosis. In most cases, the next step is to identify the infesting organisms. Most known ectoparasites are visible to the naked eye (or with simple magnification); therefore, identification usually relies on visual inspection. Astute clinicians, pathologists, entomologists, and many other professionals can usually identify undamaged ectoparasites to the level of genus. There are no blood tests (e.g., serologic studies), PCR, or culture techniques for most types of ectoparasite infestations in the clinical setting.

Even when a creature is obtained, additional steps are necessary before it can be construed as an infesting organism. The organism should be a recognized ectoparasite [i.e., known to infest humans (see Chap. 13)] and should be capable of producing the dysesthesia felt by patients. And there should be biological plausibility. For example, finding a dead housefly on one's bedspread should not be construed as evidence of myiasis. Similarly, the presence of louse nits on the scalp does not explain pruritus on the soles of the feet. If the organisms are known to infest humans and cause abnormal sensations, then the clinician and patient can work together to develop a sound therapeutic strategy. In addition, the clinician should ensure that the patient is educated on their specific condition: how it will be treated; how to prevent recurrences; and whether the patient's contacts (e.g., household members, intimate partners, or others) need to be engaged as well.

8   Common (True) Infestations

During the physical examination, look for evidence of ectoparasitic infestation with scabies or lice (head lice, body lice, and pubic/crab lice). Also, look for evidence of:

1. Bites from other anthropophilic insects (e.g., mosquitoes, bed bugs).
2. Accidental/opportunistic feeding by non-scabetic mites (e.g., bird mites, chiggers).
3. Other arthropod-associated bites (e.g., ticks).
4. Or arthropod-related envenomations (e.g., scorpions and some caterpillars), and sometimes even nematodes.

The two most common (true) ectoparasitic infestations in North America and Europe are scabies (*Sarcoptes scabiei* var. *hominis*) and head lice (*Pediculus humanus* var. *capitis*). Both species can fully sustain themselves on a person and can go through complete life cycles for generations without leaving their host. Furthermore, these ectoparasitic infestations are easily transmitted from one person to another. Once on a new host, they can establish another self-sustaining and self-replicating population that can survive for generations or send its emissaries to infest other people.

Only one other ectoparasite, the pubic (or crab) louse (*Pthirus pubis*), can accomplish this. Nowadays, pubic lice are relatively uncommon—and are becoming more rare. The trend in personal grooming for people to trim, clip, or shave their pubic hair has caused a remarkable (but widely appreciated) decline in the prevalence of this insect.

Many ectoparasites can feed on human skin; some feed on blood (e.g., ticks or bed bugs), and some feed on non-sanguinous tissue fluids. But they do not (and cannot) reside on humans, and none can conduct the full range of physiologic and reproductive tasks that are part of a life cycle on their temporary hosts. These are completed off their hosts.

The next most common ectoparasites in North America and Europe are hard ticks and bed bugs (*Cimex* spp.). Taxonomically, hard ticks constitute the family, Ixodidae, with approximately 10 genera and 650 species. Relatively few species of hard ticks are known to feed on humans. In North America, perhaps only a dozen species of ticks have medical importance for humans. Most of these are capable of transmitting microbial pathogens to humans. Tick-borne pathogens come from across a vast taxonomic spectrum such as bacteria, viruses, and protozoa (but not fungi or helminths, as far as we know).

When ticks attach themselves to humans, they usually do so singly, unlike small groups of bed bugs that troupe out together from their dwelling space to visit a host. Ticks feed individually as a rule. On humans, they take a blood meal (up to 2–3 days). Once engorged, they drop off their human hosts (hopefully outdoors) to continue their life cycles, e.g., larva-to-nymph-to-adult-to-egg. Adult female ticks require a blood meal to lay their eggs, but they do not engage in any reproductive processes while on a person. Ticks neither copulate nor lay eggs while on a person, and they

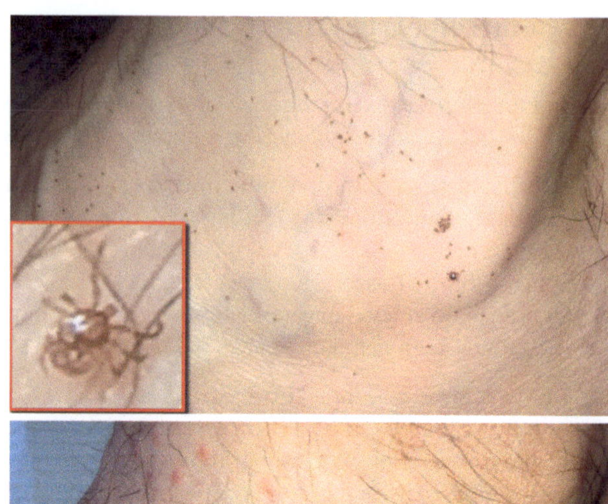

Fig. 7 A true infestation that produced primary lesions. During one autumn walk in a forested area in Central Maryland, part of the US mid-Atlantic states, the author noticed hundreds of tiny ticks crawling rapidly over his skin and clothing. A close examination showed that these were lone star ticks, *Amblyomma americanum*, which lay eggs in masses of several thousand eggs. The eggs hatch simultaneously, releasing thousands of speedy, vigorous larvae. My wife counted roughly 375 bite papules on my left leg from knee to toes. I probably had >1000 total bites

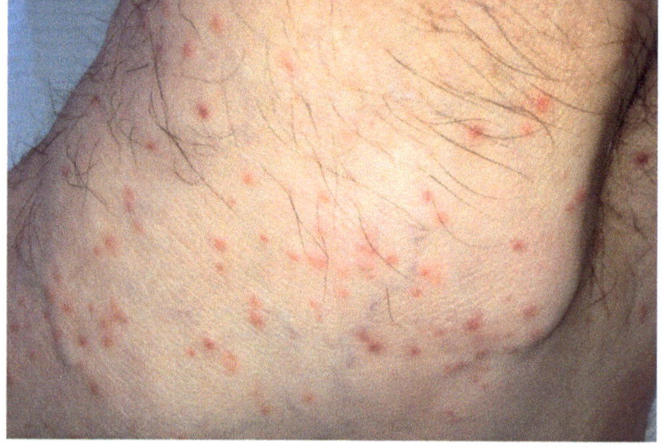

certainly do not lay eggs within a person's skin, no matter the wild tales one might hear.

On rare occasions, large numbers of hard ticks can attack or infest a single individual. For example, I was once swarmed by more than a thousand very aggressive ticks. Literally, Fig. 7 shows my ankle with approximately 1000 larvae of the lone star tick, *Amblyomma americanum*. While on a hike in Central Maryland, I unknowingly stepped on a lone star egg mass. A lone star egg mass consists of hundreds, if not thousands, of eggs. The presence of my warmth and my expired carbon dioxide (CO_2) led to a mass hatching of these aggressive little critters. They swarmed over me in a matter of seconds and proceeded to bite. At one point in time, my wife counted nearly 400 ticks on my left leg from knee to toes. Now, that is an infestation!

Other infestations encountered less frequently in North America are body lice and pubic lice. We see uncommon, imported cases of furuncular myiasis, tungiasis, and cutaneous larva migrans. Even more rare are cases of endemic myiasis, where fly larvae that normally feed on wild animals or farm animals wind up accidentally on humans and become opportunistic feeders. Figure 8 and Table 6 illustrate and describe papular urticaria.

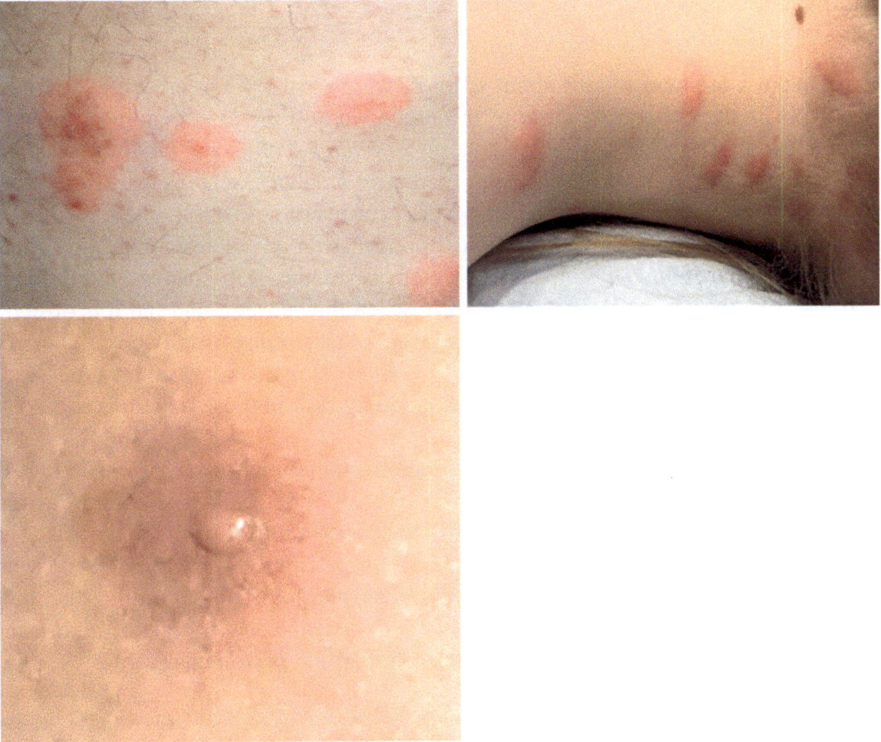

Fig. 8 Primary lesions seen with papular urticaria (common insect bite reactions)

Table 6 Signs and symptoms of papular urticaria (insect bites)

Identifier	Signs and symptoms
Erythematous and edematous papules or plaques	Slightly elevated, juicy/swollen, red bumps
	Lesions are itchy; skin between lesions is not

Observations

Many have a central punctum where the insects' mouthparts entered the skin. This punctum is usually near the center of the lesion. It may be a minute, slightly elevated papule (<1 mm diameter) or a tiny speck of dried blood)

A focal crop of lesions appears to be roughly the same age and appears to be resolving at the same pace

Note

The above features describe the primary lesion of papular urticaria. Because these lesions are itchy, often intensely so, repeated or vigorous scratching may create secondary changes that alter their appearance

In people who develop severe insect bite reactions, one or more lesions may develop small blisters. It is rare for all lesions to develop blisters. If it appears that all suspected bites evolve into blisters, consider a different diagnosis

The bites of some mites (e.g., furniture or straw mites, *Pyemotes* spp.) may develop a "comet sign." this is a streak that is the same color as the primary lesion that projects in a small paisley-shaped curve from the main bite site

Scabies

Let us take a moment to talk about scabies, which usually causes severe pruritus. "Scabies" is the colloquial name of the organism and the skin condition. It is an ectoparasitic infestation caused by the mite *Sarcoptes scabiei* var. *hominis*. In emergency rooms and overwhelmingly busy pediatric practices, scabies is often the first proposed explanation for generalized severe pruritus of new onset.

The classic teaching about the clinical presentation of scabies is that the telltale signs are burrows in the skin and involvement of the web spaces of the hands (Fig. 9). Yes, these two features are helpful when they are present, but it is difficult to find pathognomonic burrows. The simple act of scratching usually turns a burrow into a red excoriated bit of skin. Furthermore, many skin disorders involve the hands and dorsal web spaces, so the presence of an itchy rash in those locations is consistent with scabies, but not diagnostic.

I believe that there are two more reliable presentations of scabies. The first is symmetric involvement of the volar wrists (the soft side or palm side of the wrist) (Fig. 10a). If the wrists are not involved, then I am skeptical of the diagnosis. The second reliable presentation is found only in males. Nearly every male over 1 year of age who has scabies will develop nodules (or nodular scabies) on their genitals (Fig. 10b). If the male genitalia (i.e., the penis, particularly the glans, and/or the scrotum) lack these nodules, then it is unlikely that the person has scabies.

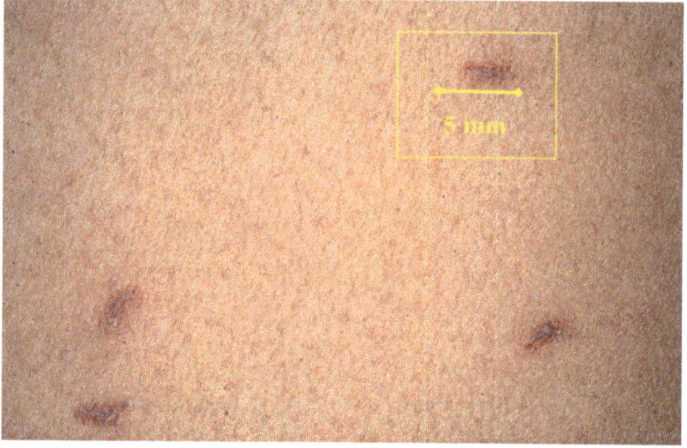

Fig. 9 Scabies burrows. Scabies burrows are the linear, tunnellike homes of scabies mites. Burrows are in the stratum corneum, the uppermost layer of the epidermis. They are rarely as well formed and intact as the ones in this figure. Because scabies is so itchy, people scratch areas of active infestation, rendering the burrows unrecognizable as such and appearing like the lesions in Fig. 10a. Burrows are usually only 2–5 mm long. Patients with CI/DI often misidentify other linear findings on the skin, such as dermatographism, as burrows. Dermatographic lines are a form of pressure urticaria that are usually several centimeters long. Urticaria (hives) are situated much deeper in the skin, do not involve the epidermis, and usually resolve in 10–30 min. CI/DI patients may describe dermatographic urticaria as evidence of "deep scabies" that can vanish even deeper into the body, saying, "And that's why they're so hard to treat"

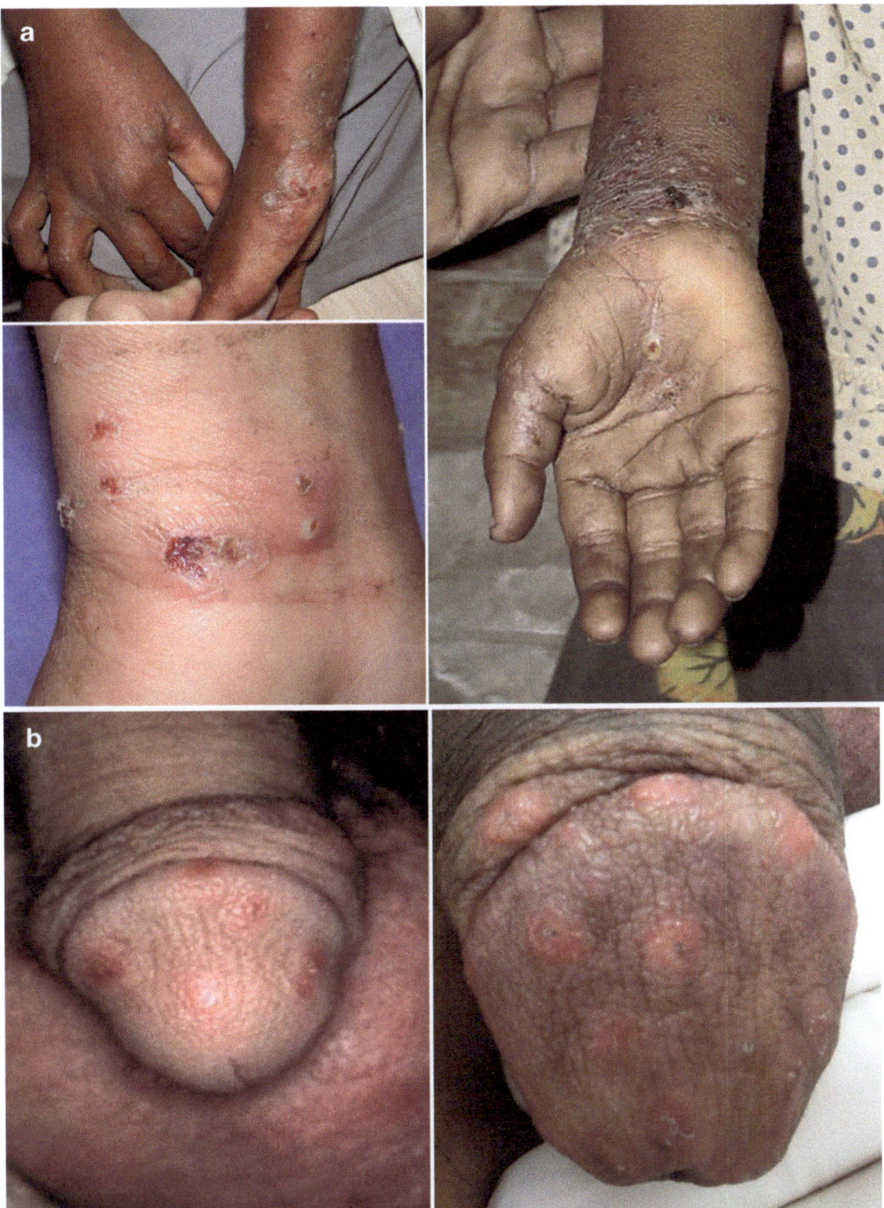

Fig. 10 (**a**) Common sites for scabies infestations: the volar wrist. Textbooks describe the web spaces between fingers as a classic location for scabies. That is true—but it is even more common for a person with scabies to have lesions on the volar (palm side) of the wrist. (**b**) Common sites for scabies infestations: male genitals. Nodular scabies (NS) on male genitalia. Both images show nodules on the glans penis; however, nodules are common on the penile shaft and scrotum. There may be fewer than a dozen total nodules. NS are often asymptomatic (not itchy) and do not resemble the scabies eruption elsewhere on their bodies. A person with NS may not realize that the genital lesions are part of the scabies that is so noticeable elsewhere on their bodies. Men with scabies nearly always have genital lesion, which makes this an important part of the physical examination for men with CI/DI

Table 7 Summary of the 2020 IACS consensus criteria for the diagnosis of scabies [5]

A. Confirmed scabies	At least one of: 1. Mites, eggs, or feces on light microscopy of skin scrapings (Appendix A) 2. Mites, eggs, or feces visualized on an individual using a high-powered imaging device 3. Mite visualized on an individual using dermoscopy (Appendix B)
Clinical scabies	At least one of: 1. Scabies burrows 2. Typical lesions affecting male genitalia 3. Typical lesions in a typical distribution and two history features
Suspected scabies	Either one or two: 1. Typical lesions in a typical distribution and one history feature 2. Atypical lesions or atypical distribution and two history features
History features	1. Itch 2. Positive contact history

Diagnosis can be made at one of the three levels (A, B, or C). A diagnosis of clinical or suspected scabies should only be made if other differential diagnoses are considered less likely than scabies
Information from the 2020 International Alliance for the Control of Scabies Consensus Criteria for the Diagnosis of Scabies [6]

Wrists, dorsal aspect of finger web spaces, ankles, soles, and male genitalia are high-yield locations when looking for scabies, but a proper examination requires a complete skin examination as well. A full body inspection allows the detection of other skin conditions that might explain the patient's dysesthesias.

The International Alliance for the Control of Scabies (IACS) recently published guidelines on the clinical diagnosis of scabies and how to do a diagnostic scabies preparation aka. scabies prep or scabies scraping (Table 7 and Appendices A and B) [5, 6]. This procedure enables the clinician to collect scant amounts of the most superficial layer of the epidermis (the stratum corneum) to look for the diagnostic presence of scabies mites.

Head Lice

The diagnosis of head lice infestation is best made by identifying live, freely moving lice on a patient's scalp. Nevertheless, in some situations, a firm diagnosis of head lice is possible even without directly observing the organism. For example, detecting nits *against* the scalp can indicate an active infestation of lice, even if no adults are detected. Nits are louse eggs that are cemented onto a substrate, usually a hair shaft or textile fibers. Nits on scalp hair indicate an infestation with head lice, but there are nuances to interpreting the presence of nits.

The most common situation is when one finds the nits of head lice (*Pediculus humanus* var. *capitis*) on scalp hair. If nits on hair shafts are within 0.5 cm (¼ inch) of the scalp, one can surmise that live adults are on the scalp. Gravid head lice attach their eggs on hair shafts using a biological cement collar. Once cemented to a hair, the nit is fixed in place. As the hair grows, the nit will continue to drift further away from the scalp. The more the distance, the greater the length of time to the attachment. Louse eggs hatch roughly 6–10 days after attachment, which is roughly enough time for hair to grow about 0.5 cm (hair grows at a rate of ~0.3 mm/day).

The diagnosis of head lice infestation can also be made by detecting a proxy for adult organisms [8]. For example, the presence of nits on hair shafts close to the scalp shows a current (or very recent) infestation with live adults. In this situation, one should continue to examine the scalp for viable adults. Microscopic examination of a nit that is <5 mm from the scalp may show an embryonic louse, distinguished by characteristic dark-red eyespots seen through the chorion (shell). However, if *all* nits are more than a centimeter from the scalp, that points to an earlier infestation, but it does not help determine if there is a currently active infestation. Nits that are >1 cm from the scalp do not confirm an active case of head lice. A closer examination of those nits shows either that they are empty (i.e., already hatched) or, if not empty, then one is probably looking at a dead embryo that will not hatch. If an embryo is found, it will not be alive but be retained in the nit as the hair grows out. If *all* nits are further than 0.7 cm from the scalp, it suggests evidence of a prior infestation.

Pseudo-Nits

Several minor scalp conditions can generate harmless, minute, keratotic nodules along a hair shaft that superficially resemble nits. They are called, appropriately enough, pseudo-nits and are not associated with head lice or any other infestation [9, 10]. Conditions in which pseudo-nits can occur include seborrheic dermatitis, tinea amiantacea, accumulations of haircare products, and peripilar hair casts. The simplest way to distinguish true nits from pseudo-nits is to see if the purported nit can be moved along the hair shaft. Pinch the tiny nodule between the index finger and thumb, and then try to slide it up and down the hair shaft. True nits remain fixed in place and cannot slide along the hair shaft; pseudo-nits, on the other hand, can be grasped and made to slide freely along the hair shaft.

I have seen several patients with pseudo-nits who have been told by physicians, school nurses, and other health professionals that these are nits and that the patients have head lice. In the case of schoolchildren, there have been stressful events where the child has not been allowed to return to school, and another case in which the mother felt so humiliated that her son had head lice she disenrolled him from public school and began homeschooling him. She and her son became hermits, rarely venturing out of the house, wearing stocking caps when they did for fear that they might infest others. Of note, their scalps were not itchy; they had no insect bites on the scalp, neck, or around the ears; and no one had ever seen an adult louse. The diagnosis of pseudo-nits was made without difficulty, and we demonstrated the features with a microscope and textbook photos of nits and pseudo-nits.

Bed bugs

When a physician suggests that a child's itchy eruption might be due to bed bug feeding, parents are often surprised, shocked, disbelieving, and offended. The intimation that bed bugs are present in one's home can create intense feelings such as revulsion, bewilderment, defensiveness, shame, and anger. Until the presence of bed bugs in a home is proven, many people will not accept this as a possible explanation. Because reactions to bed bug bites vary among people, one cannot conclude with certainty that particular lesions were caused by bed bugs. If bed bugs are suspected, then efforts should be made to confirm that impression. This can be done by family members or by pest control professionals.

People often construe the presence of bed bugs as an indictment of housekeeping and hygiene. There is deep social anxiety and stigma that a diagnosis such as this will induce others to judge them as maintaining a filthy household (or that they are dirty people). But bed bugs are indifferent to cleanliness; they simply want to feed on warm bodies, rich or poor, clean or dirty, irrespective of class or cleanliness. Of course, when a child has bed bug bites, that entails the child has been (and perhaps still is) in an environment where bed bugs are present. When parents reject the possibility of bed bugs in the home (or elsewhere in the child's living space), this can delay treatment and subject the child to continued exposure to the biting arthropods. Many families are reluctant to face the cost and inconvenience of engaging an inspector, hiring a professional exterminator, or confronting a landlord. Ironically, some true DI patients are quite the opposite; they bring in inspectors and exterminators on a regular basis and are baffled when the professionals report that there is no evidence of a household infestation.

Several colleagues who work as public health entomologists (for example, at universities or state-level public health or agriculture departments) tell me that they receive many inquiries about bed bugs (and many panicked calls for immediate rescue from bed bugs). Some DI patients call several times each week. Consequently, most local (county and state) public health departments have valuable web pages, handouts, guidelines, etc. on bed bugs in the home. Their messages are clear:

"… if poorly managed, significant financial costs can be incurred in eliminating bed bugs from a residence. If managed correctly it can be quick and cheap. Many jurisdictions now have bed bugs laws that help garner cooperation among parties if there is disagreement. These can be used to good effect. No personal effects, furniture, or articles should be abandoned or thrown out because of bed bugs. Contrary to popular myths perpetuated by social and commercial media, bed bugs are weak, delicate, shy, medically harmless little insects that are easily killed. They transmit anxiety resulting in people dissolving into blind panic and fear. They benefit from this dysfunction and thrive as a result. When people become pragmatic and systematically work together in managing the insects, minus the finger-pointing, bed bugs are always successfully exterminated" (Personal correspondence with Gale Ridge, PhD, The Connecticut Agricultural Experiment Station).

9 Additional Information Arising During the Medical History

Travel

Patients often identify foreign travel as the origin of unwanted sensations. They are perhaps aware of the public health topics and return home with the belief that they became infested while abroad. They may suspect some sort of globally common infestation, such as bed bugs or scabies, or that they now have a disease or an infestation that might be common in their travel destination but unusual in the United States. The patient may further conclude that the rarity of their particular disorder explains why doctors at home have such difficulty in finding the mysterious creatures. Knowledge gaps in medical science, they believe, help explain the medical community's difficulty in finding their unusually mysterious creatures. Those conditions are relatively easy to confirm with laboratory tests (blood smears, serologic studies, examination of stool for ova and parasites).

Pets and Animal Contact

Chapter 4 by Professor Nancy Hinkle, PhD., addresses the flawed logic of blaming one's pets or other animal contacts as the source of an enduring infestation. Yes, several species of mites that ordinarily associate with common pets, such as dogs, cats, and birds (and bird nests), are capable of biting humans, but none of these actually infest humans. In other words, these mites occasionally bite people, but they are unable to live on (or in) humans. They certainly cannot invade human skin, reproduce in human bodies, or maintain their life cycle after feeding on people. Several websites and social media platforms disseminate factually and biologically incorrect information that these mites cause unexplained dysesthesias in the types of patients we are discussing in this volume. When people believe the online information and ignore information provided by their veterinarians, they may seek to treat or abandon or destroy their once-beloved pets.

Possible Causes of the Patient Dysesthesias

Another standard question is what the lesions looked like before one started scratching. The dermatologist wants to see primary lesions, those that are untouched, unaltered by kicking, scratching, rubbing, or anything else that changes the appearance of the lesion.

Overbathing Leads to Dry, Itchy Skin

It is important to take a good personal hygiene history, which may help explain the cause of generalized itchy skin. Probably, the most common cause of itchy skin is dry skin—and the most common cause of dry skin is overly zealous hygiene habits. It is understandable that a person with CI may become particularly focused on personal hygiene. The problem is that the pursuit of hyper-cleanliness leads to frequent exposure of one's skin to water, especially hot water, and to soaps and cleansers. Worried patients often shower or bathe three or more times a day. Exposure to water, in general, can deplete the skin of its natural lipids and prolonged exposure. Exposure to hot water is even more drying, and scrubbing one's skin with harsh soaps and coarse washcloths will dry out the skin even further. These overly vigorous hygiene habits can cause dry, chafed, irritated skin that is profoundly itchy or even painful. In other words, efforts to sanitize one's skin only make it more inflamed and uncomfortable.

When skin gets dry or xerotic, in the parlance of dermatologists, the stratum corneum often sheds minute white flakes that are similar in size and color to flakes of scalp dandruff. Patients may perceive the barely perceptible white flakes as creatures trying to exit the skin. The patient may attempt to remove the white flakes with their fingernails or common household implements. This typically leads to excoriations and erosions, followed by serous or hemorrhagic crusts. Patients often perceive the scratches and crusts as tracks produced by the infesting organisms.

Diagnostic Tests

To confirm the diagnosis of an ectoparasite infestation, we try to obtain laboratory evidence by examining skin scrapings or strands of hair aided by a microscope. Another method is to perform a skin biopsy. This is a quick, simple surgical procedure to obtain a small skin specimen, which is sent to an anatomic pathology laboratory. There, the pathologist runs a histopathological examination looking for organisms, their residua, or inflammatory patterns such as tissue eosinophilia in the superficial dermis that are characteristics of arthropod infestations. Alternatively, the pathologist may provide the diagnosis of another skin condition, one that explains the dysesthesias and is unrelated to infestations. Table 8 lists tests that clinicians may consider ordering during the evaluation of a patient with unexplained dysesthesias. Some of the tests are part of an advanced evaluation for patients with widespread, unexplained, intractable pruritus; other tests are intended for patients who have CI/DI with less severe dysesthesias.

Table 8 Tests to consider ordering as part of workup for unexplained intractable pruritus

Order only if and when clinically appropriate. In each CI patient, only a few of these tests will be relevant
• Scabies preparations (see Table 7 and Appendices A and B)
• Fungal potassium hydroxide (KOH) preparation
CBC with differential (to look for eosinophils
• IgE levels
• ESR and CRP are nonspecific indicators of systemic inflammation. If elevated, they do not point to any specific disease as the cause of the inflammation. If the results are normal, that provides a level of reassurance that any internal or systemic inflammation, if present, is not severe
• ANA (antinuclear antibody panel for conditions such as systemic lupus erythematosus)
• RPR (a screening test for syphilis; many inflammatory conditions produce false-positive results)
• HIV
• Bacterial culture of lesions and nostrils to assess the possibility that skin lesions may be due to secondary staphylococcal infection. In the United States, if patients who present for medical care with suspected spider bites, probably >50% have staphylococcal infections, especially *Staphylococcal furunculosis*. Many of these patients are nasal MRSA carriers
• Thyroid studies (TSH alone; no need for full thyroid panel; both hypothyroidism and hyperthyroidism can cause generalized pruritus)
• Liver function panel (total bilirubin, aminotransferases, alkaline phosphatase)
• Renal function panel (creatinine). A phase of end-stage kidney disease, uremia, can cause severe pruritus. It would be exceedingly rare that a patient with uremic pruritus would not already carry the diagnosis of ESKD
• Pregnancy test. Cholestasis of pregnancy is extremely itchy, but it usually occurs in the third trimester. At this point, most pregnant women will know that they are pregnant
• Urine drug screen for cocaine, amphetamines, opioids, and cannabinoids
• Stool ova and parasites × 3. Examines a person's feces for evidence of intra-intestinal parasites, such as nematodes (hookworm, whipworm, *Ascaris*) and a variety of protozoal organisms. Some nematode parasites may cause itchiness during tissue phases of an infection, but rarely do so when they have reached their intestinal destination
• Chest radiograph. Pruritus is a well-described, although uncommon, symptom of Hodgkin disease
Helminth serologies, only if indicated, e.g., lymphatic filariasis, schistosomiasis, and onchocerciasis
• Electrocardiogram. Recommended by some authorities if the patient's DI will be treated with pimozide. High-dose pimozide can induce Q-T prolongation
• MRI of head

"Sample Sign," Formerly Called Ziplock Bag or Matchbox Sign

This phenomenon of excessively collecting specimens by patients was once called the "matchbox sign" or "ziplock bag sign," but nowadays, it is more appropriate to call this the "specimen sign." Patients often carefully collect material evidence of the suspected infestation. They may show up at the appointment with specimens of suspected creatures they have collected from their homes, clothes, person, or bodily excretions. Material collected from household floors, carpets, and furniture usually consists of amorphous dust, lint, textile fibers, and other inert debris. On occasion, patient-collected materials contain animate (or remnants of formerly animate) items, such as non-infesting insects, mites, and other arthropods, or miscellaneous invertebrates. These are usually an innocuous part of ordinary domestic microfauna. Intact organisms are often easily identified, but more often the material consists of desiccated fragments. Sometimes, the materials have been assembled in neatly labeled, clear plastic specimen bags, meticulously cataloged by date, body site, geographic location, and circumstances of the "capture." Other patients may bring in a haphazard jumble of containers (e.g., plastic bags, food jars, medication bottles, wads of cellophane tape), each with varying amounts of dry material. Some patients deliver bottles or jars containing wet materials, usually preserved in rubbing alcohol or something similar. Very often, patients will present a diary or logbook that minutely details every episode of the unwanted sensations.

Be careful not to overinterpret the presence of unidentified arthropod legs, wings, etc. as evidence of a true infestation. If necessary, one may choose to engage an entomologist to assist in identifying fragments. A skilled entomologist using only arthropod fragments can identify species level.

How Does One Manage Specimens and Materials Brought by Patients, e.g., Sample Sign?

Patients with DI are likely to bring specimens with them as proof of their infestations. There are several ways to manage these materials and questions to ask. When patients arrive with bags of specimens, I ask them to select two or three specimens that they are reasonably confident will have a creature in them. I bring a two-headed microscope into the exam room and ask the patient to join me, so we can examine the selected materials together. This helps with patient rapport by showing that I genuinely want to determine if there is an infestation. It also assures the patient that I am open and honest with them—and this sets a good tone for any sort of therapeutic contract we might create. In addition, I look through any photographs the patient might have brought in.

Many patients purchase their own microscopes for examining suspicious materials at home. If a patient feels the need to submit their own samples, then their choices might include a private commercial entomology or parasitology laboratory; academic laboratories (at a university); or a state or local agricultural extension office. However, expertise, turnaround time, and detailed replies are not guaranteed by other providers. Many entomology consultants have developed an interest in this field, but if these duties are not part of their core job responsibilities, they may regard examining bags and bags of poorly preserved insect parts to be an especially burdensome task, especially when DI patients send materials in bulk quantities.

If patients want to conduct their own investigations, I recommend they coordinate with a reliable private or state entomology laboratory. This is best done by collaborating with the patient's primary physician.

In some situations, a prudent clinician may want to ensure a chain of custody. Ideally, one clinician will obtain the samples or biopsy specimens. That person, not the patient, will package and send the specimens. Within the routine or community healthcare system, it may be simplest to send the materials to a pathology service that has experienced dermatopathologists on staff.

If patients feel a need to send their own samples, then their choices are to a private lab (such as mine) or an academic institution such as a university extension service. These services provide patients with thorough, informative, and nonjudgmental reports. The patient must understand that "send-out labs" employ or are operated by entomologists or parasitologists who are scientists, but who are not necessarily physicians. They can describe and identify specimens and add useful information from an environmental health perspective, but this information must be sent back to the attending physician so as to assist in the differential diagnosis.

If there are specimens that need a more detailed analysis and/or a second opinion, I may send them to the laboratory or to Dr. Richard Pollack, a public health entomologist at the Department of Environmental Health and Safety, Harvard University, Needham, Massachusetts. Dr. Pollack has a private consulting firm, Identify US, that will examine specimens submitted by patients. He can be reached online at https://identify.us.com/ or rich@identify.us.com.

Examples of written reports:

"Dear Ms. Jones,

Thank you for submitting specimen bags (by mail) and digital images (online) for evaluation. I have scrutinized the digital images. Nothing in those images appears at all consistent with an insect, tick, worm, or other creature, either in whole or part. Instead, the images appear to show bits of debris. Whereas it is possible that these inanimate objects caused some amount of discomfort, they are not evidence of anything that would cause a bite, sting, or infestation.

The samples that you sent include about two dozen objects in a clear plastic zipper-lock specimen bag. I have taken great care to examine each object microscopically. I sought evidence of arthropods (insects, mites, ticks, and related forms) as well as of other animals. Nothing in the sample was consistent with such a creature, either in whole or part. Instead, the objects were consistent with bits of dried plant leaves (the leaf veins were clearly evident); fabric fibers (some organic, others synthetic); dried skin flakes and scabs; and other debris. Because none of these objects had hard or sharp edges, it would seem unlikely that they would be the cause of any bite-like sensations. The good news is that there is no evidence here of a biting, stinging, or infesting creature. Based on this sample, I do not see anything that is capable of infesting a person.

I hope this reply is timely and of value. I encourage you to share these findings with your doctor, who will consider other explanations for your discomfort. I hope this evaluation is helpful. Please do let me know if you have any further questions or comments. I wish you well.

Sincerely,"

When submitted materials contain arthropods (or arthropod parts) of a non-infesting species, a letter will be like this:

> "Thank you for sending the specimens. I examined them using a microscope and other optical instruments. I was able to identify specimens as [*Genus species*], which is the scientific name for a type of [aphid or spider mite or mealy bug or scale insect]. Members of this species are commonly found on house plants, especially plants such as [common name of household plant]. They feed only on plant material and are not known to bite, infest, attack, or live on humans or other animals. I have attached several close-up photographs of the specimens. Please inform your physician of my findings."

There was a time when I examined nearly all materials that patients brought with them. Overall, the yield was negligible; few specimen containers included arthropods (or arthropod parts). The examination process helped me understand the sorts of things that patients might regard as invasive pathogens. Nowadays, I examine 2–3 "high-yield" specimens per visit (and code the microscopic exam as 87,220) but encourage patients to send specimens to a reliable private, state, or university entomology laboratory. Private labs will charge the patient a modest fee for an examination, while state and university labs are usually free. I also arrange for the laboratory report, which is usually thorough, to be shared with me. When warranted, I will provide patients with specimen containers (e.g., urine cups), glass microscopy slides, transparent tape (non-frosted), slide cassettes, ziplock bags, and/or adhesive labels with clear instructions on how to obtain samples.

I believe that there is a finite number of specimens that the medical, academic, or entomology communities should accession, process, examine, and report. In other words, there should be a limit on the number of packets of debris that a patient submits for analysis. Fully delusional patients could probably collect and submit materials ad infinitum. During a patient's initial visit, I review options regarding the examination of materials and recommend ways for the patient to collect them in the future.

10   Skin Biopsies

See Box 1.

> **Box 1. Billing**
> For the purpose of disease classification for patients with CI, I use two ICD-10 codes: R20.8 (other disturbances of skin sensation) or Z11.9 (encounter for screening for infectious and parasitic diseases, unspecified) or both. If it is clear that the patient is delusional, one may use F22.0 (delusional disorders). Before a reader uses these codes for billing purposes, it is advisable to check with a billing authority.

Roughly half the patients who ultimately receive a diagnosis of DI undergo a skin biopsy during the evaluation. These biopsies are usually performed at a dermatologist's office and are often at the patient's request, even if the dermatologist explains that the histopathological examination is unlikely to find an infesting organism.

Two words of caution when considering a skin biopsy:

1. Take biopsy tissue from more than one site.
2. Take biopsy tissue from intact skin (i.e., without excoriations, erosions, or crusts).

In the Centers for Disease Control (CDC)-funded study of patients with undefined dermopathy (Morgellons disease and similar disorders along the DI spectrum), the investigators examined 115 case-patients. They collected detailed epidemiologic data, analyzed skin samples, performed clinical evaluations, and conducted geospatial analyses on all the patients. They employed conventional light microscopy, electron microscopy, and various spectrographic methods, but no organisms were found in any specimens [4]. In nearly all cases, the only foreign substances detected in skin samples were ordinary textile fibers that were trapped in the serum crust of the raw, excoriated, or ulcerated skin.

Discernment is necessary when selecting sites to biopsy. Find fresh primary lesions; avoid secondary lesions on heavily excoriated, eroded, crusted, lichenified, or secondarily infected. Also, try to select lesions that represent typical sites for these are more likely to give meaningful information, rather than a large, dramatic, or alarming lesion.

Furthermore, I recommend taking skin biopsy specimens from two sites. Otherwise, when the surgical pathology report indicates that no pathogens or parasites were found, the patient may wonder about a possible sampling error and they may follow up with a request for a second biopsy.

11   After the Examination Is Complete, What Are the Patient's Goals?

Of note, patients often have different goals:

1. Some are focused on learning the cause of the abnormal sensations. The sensations are a mystery, but they are not terribly unpleasant, nor do they hamper daily life. It is a nagging curiosity that drives the need to know their cause. Once things are explained, pursuing treatment is a lesser factor. "I just want to make sure it's not black mold on my walls that is causing this."
2. Others are worried that the sensations indicate a serious disease (e.g., the return of a previously treated cancer). "I was itchy all over when my Hodgkin's disease started. I want to make sure that this new itchiness isn't my Hodgkin's coming back."

3. Some worry that they might transmit the itchy disorder to other people (family members, friends, work colleagues, acquaintances). They want to know if their condition is contagious. "The biggest joy in my life is when the grandchildren visit. We haven't had them over since this weird itching started two years. I've been so afraid that the grandkids might catch this … whatever it is."
4. Some find that the sensations interfere with daily life. They simply want the sensations to stop. They are healthy otherwise and are not particularly worried about the cause itself. "I'm in the public eye all day long. I just can't keep scratching and rubbing and swatting in front of people. This has got to stop now."
5. Some are surprisingly indifferent whether or not the sensations continue, as long as they are reassured about the other matters.

12 When the Evaluation Does Not Support the Diagnosis of an Infestation

How should the clinician proceed when the medical history and physical examination do *not* support a diagnosis of an infestation? If the evaluation points to a different diagnosis, then medical management can shift away from CI. At this point, many patients are relieved to learn that their provider was unable to find evidence of an infestation. If the clinician has identified an unrelated, "non-infestation" condition that explains the sensations, patients may also accept this as good news.

From a practical standpoint, when the dysesthesia is fully explained as a "non-infestation," the patient and clinician will likely infer that there is no infestation. This is somewhat of a logical shaky ground: the presence of an itchy "non-infestation" condition may explain the dysesthesias, but it does not negate the possibility that an infestation exists. Occam's razor points in one direction (i.e., there is no infestation), but rules of logic prevent one from stating, "There is no infestation," as an absolute truth.

Ruling In or Ruling Out an Infestation

Many patients are thrilled to learn that the medical history, physical examination, and laboratory tests do not show an infestation. And they are thrilled if we add that we have been able to identify what is causing their unwanted sensations. It would be nice to add that we have ruled out the possibility that the patient has an infestation " … and, by the way, we've proven that you don't have an infestation." But the rules of logic prevent us from adding this bit of reassurance. The colloquial way to state this is, "You can't prove a negative."

An Unexpected Response

On the other hand, a subset of patients will respond differently. They may be disappointed or surprised to learn of an alternative different explanation; they may be angry with or feel betrayed by clinicians who cannot confirm their suspicion of an infestation. If the clinician suggests another explanation to account for the abnormal sensations, these patients may be totally skeptical or simply reject the idea peremptorily.

There is also a large subset of patients for whom the clinician cannot find a pathophysiologically plausible explanation. The patient is adamant about an infestation being the sole cause of the sensations, but there is no evidence for that.

This introduces the third criterion in making the diagnosis of DI: the patient's unwavering conviction that the sensations are caused by an infestation. Despite a lack of evidence to support the patient's belief, patients with DI are unwilling to accept an alternative explanation.

In my opinion, there is a corollary to Criterion #2 ("The patient has an unwavering conviction that the sensations are caused by an infestation"). The corollary has not been included as an essential criterion for the diagnosis of DI, but I find that this feature helps gauge the depth of a DI patient's delusion. Brownstone, Talasila, and Koo (Chap. 3) refer to this as "insight." The amount of insight can serve as a crystal ball to anticipate how the patient's treatment course will ultimately turn out. A complete absence of insight heralds a dismal, downward spiral in which the patient's health (physically, mentally, emotionally, financially, socially, and occupationally) descends.

Fantastical Shape-Shifting

Patients often rationalize that physicians cannot confirm their infestation because the patient's particular strain of organisms can hide on or in the human body in ways that evade detection. Patients regularly hypothesize that "their creature" can change its appearance almost at will.

The explanations call to mind an insect with a complete (holometabolous) life cycle, in which no two stages can be recognized a priori as related forms of life. However, unlike a butterfly whose full metamorphosis has four well-understood, sequential, and predictable stages (egg to caterpillar/larva to chrysalis/pupa to winged butterfly/adult), many DI patients believe that their parasite can change into phantasmagorical different forms of life. For example, a patient may report that "their creatures" usually live on the skin's surface (where they create bloody crusts) or within the skin (where they produce amorphous multicolored fibers). The infesting creatures resemble the mythical Greek God, Proteus, who could shape-shift at will to avoid capture. In general, creatures actively propel themselves by burrowing through the skin; crawling or hopping on the skin; or flying purposefully from one body part to another. But when subject to an examination, creatures can change shape and habits, pass through the skin, and enter the bloodstream, where they generate wings to help them swim penguin-like through arteries and veins. Patients postulate that their creatures can escape the body through one or more routes, perhaps exiting directly through the skin or maybe out a body orifice via bodily excretions (e.g., nasal mucus, urine, stool, saliva, or tears).

13 Consequences of Empiric Treatment for a Presumptive Diagnosis

When a DI patient receives presumptive treatment for a disorder that is neither confirmed nor actually exists, the situation is distorted in two ways. First, the patient will likely interpret the act of a doctor prescribing an antiparasitic medication to mean that the doctor found convincing evidence of an infestation. At subsequent visits to other providers, the patient will likely report, "I've seen other physicians who have all diagnosed me with scabies and treated me with ivermectin or permethrin or both. The treatment worked for a few days, but now I'm itchy again and need more medicine. I think I need a higher dose and to take it for a longer time. That's why I'm here to see you."

A conundrum arises: when the new provider explains that he is unable to find evidence of an infestation, the CI patient might reply, "Of course, you can't find any evidence. The treatment that I took last week shut the creatures down. It seemed to kill most of them, but some must be resistant and didn't die—or somehow evaded the treatment. But those survivors are now starting to repopulate me, and I can feel more of them. There might not be enough for you to find them easily, but believe me, the last three doctors all diagnosed scabies ...."

The unwanted sensations inevitably return, usually within a week, which the DI patient interprets as evidence of persistent infestation due to resistant organisms. The antiparasitic medicine, they reason, may have killed most of the offending creatures, but several particularly vigorous organisms survived the treatment and have now repopulated the patient with a drug-resistant strain of organisms. Now, the fixed delusion has returned, bolstered by the beliefs that:

1. One or more doctors have already confirmed the infestation.
2. The prescribed antiparasitic medication(s) gave temporary relief from the unwanted sensations.
3. Their particular organisms have biological superpowers. The organisms are drug-resistant and seem to recrudesce (bounce back) within a few days of any treatment.

This places us in an unsettling situation where the patient believes that there is proof of an infestation and that their particular infestation is resistant to treatment. When the delusions return, the patient typically resumes pursuit of treatment and will undoubtedly see a clinician, at some point, who will inform the patient that there is no evidence of infestation. *Plus ça change, plus c'est la même chose.*

14 Biological Plausibility

Many DI patients propose that the inability of a particular physician (or the entire medical community) to confirm the infestation can be explained by Western medicine's unfamiliarity with "exotic infection." DI patients often conclude that their particular "infesting organism" must be unknown to science.

Several of my DI patients bring up the example of Barry Marshall, the Australian physician who identified *Helicobacter pylori* as the major cause of gastritis and peptic ulcer disease. Marshall's hypothesis that these chronic diseases had an infectious cause was radical. It was met with the usual skepticism that arises in response to any proposed major paradigm shift [11]. But Marshall's work was replicated, and his conclusions were shown to be valid. My patients liken their situation to that of the patients of Dr. Marshall. They believe that the established medical community is close-minded and unwilling to recognize the validity of the discoveries that they have made. (Several of my patients believe that their situation will lead to the awarding of a Nobel Prize, in the same manner that Marshall was awarded the 2005 Nobel Prize in Physiology or Medicine for his iconoclastic research.)

Such conversations are often frustrating and fruitless for all parties. The Barry Marshall analogy is specious on a fundamental level. His postulated association between *H. pylori* and gastritis generated well-constructed hypotheses, well-designed studies, well-analyzed results, well-replicated results, and well-reviewed publications. Yes, physicians and scientists acknowledge that there are an infinite number of gaps in our current biomedical knowledge. However, replacing speculation with substance requires certain cognitive and methodological steps that have not been fulfilled.

Among the most distinctive signs that a patient is detached from reality are the biologically implausible descriptions and explanations of the supposed attacking organisms. Patients often describe a creature that is a fanciful amalgamation of mutable features, behaviors, and life cycles. The patient believes that the infesting organisms are composed of parts from taxonomically unrelated organisms, such as fungal hyphae that develop wings and allow them to swim through blood vessels. The organisms, it seems, can shape-shift at will to help evade capture or detection. Indeed, the creature may be new to medical science; knowledge gaps and hubris within the medical community, they believe, help explain the undeserved skepticism regarding the organism causing their particular infestation.

Now I want to return to the cognitive framework that I described earlier. Start by looking for evidence of an infestation or other environmental assault. Next, look for evidence of another dermatologic explanation for their dysesthesias. Finally, determine if there is a biomedical or neuropsychiatric cause for the sensations. If the patient remains unrelentingly convinced that an infestation is the sole possible cause and flatly rejects any other possible explanations, I run through the list of beliefs and behaviors presented in Table 9 to conclude that the patient is delusional (Table 9). The chapter in this book written by Brownstone, Talasila, and Koo (Chap. 3) discusses the depth (or conviction) of the delusions and how to design a treatment plan commensurate with that depth.

Table 9 Beliefs and behaviors to conclude that the patient is delusional

In addition to the diagnostic criteria discussed at the beginning of this chapter, I take note of the following:

- Adamant insistence that an infestation is the cause (at the start of the initial evaluation)
- Unwavering insistence that an infestation is the cause (throughout the evaluation)
- Denying or rejecting possibility of a different (non-infestation) explanation for the sensations
- Reminds physician that "science doesn't know everything." This is true, of course. (And good scientists know that best of all. That is why they are in science … to try to fill in some of the vast gaps in our collective knowledge). But science does know, for example, that some creatures and life cycles are biologically *implausible* to the point of being biologically *impossible*. Science knows that there has never been a winged insect that swims in the bloodstream—or a nematode that reproduces by wind-blown spores—or a scabies mite that spreads from person to person by fecal-oral contamination. Several patients cite the story of Dr. Barry Marshall, the iconoclastic gastroenterologist who first proposed that stomach ulcers had a bacterial cause. His ideas were scorned for years until other researchers replicated his studies and confirmed his findings. His work led to a true paradigm shifte and he was soon awarded the Nobel Prize in Medicine/Physiology. He had a testable hypothesis that was indeed tested. Explanations offered by DI patients and supporting websites are rarely testable, verifiable, or falsifiable.
- Arguing that their particular creature is new to science. Yes, I suppose that can happen, but the new species should fit somewhere in the known biological universe. It cannot have a set of attributes that have never before been observed.
- Arguing that their doctors have been unable to find or detect creatures from the collected samples because medical and science training in the United States, Europe, Canada, etc., ignores organisms, pathogens, and diseases from non-Western countries. Yes, there is some truth to that statement, although the doctors they have seen so far are ones who do know organisms, pathogens, and diseases that are from other countries. Even if the Western-trained physicians are unfamiliar with specific "rare and exotic tropical diseases", it is unlikely that they did not know those diseases, and a pathologists who examine biopsy material of diseases tissue might not be able to identify the cause with certainty but they will be able to discern that the tissue is diseased or otherwise abnormal.
- Believing that their organisms have developed resistance to topical and oral antiparasitic medications such as permethrin and ivermectin
- Reporting near-immediate improvement when taking an antiparasitic medication, long before one's intestines can absorb the medicine. Several patients have told me that they started feeling better within seconds to minutes of placing an ivermectin table on their tongue, often before swallowing the tablet. Or that as they applied permethrin to their skin, they could feel (but not see) creatures crawling (or running) across their skin to take refuge on skin surfaces that had not yet been treated.
- Using permethrin more than twice weekly for 1 month and wishing to continue doing so to prevent a recrudescence of their organisms
- Reporting that products not known to be effective against certain organisms have been very effective for them
- Uses veterinary antiparasitic medications
- Buys bulk medications for self-treatment
- Buys antiparasitic medications over the internet
- Biological implausibility

(continued)

Table 9 (continued)

- That other doctors old the patients that they know exactly what the patient has. The story often continues that the doctor prescribed a medication that truly work. However, the creatures have returned and the patient wants me to prescribe higher doses of that medication (or something more powerful) because some of their creatures have survived the infallible medicine and are now resistant to usual doses. (The medications that the "expert doctor" has prescribed is usually something innocuous like ketoconazole shampoo.)
- Buying a microscope for home use to examine materials they have collected
- Taking a series of photographs through the microscope or other forms of magnification and identifying many bits of formless schmutz as arthropods, helminths, or previously unknown creatures
- Fear that household members/guests will become infested
- Maintaining a logbook with dates and times, geographic locations, location on their bodies, sketches of the creatures and their life cycles
- Removing pets from the household (allergies, OK, infestation not)
- Reporting that "their mites" which they believe came from pets or other animals, household materials, or outdoors have burrowed deep into their skin where they live and reproduce (scabies and *Demodex* are the only mites known to do that in humans)
- Belief that their creatures no longer live in the bodily tissues where they normally live in other people, but they now have evasive tactics to avoid detection, capture, or treatment. For example, their scabies mites are no longer in the epidermis but are now deeper in the skin (i.e., the dermis) or have invaded other organ systems that they are not known to invade
- Report that they have a scabies infestation of the scalp. Scabies almost never involves the scalp; when it does, the patient is usually immunosuppressed and has severe, widespread crusted scabies.
- Report that they have crusted scabies and point to intact normal skin as evidence, which they know is difficult to treat and may require multiple treatments
- Report that their creatures lay eggs within their skin
- Of the ectoparasites discussed in this book, only scabies can lay eggs in the skin. Two types of lice (head lice and pubic lice) lay their eggs while *on* a person. Both lay their eggs on hair shafts. On rare occasions, opportunistic flies will lay eggs on an open, necrotic wound. The chigoe flea, *Tunga penetrans*, gravid females will extrude eggs from pores
- History of discarding or destroying household items
- Moving from their homes permanently—or developing a peripatetic lifestyle by moving from hotel to hotel. They experience some relief upon arriving at each new place. Nevertheless, within a few days at each new hotel, their symptoms return, or the creatures reappear

Responses that suggest something else may be going on:

- Inability to describe how the purported organism disturbs patient
- Fixed belief that infestation is the cause. Will not consider other organ systems as possible cause, e.g., hyperthyroidism
- Report that organisms vanish rapidly upon using certain medications. A focus on collecting specimens and maintaining a diary/logbook. Discarding/destroying belongings
- More willing to try internet-touted nontraditional therapies. Patients refuse to even consider a psychiatrist/psychologist

In most cases, while we wait for laboratory test results and biopsy reports to come back, I try to characterize the unwanted sensation so that we can develop an acceptable treatment regimen to mitigate that sensation. The most common situation is pruritus, which is due to an irritant dermatitis, not due to a true infestation. Many patients become overly vigorous in soaping and scrubbing their skin, or from applying astringents, household cleaners, and insecticides to their skin. At this point, it is important to palliate the uncomfortable sensations and help relieve the patient's misery. This usually consists of encouraging gentle skin care and hygiene practices (to avoid drying out the skin); recommending moisturizers, topical anti-pruritic agents, and occasionally mild topical corticosteroid products; and advising the patient to obtain training on biofeedback therapy. I also ask for permission to discuss the condition with the patient's other health providers and with family members.

On the other hand, many of my patients never describe an itching, burning, stinging, or any other nociceptive property. Instead, they are bothered by the perceived presence of something in or on their skin, not by any particularly distressing sensation itself. I ask the question several ways and several times during the visit, yet many describe only the distress of "knowing" that something is in or on their skin. Furthermore, most of my patients do not describe a health consequence that concerns them. They are not worried about a disease state or particular illness. Most are simply bothered mainly by the "awareness" or "knowledge" that something exists in or on the skin.

In my opinion, many people with DI, as bright and capable as they are in most facets of their life, seem relatively untroubled by the sensations, but are bothered mainly by the presence of an infestation, not by the consequences of an infestation. They have often been extraordinarily inarticulate and unable to state whether or not they experienced any physical discomfort, such as pain or itch. The value of these questions is that the answers serve as data points as I try to assess the depth of the delusion.

Another valuable litmus test is to offer a course of psychoactive medication (e.g., an SSRI or an atypical antipsychotic). My position is that rational individuals, even while firmly suspecting an infestation, are amenable to trying a psychotropic medication on the chance that it might help. On the other hand, delusional individuals are so firmly fixed on the certainty of the infestation that they are unwilling to try a psychotropic medication. DI patients regard a suggestion to try a psychoactive medication (or almost any medication other than an insecticidal agent) as anathema—and they typically respond with bitterness and anger. This response is another data point that indicates how much insight, if any, the patient has with their condition.

DI Evolution

It is not uncommon for DI to have a trigger event. Patients naturally become anxious about their unexplained discomfort. Visits to their primary physicians, dermatologists, and other medical professionals provide no answers, and they start to drift through doctors, "doctor shopping." Many try nontraditional therapies; many discard, destroy, and replace belongings. Many misuse and abuse pesticides and medications. "Misery loves company," so many develop strong emotional ties with others. These are often with close friends and family members they trust. It results in mutual support partnerships called *folie à deux*. Since in their minds they have been failed by the medical profession, all are left to their own imaginations. This results in elaborate descriptions of their parasites with a great deal of effort in seeking confirmation and vindication (not relief from suffering). "I am invincible, I am right!"

Effects of CI on the Person and Their World

During the evaluation of patients with possible DI, clinicians should be aware of the devastating consequences of this condition on individuals and their world. If possible, I ask questions to help me understand how the condition of CI/DI is affecting the individual and their families. Table 10 lists questions I often ask these patients. The answers to these questions might not help get the patient on proper therapy, but the answers can help providers and family members assess the trajectory of the DI patient's future.

Many DI patients make a "therapeutic contract" with their providers. This is an arrangement where a patient agrees to try a medication (usually an antidepressant, antipsychotic, or atypical antipsychotic medication) if a thorough evaluation shows no evidence of infestation and reveals no other biologically plausible cause for the dysesthesias. I find that a set of DI patients continue to reject any possible explanation other than the presence of an infestation. If the characteristics listed in Table 9 indicate that an individual has no insight into the problem and will not try the recommended medications, I worry about the eventual outcome. In their efforts to obtain relief from their misery, they wind up battling those who love them, work with them, socialize with them, and care for them. At this point, their personal lives, family support, and employment collapse around them.

My cohort of patients who fall in this category have generally been well-educated, high-functioning individuals, who have had steady employment, friendships, homes, and families. This group has included attorneys, schoolteachers, a licensed therapist, a lobbyist, a congressional staffer, a biomedical scientist, two physicians, a retired bank president, and a college dean at a major East Coast university. After a few years with DI, they are often unable to continue working, stop receiving house guests, end all social contacts, and push family members away. Their days are spent beseeching or battling family members, landlords, pest management professionals, public health scientists, and physicians. Many wind up living alone and no longer gather with family members, even at the usual major family events. They spend hours cleaning and washing their bodies, their clothes, and their homes. They try increasingly bizarre methods to rid themselves of their infestation. They buy new clothing and bed linens regularly, even daily. They rid their homes of objects and dear possessions, fearing that those objects may be the source of the infestation.

Table 10 Additional questions for the psychological and behavioral history

Patient's assessment of the situation	• What do you think is going on?
	• Do you think you acquired this from pets?
	• Do you think you acquired this from travel?
	• Have you collected specimens?
	• How do you collect them?
	• Have you stored them?
	• Do you have your own microscope or magnification device?
	• Has anyone examined the specimens?
	• Have you had anyone else provide results?
	• Do you maintain a diary or logbook?
	• Do you have the problem when you travel?
	• When you stay outside the home, is there a problem?
Patient's assessment of their well-being	• How has this affected you?
	• Your health?
	• Your sleep?
	• Work or school?
	• Family relations?
	• Social activities, sports, and recreation?
	• Physical activity?
	• Clothing choices?
	• Physical intimacy?
	• Bed partner?
	• What do family members think is going on?
	• Do you believe you are contagious?
	• Anyone else at home have this?
	• Are you fearful to hold or touch people, even when no one else gets this?
	• Taken efforts to isolate self or to prevent possible spread? • Attempts to move domicile or go from hotel to hotel?
	• Discard and abandon items?

Some leave their homes and move from motel to motel, hoping to find relief. And if the new environment provides some relief, that improvement dissipates within a few days, prompting another move.

The college dean, who was mentioned at the beginning of the previous paragraph, hired pest management professionals to inspect her home almost monthly. She had her home tented and fumigated three times. She sealed most possessions in plastic bags and bought new clothes and bed linens several times a week. The city health inspectors stopped responding to her complaints of bugs everywhere. She needed more money to migrate from motel to motel, so she soon sold her stately home and depleted her retirement accounts. She declined invitations to move into the homes of family members, fearing that she would carry her infestation into those homes. She stopped coming to see me or her other physicians. Family members soon lost track of her. But they and I still worry about her.

15 Conclusion

Working with patients who are concerned about a possible infestation can be exceptionally rewarding. There is a great swirl of anxiety, guilt, uncertainty, fear, self-loathing, and other negative emotions associated with a concern of infestation. Not only does the purported infestation cause unpleasant (and often debilitating) sensations, but it also makes people worry that they might infest others, especially the loved ones with whom they live.

Many health providers and other professionals who interact with these patients have been able to identify and treat true infestations of the individuals or their homes. Many of us have been able to assuage the person's anxieties by identifying a non-infestation cause of the unwanted sensations.

There are times when the CI patient has the same set of anxiety, guilt, fear, and other negative emotions as the result of a fixed and unshakeable delusion that seemingly cannot be mitigated by thorough medical examinations, tests, and investigations. The unifocal psychosis that we now call delusional infestation has an astonishingly similar set of features in nearly all patients who have it. It has been argued that communication across the internet has influenced DI patients, leading to a convergence of the medical and psychological aspects of the disorder. While that degree of communication has probably influenced many individuals with DI, similarities in patient presentations were noted decades ago.

In this Chapter, I outlined an approach for health providers, dermatologists in particular, to examine patients with CI, who might in fact have DI. Clinicians certainly want to correctly identify cases in which there is a true infestation or there is a dermatologic condition that induces dysesthesias that may lead a person to believe that they are infested or to identify the dozens of other causes that may induce such dysesthesias. But there comes a time in that evaluation when one must consider that the patient's concern for infestation is, in fact, delusional. It is not a fabrication for any self-gain; it can be a devastating disorder that causes their world to collapse around them.

In order to confidently make the diagnosis of DI, one must rule out other non-delusional causes of the unwanted sensations. But to make the diagnosis of DI, the provider must be cautious in the evaluation and confident in the conclusions. I hope that this chapter will help clinicians as they work with such patients.

Appendix A: The Scabies Prep: Microscopy for the Diagnosis of Scabies

International Alliance for the Control of Scabies Consensus Criteria for the Diagnosis of Scabies [6]

In individuals with common scabies, material for microscopy should be collected from suspected burrows. Dermoscopy or an ink test may help to identify a burrow (see "Burrows" in the main manuscript) [12, 13]. If no definitive burrows are seen, samples taken from newly developed papules or vesicles around the fingers, hands, and wrists may also demonstrate evidence of mites. It is less common for mites to be found from skin lesions on the trunk, as these lesions most commonly develop from hypersensitivity reaction. Collecting material from multiple sites improves sensitivity, noting that most individuals with common scabies have less than 15 mites on the body, and most lesions will not have mites within them [14]. This can be a time-intensive process.

With all of these techniques, the material examined should include the most recent (advancing) end of the burrow where the scabies mite is located. This can sometimes be visually detected by the presence of a small papule. The main methods for obtaining material are the following:

(a) Skin scraping: direct scraping of skin to remove the *stratum corneum* over a suspected burrow or papule (see videos at https://www.controlscabies.org/resources/media-library/videos). A scalpel blade (e.g., #10 or #15) is recommended, used either as the blade alone or in a scalpel handle. Other sizes, round-edged blades, or a vaccinostyle can also be effective. A curette can also be used and may be more acceptable to children [15]. The blade is held at right angles to the skin and scraped along the surface of the burrow with light-to-medium pressure. The region should be scraped down to the base of the *stratum corneum*, where the mite resides, or until the first punctate droplets of blood are seen indicating that the blade has traversed the epidermis and has exposed dermal capillaries. A gentle, superficial slicing motion, with the blade held parallel to the skin surface, can also be effective. Application of a minute amount of mineral oil to the scalpel or the skin lesion allows the skin scrapings to adhere to the blade.

(b) Needle or tweezer extraction: using a needle or fine-tipped tweezers, scraped along the course of a burrow but including the distal segment. In both methods (a and b), the examiner should carefully place the skin material on a standard glass microscope slide, add one to two drops of potassium hydroxide (KOH) at 10–20% concentration, and then apply a coverslip. The material from several sites of the same individual may be placed on a single slide. After waiting several minutes to allow the KOH to soften the keratinized epidermal cells and increase their translucency, the specimen is examined with standard light microscopy. Usually, the slide is scanned initially at low magnification (4–10 ×)

and then more closely at higher magnification (20–40 ×). Adjusting the microscope condenser, if available, may increase the refractility, which makes it easier to discern the mites. Some scabies mite parts are also more readily visible with polarized light microscopy [16].

Some clinicians use mineral oil instead of KOH, noting that live mites, if present and motile, are easily detected in mineral oil, whereas KOH reduces mite movements and eventually kills them [17]. On the other hand, mineral oil does not dissolve the keratinocytes and thus nonmotile, diagnostic material (dead mites, eggs, or feces) may be obscured. Therefore, mineral oil is less useful when examining patients with common scabies but potentially more useful when examining those with crusted scabies. KOH may potentially also dissolve fecal pellets, but only if left for more than 1 h.

(c) Transparent, pressure-sensitive adhesive tape (e.g., Scotch Tape™, Cello Tape™): the tape is applied across a lesion to lift off the cornified epidermis and adherent material and then applied directly, sticky surface down, to a glass slide [18]. In a field setting, this method is often simpler than using KOH preparations [19]. Completely transparent tape must be used as "frosted tape" that will obscure the specimen. The skin and mite material are more difficult to visualize through the tape than through glass, which may lead to reduced sensitivity compared to skin scrapings or extraction methods.

Microscopy is frequently negative in patients with clinically diagnosed scabies, particularly when performed by inexperienced examiners [19, 20]. Training is also necessary to reduce false-positive test results, in particular to distinguish eggs from epidermal cell aggregates and fecal pellets from debris, environmental dirt particles, or red blood cells. If structures resembling fecal pellets are suspected, but no mites or eggs are seen, the diagnosis should be considered carefully and, where possible, further specimens or interpretation sought. Skin biopsy may also confirm the diagnosis but is not indicated unless other differential diagnoses are being considered.

Appendix B: Dermoscopy for the Diagnosis of Scabies

International Alliance for the Control of Scabies Consensus Criteria for the Diagnosis of Scabies [10]

Dermoscopy, also known as dermatoscopy, is a technique performed with handheld devices that allow 10 × magnification of the skin and does not require computer assistance. It includes a high-quality magnification lens and a transilluminating lighting system. Some devices require a liquid medium (e.g., oil, alcohol, glycerin, water) at the interface between the skin and the device's glass lens ("epiluminescence" technique). Other handheld devices, equipped with polarizing filters, do not always require the use of a liquid medium and are capable of viewing the skin with or without direct skin contact. For other skin conditions (for example, pigmented lesions such as melanoma), gel is used to reduce the distraction of the *stratum corneum* ("wet" dermoscopy). However, for diagnosis of scabies, visualization without gel is also recommended ("dry" dermoscopy) as the gel may fill the cracks of the *stratum corneum*, making features of the burrow less visible. When scabies burrows are examined by wet dermoscopy, they may show what has been called the "jet with contrail" sign, where the "delta-winged jet" represents the pigmented mouth parts and anterior legs of the mite, and the "contrail" represents the linear segment of the burrow which may contain small bubbles [21].

Dermoscopy does not allow visualization of eggs or fecal pellets. Clear differentiation of the tiny mite from excoriation, point bleeding, dirt, and other artifacts induced by scratching can be challenging, especially for less experienced examiners. For this reason, dermoscopy is more operator-dependent than high-magnification techniques described above [22]. To differentiate a scabies mite from a tiny blood crust, the brown triangular spot should be seen at the end of the scabies burrow [19]. This may be more readily appreciated using dry dermoscopy. The brown part of the mite may be less visible in individuals with darker skin types (Fitzpatrick V and VI) and in hair-bearing areas. Mites may be more readily visualized at less pigmented sites, such as the palms and soles. Use of dermoscopy in the genital region may be awkward because the examiner's head and patient's genitals are in close proximity. The diagnostic accuracy of handheld dermoscopy may be reduced in individuals with impetiginized lesions.

Dermatoscopes are now standard equipment for many dermatologists in high-income settings. The price (around US $600), however, may be prohibitive in low- and middle-income settings [4]. Newer technologies, such as handheld digital microscopes or simple lenses that attach to smartphones, may make this technique more widely available [23–25]. Further training in the use of magnified skin examination may improve the accuracy of diagnosis.

References

1. Overstreet RM, Lotz JM. Host–symbiont relationships: understanding the change from guest to pest: the Rasputin effect: when commensals and symbionts become parasitic. Nat Pub Health Emerg Coll. 2016;3:27–64. https://doi.org/10.1007/978-3-319-28170-4_2.
2. Hinkle NC. Ekbom syndrome: a delusional condition of "bugs in the skin". Curr Psychiatry Rep. 2011;13(3):178–86. https://doi.org/10.1007/s11920-011-0188-0.
3. Hinkle NC. Ekbom syndrome: the challenge of "invisible bug" infestations. Annu Rev Entomol. 2010;55:77–94. https://doi.org/10.1146/annurev.ento.54.110807.090514.
4. Pearson ML, Selby JV, Katz KA, Cantrell V, Braden CR, Parise ME, Paddock CD, Lewin-Smith MR, Kalasinsky VF, Goldstein FC, Hightower AW, Papier A, Lewis B, Motipara S, Eberhard ML, Unexplained Dermopathy Study Team. Clinical, epidemiologic, histopathologic, and molecular features of an unexplained dermopathy. PLoS One. 2012;7(1):e29908. https://doi.org/10.1371/journal.pone.0029908. Epub 2012 Jan 25. PMID: 22295070; PMCID: PMC3266263.
5. Hoverson K, Wohltmann WE, Pollack RJ, Schissel DJ. Dermestid dermatitis in a 2-year-old girl: case report and review of the literature. Pediatr Dermatol. 2015;32(6):e228–33.

6. Engelman D, Yoshizumi J, Hay RJ, Osti M, Micali G, Norton S, Walton S, Boralevi F, Bernigaud C, Bowen AC, Chang AY, Chosidow O, Estrada-Chavez G, Feldmeier H, Ishii N, Lacarrubba F, Mahé A, Maurer T, Mahdi MMA, Murdoch ME, Pariser D, Nair PA, Rehmus W, Romani L, Tilakaratne D, Tuicakau M, Walker SL, Wanat KA, Whitfeld MJ, Yotsu RR, Steer AC, Fuller LC. The 2020 international Alliance for the control of scabies consensus criteria for the diagnosis of scabies. Br J Dermatol. 2020;183(5):808–20. https://doi.org/10.1111/bjd.18943.

7. Schwartzman GH, Dekker PK, Silverstein AS, Fontecilla NM, Norton SA. Dermatologic consequences of substandard, spurious, falsely labeled, falsified, and counterfeit medications. Dermatol Clin. 2022;40(2):227–36. https://doi.org/10.1016/j.det.2021.12.008.

8. Mumcuoglu KY, Pollack RJ, Reed DL, Barker SC, Gordon S, Toloza AC, Picollo MI, Taylan-Ozkan A, Chosidow O, Habedank B, Ibarra J, Meinking TL, Vander Stichele RH. International recommendations for an effective control of head louse infestations. Int J Dermatol. 2021;60(3):272–80. PMID: 32767380.

9. Zalaudek I, Argenziano G. Dermoscopy of nits and pseudonits. N Engl J Med. 2012;367(18):1741.

10. Ruiz-Villaverde R, Galán-Gutierrez M. Hair casts (pseudonits). CMAJ. 2013;185(9):E425.

11. Kuhn TS. The structure of scientific revolutions. Chicago: University of Chicago Press; 1970. NLM ID 0325654.

12. Diab HM. Scabies incognito. J Egypt Womens Dermatol Soc. 2017;14:56–60.

13. Park JH, Kim CW, Kim SS. The diagnostic accuracy of dermoscopy for scabies. Ann Dermatol. 2012;24:194–9.

14. Mellanby K. Biology of the parasite. In: Orkin M, editor. Scabies and pedic. J.B. Lippincott; 1977.

15. Jacks SK, Lewis EA, Witman PM. The curette prep: a modification of the traditional scabies preparation. Pediatr Dermatol. 2012;29:544–5.

16. Foo CW, Florell SR, Bowen AR. Polarizable elements in scabies infestation: a clue to diagnosis. J Cutan Pathol. 2013;40:6–10.

17. Muller G, Jacobs PH, Moore NE. Scraping for human scabies. A better method for positive preparations. Arch Dermatol. 1973;107:70.

18. Katsumata K, Katsumata K. Simple method of detecting *Sarcoptes scabiei* var hominis mites among bedridden elderly patients suffering from severe scabies infestation using an adhesive-tape. Intern Med. 2006;45:857–9.

19. Walter B, Heukelbach J, Fengler G, et al. Comparison of dermoscopy, skin scraping, and the adhesive tape test for the diagnosis of scabies in a resource-poor setting. Arch Dermatol. 2011;147:468–73.

20. Thompson MJ, Engelman D, Gholam K, et al. Systematic review of the diagnosis of scabies in therapeutic trials. Clin Exp Dermatol. 2017;42:481–7.

21. Argenziano G, Fabbrocini G, Delfino M. Epiluminescence microscopy. A new approach to in vivo detection of *Sarcoptes scabiei*. Arch Dermatol. 1997;133:751–75.

22. Dupuy A, Dehen L, Bourrat E, et al. Accuracy of standard dermoscopy for diagnosing scabies. J Am Acad Dermatol. 2007;56:53–62.

23. Heukelbach J, Feldmeier H. Scabies. Lancet. 2006;367:1767–74.

24. Miller H, Trujillo-Trujillo J, Feldmeier H. In situ diagnosis of scabies using a handheld digital microscope in resource-poor settings-a proof-of-principle study in the Amazon lowland of Colombia. Trop Med Infect Dis. 2018;3:116.

25. Blum A, Giacomel J. "Tape dermatoscopy": constructing a low-cost dermatoscope using a mobile phone, immersion fluid and transparent adhesive tape. Dermatol Pract Concept. 2015;5:87–93.

Neurological Aspects of Delusional Infestation

Anne Louise Oaklander

The skin, our largest sensory organ, contains a symphony of sensory receptors and associated peripheral nerve fibers. These encode perceived stimuli and transmit electrical signals to the spinal cord and brain for processing and response. Sensory stimuli are broadly dichotomized into noxious, those suggesting impending harm, vs. innocuous, for non-threating contacts. The receptors for innocuous stimuli (e.g., Ruffini slowly adapting receptors for skin stretch, and warmth, Merkel touch cells, Pacinian corpuscles for vibration) are coupled to low-threshold mechanoreceptors connected to large, rapidly conducting A-beta myelinated axons [1]. Overall, there are more very thin, slower conducting, thinly myelinated A-delta and unmyelinated C-fiber "nociceptors" that receive and transmit messages of impending harm centrally. These signals are modulated and processed in the sensory dorsal root ganglia lying just outside the spinal cord (or the trigeminal ganglia under the brain for signals from the head and neck), then even further interpreted in the dorsal horn of the spinal cord [2, 3]. Spinal processing can trigger reflexive scratching before the brain is even aware of it or has the chance to evaluate if scratching is helpful or harmful. Nociceptive signals projecting to the brain are then interpreted as itchy or painful. They are highly salient (important) to the brain, meaning they trigger many other neurons including those involved in attention, action, emotion, and memory. Parts

Dr. Oaklander is associate professor of neurology at the Massachusetts General Hospital and Harvard Medical School in Boston and assistant in neuropathology at the Massachusetts General Hospital, where she directs the Nerve Unit. She is internationally recognized for her research on the small fiber peripheral neurons that mediate itch, pain, and many internal functions. Here, she reviews the neurological itch pathways and conditions that cause persistent itching and scratching. Specific neurological disorders lead patients with odd sensations to suspect insect infestations.

A. L. Oaklander (✉)
Massachusetts General Hospital and Harvard Medical School, Boston, MA, USA

G. E. Ridge (ed.), *The Physician's Guide to Delusional Infestation,*
https://doi.org/10.1007/978-3-031-47032-5_10

of the primitive hindbrain activate to modulate and calm incoming nociceptive signals, including the periaqueductal gray, and the rostral ventromedial medulla located on the bottom of the upper brainstem [4]. Finally, the strip of somatosensory cortex above the ears helps the brain determine the exact location of the stimulus so the appropriate muscles are activated to scratch at exactly the site of irritation to try and dislodge the cause of the itch. This can happen without reaching conscious awareness, so it is nearly impossible to control the urge to scratch an itch. Prolonged scratching of neuropathic itch in skin that has become numb is an underlying cause of self-injurious scratching.

Just like fire alarms, incoming pain and itch signals elicit widespread responses designed to defend us against apparent danger and prepare for battle and injury. "Fight or flight." In addition, other stress hormones can be released, blood pressure can rise, and the heart pumps harder. Strong emotions (e.g., fear and anxiety), learning, and memories are triggered to insure we avoid similar threats or situations in the future. These strong memories and emotions help explain why some people develop situational anxiety about potential threats, including tiny needlesticks and insect bites. These sometimes build and reinforce over time. If they become disabling, it is recommended to discuss treatment options for situational anxiety with a physician.

Given that itch and pain are closely related and share many common receptors, neurons, neurotransmitter chemicals, and pathways, the evidence suggests that itch evolved as an offshoot of pain sensation. But why have two different defensive sensory systems? Likely because different motor (muscle) responses are needed to defend against different types of threat. None of us were present then, but most sensory scientists think that itch evolved from pain specifically to protect us against any number of arthropod interactions with the skin from envenomization, physical biting, piercing, to parasitism. A different defense is required to dislodge clinging stimuli from the skin–moving towards the stimulus to scratch and dislodge it. Again, why? Because near the equator where modern humans evolved, insects and many disease causing pathogens they either carried or vectored were major evolutionary threats. Mosquitoes and the pathogens they vector, cause more human deaths than any other animal or insect on the planet (see Chap. 13, "Brief Descriptions of Some Common Medically Significant Arthropods Related to DI," by Gale E. Ridge, Lyle Buss, and Katherine Dugas). We all think of malaria but also consider the kissing bugs (Triatomine), that vector Chagas disease *Trypanosoma cruzi*, the tsetse flies (Glossina: Glossinidae) that vector sleeping sickness (Two forms: *Trypanosoma brucei gambiense* and *Trypanosoma bruci rhodesiense*), tick-borne illnesses, and the many people allergic to the venom of hornets, bees, ants, wasps and urticating caterpillars. Not to mention the nearly universal misery and health problems caused by non-fatal insect bites. The problem is that the motor response to pain–reflex and volitional withdrawal from the painful stimulus–is ineffective against tiny clinging insect threats.

Even tiny low intensity mechanical stimuli such as very light touches (such as an insect landing or bending a single hair, or the poking of a cloth fiber) can stimulate itch neurons and circuits (labeled-line theory) to cause a single or a few pain neurons to fire. This is then interpreted by the spinal cord and brain as an itch (pattern theory). In animals, longer-lasting tonic activation of itch receptors is required to induce the scratching response, but short-duration firing instead causes withdrawal [5]. The prototype is a mosquito, which can initiate itch with its initial landing and then later if it succeeds in piercing ("biting") and injecting saliva for the purpose of feeding. It can activate immune cells to release histamine and other stimulators of itch neurons. Piercing plant spines, another tiny clinging stimulus, provoke the same itch/scratch response as insects. They can contain potentially harmful chemicals that induce local inflammation and normal itching. Despite the overlap in neurons and circuits, there is increasing evidence of itch-specific cells and processing in the peripheral and central nervous systems [6]. There is also increasing awareness that non-neuronal cells in the skin, and their secretions, also contribute to neural defensive sensations. Keratinocytes, the major skin cells, express some of the same or similar receptors as nociceptive neurons [7]. Plus, firing of cutaneous sensory neurons can be triggered or augmented by injury-associated molecules released by immune cells. Histamine released into the skin, often by mast cells, is a potent stimulus of itch in inflammatory [8].

Yes, itch neurons and pathways improve chances for survival. But when itch neurons develop a damaging neurologic condition and begin firing abnormally, they send random, excess, or inappropriate alarms to the brain, each one subliminally interpreted as "an insect has landed!" Neuropathic itch involves uncoupling of the normally tight stimulus-response curve, and it is a major contributor to otherwise-unexplained itch. Understandably, the unaffected spinal cord and brain still initiate the same protective responses that result in scratching. Normally, scratching quickly removes mechanical causes of itch (e.g., insects) and the itch/scratch cycle ends unless a noxious chemical has entered the skin to continue the irritation and inflammation. However, damaged itch neurons do not calm normally. Even after scratching, the itchiness persists or resumes almost immediately, triggering another round of scratching. And as we all know, scratching an itchy area will irritate and inflame the skin, perpetuating the itch/scratch cycle.

Another unique factor about neuropathic itch is that it enables scratching to the point of self-injury. Scratching creates a wider area of mechanical stimulation that inhibits firing of the itch neurons within it. Normally, reflexive scratching is brief. It stops as soon as it transitions to becoming painful to protect our skin from self-injury. However, if the receptive neurons themselves are damaged or the spinal chord or brain itch-inhibiting circuits are not operating normally, pain may not be perceived (sensory loss). So, people with unexplained itch should be queried about reduced or distorted sensation and unexplained pain or odd sensations at the location of their itch. They should be asked if they have a history of peripheral neuropathy, shingles (herpes zoster), pinched spinal nerve roots, or spinal cord or brain injuries. They may benefit from a neurological exam to look for these conditions. Anyone, even a layperson, can touch the affected itchy area lightly with a pin and compare the sensation to an unaffected area to see if there are hints of abnormalities. If so, they should be referred to a nerve specialist or neurologist for diagnosis and treatment advice.

Most patients with previously unexplained itch are very relieved when a neurological source is found, and in those with clear thinking, this usually ends initial concerns about insect infestation. Seeing skin damage from scratching, e.g., bleeding, is a potent deterrent to further scratching. However, visual deterrence may not suffice if the itchy part isn't easily seen, or in children or individuals with cognitive or visual impairment. But in all of us, reflexive scratching continues during sleep or when we are inattentive so we may not notice scratch injuries, or these may develop slowly or be interpreted as further evidence of insect infestation or skin disease. Wearing mittens to bed at night can help the rare patients with nocturnal scratching injuries. Many patients with uncontrolled itching and scratching do not understand that their scratching is the cause of their skin lesions rather than vice versa, and instead attribute them to an undiagnosed medical condition. Unseen insects are an obvious and natural inference.

The clinician's role is to identify the source of chronic itching, whether insects, ongoing skin inflammation, or neurological. So, patients are often first referred to dermatologists. If the skin is cleared, it is important to consider fewer common causes, which may require referral to an allergist, neurologist, rheumatologist, or other. Kidney, thyroid, and liver damage are other causes of unexplained widespread itchiness of normal skin. If physicians offer patients no explanation, or patients lack confidence in the medical explanations provided, they usually fall back on their own less informed theories, which can incorporate non-factual information perpetuated by the Internet. Some patients have anxiety or thought disorders that impair their ability to accept rational explanations. But too often, no rational explanation is offered by the clinician, or it may not be communicated adequately and reinforced in a trusted medical relationship. If a clinician does not know what the cause of unexplained itch is, it is responsible to refer the patient to specialists while continuing to support and reassure them about the absence of insects.

1 Neurological Conditions

Neuropathic itch is caused by the same disorders of the central and peripheral nervous system that cause neuropathic pain. These include sensory polyneuropathy, pinched spinal nerve root, herpes zoster (shingles), and any type of brain injury that damages itch pathways and circuits. Studying patients with facial neuropathic itching and scratching (historically known as trigeminal trophic syndrome) caused by a stroke has helped map the brain pathways for itch transmission and processing through observing killed neurons and other cells in specific brain areas. For example, neuropathic facial itching and pathological scratching in Wallenberg's stroke syndrome have been mapped to the damage in the trigeminal tract and nucleus of the brain stem (Fig. 2) [9]. Multiple sclerosis rarely causes similar itching and scratching [10]. It matters more exactly where the brain injury is located rather than what caused it. Focal areas of neuropathic itch appear to be more common on the face and neck, regardless of the cause. Indirect evidence suggests that there are more itch receptors in the head and face, presumably because the mucous membranes there particularly attract insects. For instance, patients with shingles affecting the face and head (the second most common location) are more likely than those with shingles on the torso (the most common location) to report neuropathic itch (Fig. 1) [11].

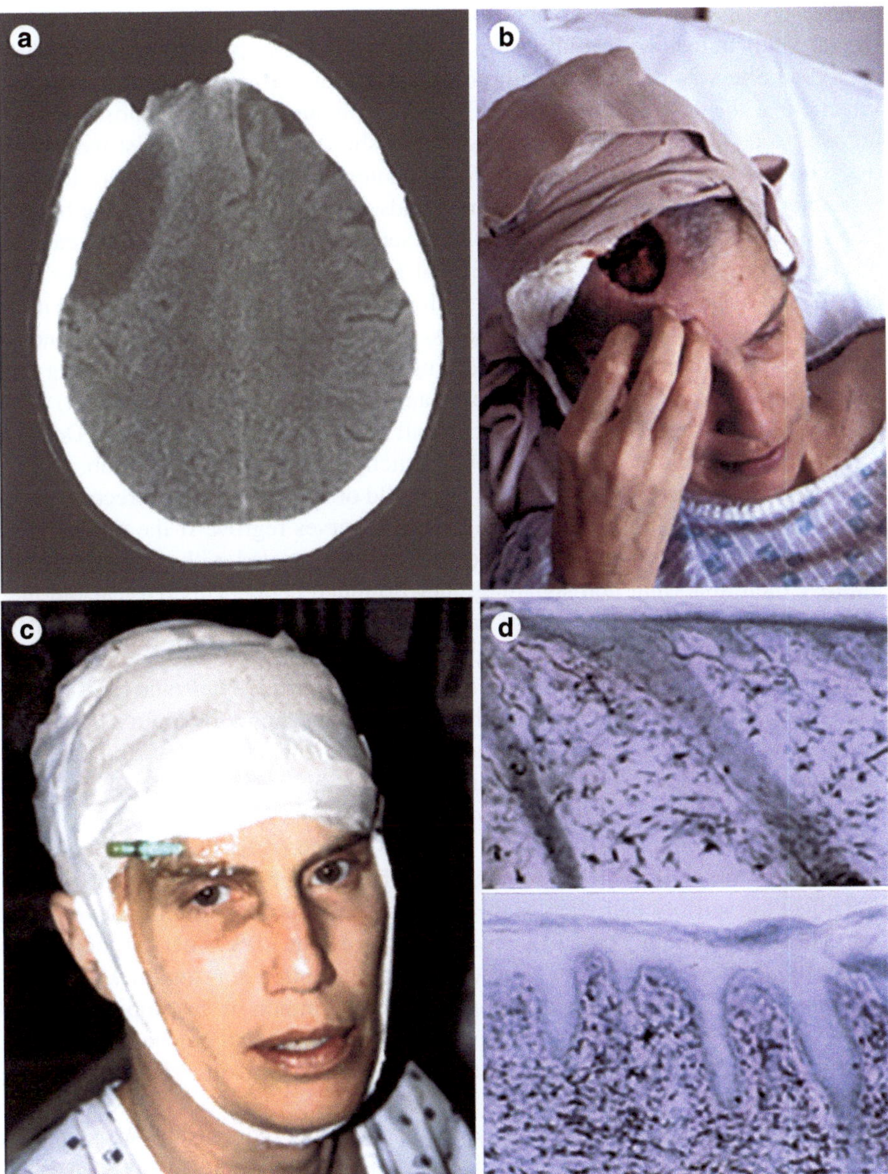

Fig. 1 Patient with severe post-herpetic itch following shingles in region above the right eye. (**a**) CT scan showing right frontal skull defect. (**b**) Patient involuntarily reaching to scratch. (**c**) Patient with head bandage and green catheter. (**d**) Skin biopsies; top panel shows healthy small fiber neurons, bottom panel shows loss of almost all nociceptors/small fiber neurons at site of itch

This dramatic and widely reported case of mine has markedly increased medical and public awareness about neuropathic itch. A woman in her 30s who had shingles around and above her right eye developed a severe neuropathic itch, and began to scratch the area uncontrollably. The skin became infected, which spread to the skull (osteomyelitis) and within a year she had scratched through her skull and into her brain as revealed by a CT scan at hospital admission. It showed a 4 cm × 6 cm right frontal skull defect with midline shift and an infected epidural fluid collection (Fig. 1a). She was diagnosed with an epidural brain abscess with softening and infection of the frontal lobe. Later after her brain surgery and diagnosis of post-herpetic itch, when I lifted the bandage back, she still could not resist the urge to reach up and scratch (Fig. 1b). Her life-threatening scratching was controlled by a protective head bandage, restraining her hands at night, and administering bupiva-caine, a sodium channel blocker that blocks sensory nerve action through a catheter inserted into the effected area. Her frontal lobe brain damage impaired her self-control and made it nearly impossible for her to resist the urge to scratch, and it left her with epilepsy and left body weakness. She later wore a protective helmet at night to prevent her from scratching during sleep. Special skin biopsies (Fig. 1d) taken from areas of normal scalp (upper panel) and itchy scalp (lower panel) demonstrated loss of almost all nociceptors/small fiber neurons at site of itch. Detailed sensory testing confirmed that this woman's itchy skin was virtually numb.

Most patients with neuropathic itch have mild or no self-injury and recover gradually from their itching and scratching as their nerves regrow. If the neurological causes for their situation are explained, and they are treated with medications that improve their abnormal itch signaling, they do much better. This 63-year-old man had a central cause of his focal neuropathic itch, pain, and scratch-induced painless facial ulcers. He developed a condition historically known as "trigeminal trophic system" (TTS) itch and pain as part of a Wallenberg's stroke syndrome (dorsolateral medullary infarction) (Fig. 2). Here, the causal lesions are in the brain's trigeminal nucleus and tract that transmit the main sensory innovation from the face [9]. Multiple sclerosis lesions or other brain injuries in this area can cause similar one-sided facial itching and subsequent scratching [10].

Neuropathic itchy areas below the neck are less common than on the face and frontal scalp. We and others have shown that different types of neurological disorder that affect the sensory pathways in the spinal cord can produce similar neuropathic itch. The most common cause of itchy patches below the neck is nerve roots getting "pinched" as they exit the spine. These itchy patches often develop thick and darkened skin from the repeated scratching, are more common on the upper torso and arms, where they are called "notalgia paresthetica." As with all pinched nerve roots, arthritis is the most common cause and surgery may be required for continuing severe symptoms.

Less often, the cause of itching is within the spinal cord itself. We studied a patient with neuropathic itch from a cavernous hemangioma (a non-cancerous blood vessel tumor) in the sensory area of his dorsal spinal cord and identified specific pathologic features of cavernous hemangiomas that we believe lead to excess firing of the itch projection neurons in lamina of one of the dorsal horns of the

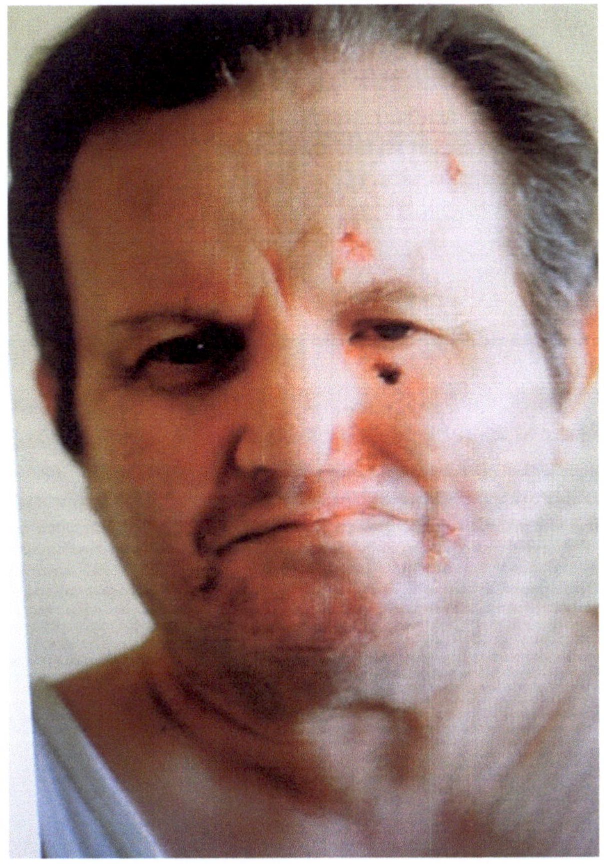

Fig. 2 Patient with Wallenberg's syndrome caused by a vertebral artery occlusion [12]

spinal cord [12]. We studied a corresponding rat model of neuropathic itching and scratching created by Robert Yezierki's, University of Florida lab [13]. We injected neurotoxic quisqualic acid into the spinal cords of rats. Some injected rats began to scratch while others did not. We compared which areas of the spinal cord the toxin had been injected, in rats that scratched versus those that had not; and it implicated this same dorsal itch-processing area of the spinal cord. Like human patients, skin biopsies confirmed that in the itchy rats. Almost but not quite 100% of the skin neurons had degenerated as a result of the injections. It permitted affected rats to apparently painlessly scratch through their skin to the flesh below as seen with the patient in Fig. 1.

A more common itch-associated peripheral neurological condition is small fiber polyneuropathy (SFN). This body-wide condition causes more widespread neuropathic itch on both sides of the body. For example, most often patients begin to itch and scratch both feet, where they may also have chronic pain, sensory loss, and other signs of neuropathy such as swelling and redness, reflecting neurogenic inflammation from excess firing of their C fibers [14]. If untreated, neuropathy symptoms can progress up the legs to affect the hands and arms, and occasionally much of the body. Occasionally patients have "non-length dependent" neuropathy, where symptoms develop randomly in patches on both sides of the torso or face or the entire body. Non-length dependent sensory neuropathies are almost always dysimmune and should be carefully evaluated by a nerve expert sooner rather than later, as very rarely, these can indicate an underlying cancer-related cause.

Concerns about potential neuropathy require neurological evaluation for diagnosis, then blood testing to identify the underlying medical cause of nerve damage so it can be treated effectively. Diabetes is overall the most common cause of neuropathy in most developed countries. Blood tests can be used to identify another dozen causes that require targeted treatment [15]. In young otherwise healthy adults, immune attack on sensory neurons is increasingly recognized, particularly in females. Very young children or families with neuropathic itch conditions may have a genetic cause of neuropathy. These require genetic testing, which sometimes identifies a highly effective precision treatment, such as for sodium channelopathies [14]. Immunotherapy may be indicted for patients with evidence of immune cause of their nerve damage, among which Sjögren's syndrome appears particularly common [16]. Evaluation by an experienced neurologist or nerve specialist is typically required to tease this out and identify the appropriate targeted therapy. Even after detailed evaluation, the cause remains uncertain in approximately half of patients with confirmed SFN (idiopathic SFN), and such patients require treatments aimed at reducing their most disabling symptoms including neuropathic itch in addition to disease-specific treatments.

2　Treatment Options

The usual treatment for normal itch and to reduce inflammation, is to use antihistamines. Unfortunately, these do not typically work for neuropathic itch, which is neuronally driven and more difficult to manage. Because it is difficult to diagnose and is not a standard topic for either dermatologists or neurologists, there are no high-quality trials of potential therapies. Clinical experience and case reports such as ours have established that medications that reduce neuronal excitability benefit many patients. Local anesthetics, that temporarily block all sensation in a small area, are a primary treatment. For small itchy patches, I always consider whether local anesthetic patches or creams applied under an occlusive wrap to hold them against the skin might help. These deliver the anesthetics directly into the underlying skin without spreading throughout the body. Systemic administration carries risks for more serious complications, including heart rhythm changes.

For widespread or disabling itch, I consider oral medications that dampen neuronal excitability. Paramount among them are drugs originally developed to treat epilepsy. Many reduce entry of sodium into neurons, which is a prerequisite for generation of action potentials. Efficacy of these treatments specifically for neuropathic itch has not been studied in clinical trials. They are well studied in epilepsy, which involves excess firing of brain neurons, so much is known about dosing, monitoring, and potential side effects. Sometimes an area of neuropathic itch indicates a nerve or nerve root that requires surgical decompression for effective relief and prevention of additional neurological damage ("pinched nerve or root"). Finally, shingles caused by internal reinfection of sensory ganglia by the varicella zoster virus (VZV) that produces chickenpox in unimmunized children is almost entirely preventable now by the newest shingles vaccination. This has been an extraordinary success, maintaining 97% protection, even in adults older than 70 who are most susceptible to shingles [17]. The public has not been sufficiently aware of their ability to protect themselves against neuropathic itch, pain, and other severe complications including loss of vision and hearing by simple and safe vaccination. Please consider it if you are middle aged or older.

3 Summary

The skin appears plain on the outside, but inside there is a symphony of specialized sensory receptors evolved for noxious (potentially harmful) and innocuous stimuli. Large myelinated peripheral nerve cell axons convey sensations for innocuous stimuli to the brain and send motor signals to muscles to control movement. Small diameter unmyelinated and thinly myelinated small fiber peripheral nerve cell axons receive and transmit harmful/noxious stimuli that are interpreted by brain as pain and/or itch. They release chemicals into the skin that help activate immune cells, which cause skin redness and inflammation.

Itch and pain trigger widespread responses in the spinal cord and brain including release of stress hormones, fear and anxiety emotions, memory, learning, and defensive muscle actions.

Pain sensations promote flinching and withdrawal from an external threat whereas itch promotes reaching to scratch or rub to dislodge clinging threats. Due to many shared neurons, experts believe itch evolved as an offshoot of pain to protect against any number of arthropod interactions with the skin such as envenomization, biting, or parasitism.

Damaged itch and pain neurons fire inappropriately to cause chronic neuropathic itch. This is difficult to diagnose and manage, as anti-inflammatory treatments for normal itch fail. A clue to the disorder is reduced or altered sensation (pain) in the same area. If scratching is less painful, it can continue longer.

Acknowledgments Supported in part by the U.S. National Institutes of Health NIH R01-NS093653 (Oaklander), the Mayday fund (Oaklander), and other philanthropic foundations.

References

1. Abraira VE, Ginty DD. The sensory neurons of touch. Neuron. 2013;79:618–39.
2. Foster E, Wildner H, Tudeau L, Haueter S, Ralvenius WT, Jegen M, et al. Targeted ablation, silencing, and activation establish glycinergic dorsal horn neurons as key components of a spinal gate for pain and itch. Neuron. 2015;85:1289–304.
3. Akiyama T, Nagamine M, Carstens MI, Carstens E. Behavioral model of itch, alloknesis, pain and allodynia in the lower hindlimb and correlative responses of lumbar dorsal horn neurons in the mouse. Neuroscience. 2014;266:38–46.
4. Nguyen E, Smith KM, Cramer N, Holland RA, Bleimeister IH, Flores-Felix K, et al. Medullary kappa-opioid receptor neurons inhibit pain and itch through a descending circuit. Brain. 2022;145:2586.
5. Schmelz M. How do neurons signal itch? Front Med (Lausanne). 2021;8:643006.
6. Jiang S, Wang YS, Zheng XX, Zhao SL, Wang Y, Sun L, et al. Itch-specific neurons in the ventrolateral orbital cortex selectively modulate the itch processing. Sci Adv. 2022;8:eabn4408.
7. Cabanero D, Irie T, Celorrio M, Trousdale C, Owens DM, Virley D, et al. Identification of an epidermal keratinocyte AMPA glutamate receptor involved in dermatopathies associated with sensory abnormalities. Pain Rep. 2016;1:1.
8. Andersen HH, Elberling J, Sharma N, Hauberg LE, Gazerani P, Arendt-Nielsen L. Histaminergic and nonhistaminergic elicited itch is attenuated in capsaicin-evoked areas of allodynia and hyperalgesia: A healthy volunteer study. Eur J Pain. 2017;21:1098–9.
9. Fitzek S, Baumgartner U, Marx J, Joachimski F, Axer H, Witte OW, et al. Pain and itch in Wallenberg's syndrome: anatomical–functional correlations. Suppl Clin Neurophysiol. 2006;58:187–94.
10. Yamamoto M, Yabuki S, Hayabara T, Otsuki S. Paroxysmal itching in multiple sclerosis: a report of three cases. J Neurol Neurosurg Psychiatry. 1981;44:19–22.
11. Oaklander AL, Bowsher D, Galer B, Haanpää M, Jensen MP. Herpes zoster itch: preliminary epidemiologic data. J Pain. 2003;4:338–43.
12. Dey DD, Landrum O, Oaklander AL. Central neuropathic itch from spinal-cord cavernous hemangioma: a human case, a possible animal model, and hypotheses about pathogenesis. J Pain. 2005;113(1–2):233–7. PMID 15621384.
13. Brewer KL, Lee JW, Downs H, Oaklander AL, Yezierski RP. Dermatomal scratching after intramedullary quisqualate injection: Correlation with cutaneous denervation. J Pain. 2008;9(11):999–1005. PMID 18619906.
14. Oaklander AL, Nolano M. Scientific advances in and clinical approaches to small-fiber polyneuropathy: a review. JAMA Neurol. 2019;76:1240–51.
15. Lang M, Treister R, Oaklander AL. Diagnostic value of blood tests for occult causes of initially idiopathic small-fiber polyneuropathy. J Neurol. 2016;263:2515–27.
16. Lefaucheur JP, Sene D, Oaklander AL. Primary Sjogren's syndrome. N Engl J Med. 2018;379:396.
17. Cunningham AL, Lal H, Kovac M, Chlibek R, Hwang SJ, Diez-Domingo J, et al. Efficacy of the herpes zoster subunit vaccine in adults 70 years of age or older. N Engl J Med. 2016;375:1019–32.

When Delusional Patients Are Really Infested: The Exception to the Rule

Richard J. Pollack

Persons who manifest with apparently odd and seemingly inexplicable complaints and fanciful claims of parasites and parasitic conditions are well known to and often shunned by medical and pest management professionals. The authors who collaborated in writing this book have carefully considered published evidence as well as personal training and experiences to provide invaluable guidance as to how clinicians and allied professionals might best evaluate, diagnose, and manage these too frequent and often immensely challenging problems. Whereas the majority of those who steadfastly complain about self-perceived parasitosis do likely suffer from delusions or illusions; a few complainants truly are burdened with parasites or have encountered biting, stinging, infesting, or irritating organisms.

My entomological and parasitological training and endeavors of more than four decades in this arena have provided me with considerable opportunities to encounter persons who suffered from obvious and documentable parasitic conditions. Presumed parasites were often genuine, based upon incontrovertible evidence of their presence. Other times, they were inapparent, elusive, or imagined. With surprising frequency, complainants experienced actual irritation, but misconstrued sensations from other causes as bites or from creatures burrowing in the skin. Direct contact with sand grains, metal shards, glass fibers, plant fragments, insect hairs, or diverse other foreign matter can readily mimic sensations of something walking upon, biting, or burrowing into the skin. Similarly, many people develop a sensitivity to certain ingredients in personal care and household cleaning products, cosmetics, and even to pesticides and medications that they may have used with the intention to treat or prevent pest or other problems. Hence, things are not always what they seem.

R. J. Pollack (✉)
Department of Environmental Health and Safety, Harvard University, Cambridge, MA, USA

IdentifyUS LLC, Needham, MA, USA
e-mail: rich@identify.us.com; richard_pollack@harvard.edu

G. E. Ridge (ed.), *The Physician's Guide to Delusional Infestation*,
https://doi.org/10.1007/978-3-031-47032-5_11

My perspectives have been informed by examining many tens of thousands of specimens and digital images submitted from clinicians, clinical laboratories, members of the public, and because of my own laboratory and field research efforts, as well as my personal encounters (some intentional, others accidental) with diverse ectoparasites and endoparasites. Many of the samples I have reviewed were invited submissions in response to institutionally approved research efforts on lice, ticks, and other arthropods. Another considerable proportion of samples—mainly of ticks, lice, fleas, and fly larvae—were submissions that arrived with requests to support the provisional identifications rendered by clinical laboratorians.

Samples submitted from clinicians or clinical laboratories generally tend to be discrete specimens that are usually well preserved and carefully packaged. Such specimens invariably are accompanied with information of only limited interest or utility, such as the subject's age, the date sampled, and the location on the body (or elsewhere) from where the specimen derived. Information that would provide potentially greater value is rarely available, such as the sufferer's address, travel history, vocation, avocations, or pet ownership or other animal contact. Samples from clinical settings not infrequently arrive with a note requesting assistance to rule out "delusional parasitosis."

Samples from clinical settings tend to be pleasingly intact, making it relatively straightforward to examine and identify them. Clinical workers do, however, frequently cleave samples, or they embed and section most any mass without first carefully examining the entire intact specimen. Whereas many organisms can be readily characterized by scrutinizing stained slide-mounted sections, much useful information may be lost by unnecessary and overzealous processing (Fig. 1).

Samples submitted by members of the public are sometimes also separated from debris and packaged and delivered with care. Sufferers who firmly believe they're "with parasites," however, tend to become singularly focused on demonstrating "evidence" of a parasite on or in their skin or elsewhere in their body, or within

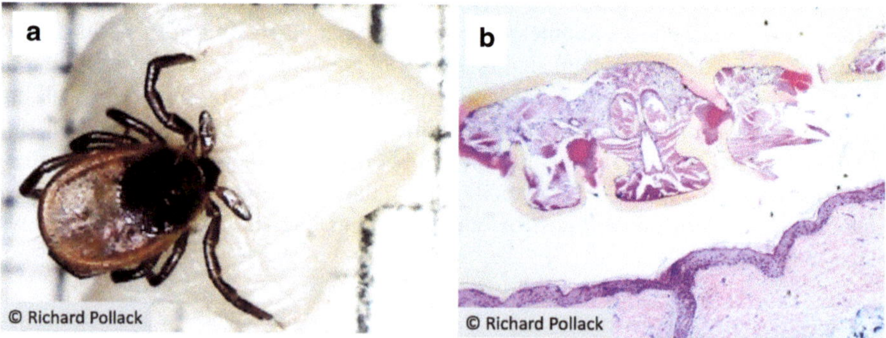

Fig. 1 Intact parasites are often more readily identified and evaluated than those that have been embedded, sectioned, and stained. (**a**) An adult female black-legged /deer tick, *Ixodes scapularis* and the biopsied skin on which its mouthparts were embedded. (**b**) An H & E-stained section of an ixodid tick. Although the sectioned organism is clearly a hard tick, features are missing that would be helpful in concluding the species of tick

bodily excretions and secretions. Such evidence may be delivered in the form of bags or boxes containing many hundreds of skin fragments, scabs, self-biopsied samples, inexpertly prepared microscope slides, floor sweepings, and even buckets of excreta. Although some samples arrive intact, others have suffered the wrath of the submitter, being crushed, torn apart, pulverized, encased in occlusive ointments, sandwiched between many layers of tape, or worse.

Persons convinced they are parasitized regularly share hugely magnified photocopied pictures, cartoonish hand drawn diagrams copied from zoological texts (or less authoritative sources), and increasingly share screen captures of "their" parasites obtained from academic or other websites. They also seem prone to submit excessively detailed and specific or overly meandering descriptions or their life's burdens. Fortunately, only a few such persons have successfully evaded building security, entered offices, and begun to disrobe to show their real or imagined parasites, or the scars or lesions that they claimed to result from a current or past infestation or bites.

Lesions that result from the bites of different arthropods may be remarkably similar in appearance, but they may vary depending on the kind of arthropod involved and the person's prior history with the close relatives of the offending organism. Clinicians should be mindful that it is rarely possible to identify the insect or tick merely by examining the appearance of the lesion. Finding and documenting the actual villain remain the gold standard.

Generally, the likelihood that a genuine parasite or pest is present is inversely proportional to the extent of time and effort the complainant exerts to "prove" this condition. Persons are far more likely than not to be considered as suffering from Delusional Infestation (DI) if they've visited many generalist and specialist clinicians in succession, have repeatedly submitted samples, have self-treated or demanded off-label treatments from their medical providers, or have become adherents to websites that proclaim the existence of invisible parasites. Over the decades, I've experienced more than a few cases where the complainant has been labeled as being delusional. Because of their preoccupation with a presumed but non-demonstrated parasite, such patients have been dismissed by their clinicians, physically removed from medical offices with the help of sheriffs, shunned by family members and acquaintances, and terminated by their employers. In desperation, too many have sought unwise and counterproductive therapies, such as by self-treating using increasingly toxic, caustic, or flammable substances.

Some persons earnestly believe they are actively infested, even though no parasite can be detected or captured. Affected individuals should not be dismissed as being mentally unstable, and reasonable efforts should be pursued to identify the true cause of the sensations, irritation, bite-like reactions, or lesions, and to capture and identify any offending creature. The proximal cause may, indeed, be some type of biting or otherwise irritating insect or mite, but it may not necessarily be infesting the person at the time of examination.

A person may be chronically infested by head lice, body lice, pubic lice, or by scabies mites. Hard ticks and tissue-infesting fly larvae may remain on (or in) a person for days or weeks. Biting flies, fleas, bed bugs, and biting mites (that usually infest rodents or birds) are more transiently in contact, usually just for a few minutes at a time. Even plant-feeding or predatory insects may occasionally bite or cause other discomfort as they sample the environment or react in self-defense. Although many of these biting, infesting, and irritating creatures may not be noticed in the act of biting, the irritation resulting from the injected saliva during feeding may manifest minutes to many days later.

Among the innumerable cases on which I've been consulted, a precious few are illustrative of apparently delusory patients who truly suffered parasitic infestations. I now present a summary of a few of these exceptions.

1 Waist Woes

A 53-year-old professor of medicine complained of anxiety, distress, and insomnia resulting from worsening itching and the ensuing excoriation of the skin around his waist. His odd and uncharacteristic behavior of sudden onset caused his colleagues to evaluate him for a drug reaction or misuse. Ruling out those explanations, he was administered topical antibiotics to treat a presumed skin infection, and topical corticosteroids for inflammation. Despite such treatment, the discomfort caused the patient to repeatedly revisit the emergency room. After several days, a line of small punctate lesions was observed around his waist. Another water-based topical antibiotic ointment was applied to each lesion, and the patient was again sent home. A few more hours of continued "torture" stimulated yet another visit to the ER where he complained of sensations of being "eaten alive." Being aware of prior stress-associated incidents burdening this patient, several of his close and increasingly concerned colleagues dismissed such complaints and encouraged a consult with a psychiatrist. For lack of other immediate options, they applied another topical antibiotic, but this time in an oil-based formulation. Within minutes, a scene characterized as "horror and chaos" transpired in the emergency room, with the patient and some of his medical attendees screaming and sobbing. The application of the occlusive ointment resulted in the emergence of 44 worm-like creatures that proceeded to drop onto the exam table.

The story and the container of formalin-fixed larvae were delivered to me some hours later for review. The specimens (Fig. 2) were readily recognizable as larvae of the tumbu fly, *Cordylobia anthropophaga*. Before revealing the identity of the villains, I asked the clinicians when the patient had been in the African tropics (the normal range of this fly). They repeatedly attested that there was no relevant travel history. After further inquiry, the patient revealed that he had, indeed, visited the Togolese Republic in West Africa a month or more earlier, and he described how he had his swimsuit and other undergarments washed in a traditional manner in the local river and dried on a rock. Female tumbu flies generally deposit their eggs in sandy locations, but they might also oviposit onto urine- and sweat-stained items of clothing. The hatchlings then seek a host to invade and in which to develop. The actions of the feeding larvae as they develop within furuncles in the skin can become

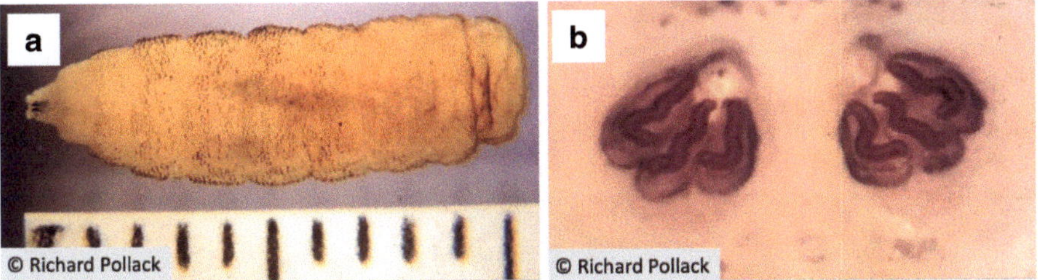

Fig. 2 Tumbu flies. (**a**) Mature larva (scale in millimeters). (**b**) Posterior spiracles of the larva, depicting the characteristic pattern for this species

exceptionally itchy and painful. The affected area of skin corresponded closely with the waist band of this patient's swimming trunks.

The emergence or extraction of the larvae effectively ends the infestation, and the lingering irritation soon abates. Although the larval developmental sites in the skin should be monitored for healing and secondary infection (generally caused by scratching by the patient), no other efforts (other than, perhaps counseling) are justified.

2 Food for Thought

A 60-year-old business executive repeatedly visited his dentist and general practitioner for several months, complaining of intermittent and worsening tooth or jaw pain, an odd mouth odor and taste, and sensations of "movement" in his mouth. Neither practitioner identified a physical cause for the complaints. Finding nothing obviously amiss other than the repeated complaints of inexplicable sensations of movement in the mouth, the primary care physician penned "delusional infestation" on the medical record, prescribed anti-anxiety medication, and referred the patient for a psychiatric consult.

The patient subsequently returned to his dentist, this time with necrotic lesions extending deep into his gum and into the underlying bone. Biopsies of the lesions taken by an oral surgeon revealed small foreign masses that the histopathologist thought suggestive of arthropod integument. Though the stained sections were not particularly informative, a fragment remaining in the paraffin block was available for examination. By transilluminating the block, a portion of the head of a tiny adult fly was evident (Fig. 3). The fly head appeared remarkably consistent with that of the cheese skipper fly *Piophila casei* (L), an occasional cause of myiasis in people.

Larvae of this fly are added intentionally to help ferment and enhance the maturation of Pecorino or sheep's milk cheese. This product, casu marzu ("rotten cheese"), is prohibited in the United States and the European Union but is a delicacy on the island of Sardinia in the Mediterranean. This cheese is often eaten while the larvae are alive and moving.

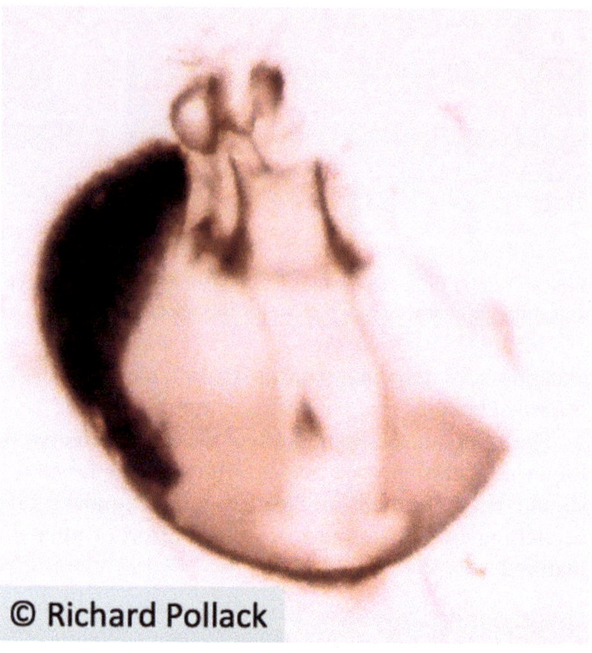

Fig. 3 Cheese skipper. Portion of the adult fly head embedded in the tissue block

After extensive questioning, the patient finally admitted he had, indeed, enjoyed this delicacy on many occasions, along with an appropriate wine pairing. He did not, however, reveal where he obtained or consumed the "living" cheese. Apparently, one or more of the larvae may have invaded his gum and jaw where it competed development but became entrapped and unable to escape. Interestingly, the patient was so unconcerned about this finding that he planned to enjoy more of the same kind of cheese at the next opportunity.

3 Security Concerns

A well-respected and otherwise vigorous and healthy member of Congress complained for several months about suffering diffuse biting sensations, mainly experienced during the evening while at home. Her incessant scratching while at work, and the resulting skin damage, caused considerable concern among her peers and staff members, by her doctors, as well as by her family members who all resided elsewhere. Her colleagues urged her to seek medical evaluation and to take time off for a vacation. The ensuing distraction provided an opportunity for some of her colleagues to propose that she was unfit for office, or at least should be removed from high security committee obligations.

The patient expressed absolute certainty that she was being bitten, but she could not see any creature in her well-appointed and pristinely maintained home. A small army of "exterminators" visited without finding a cause for her complaints, and she disallowed the application of any pesticide in the absence of a demonstrable pest. In parallel, she consulted with her general practitioner and several specialists in infectious diseases and dermatology. Whereas the skin was found to be markedly excoriated, neither a causative agent nor a discrete bite-like lesion was discovered. The patient had no pets and no known animal exposure. Her complaints only worsened with time, despite prescribed doses of anti-inflammatory medications, antihistamines, and various soothing salves. Her primary care physician recorded "delusional parasitosis" (infestation) on her medical record and referred her to a psychiatrist who prescribed an antipsychotic agent that the patient refused to fill.

The dermatologist continued to suspect a biting arthropod and requested assistance in reviewing vacuum bag samples from the home. I requested these be limited to separate vacuum bag samples—each with the contents of just about 1 min of cleaning—from the bed, the favored chairs, the sofas, and the carpeting in the residence. Instead, a massive box arrived containing dozens of overfilled vacuum bags apparently from overzealous efforts to completely erode the rugs and upholstered furniture throughout the home. Although such efforts often are exercises in futility, I agreed to scrutinize a sample from each bag.

The first of many samples were unremarkable, other than the amount of damage seemingly caused by the beater brush to fine expensive fabrics. The sample obtained from the patient's favorite chair, however, was teeming with thousands of northern fowl mites, *Ornithonyssus sylviarum* (Fig. 4). This chair was found to

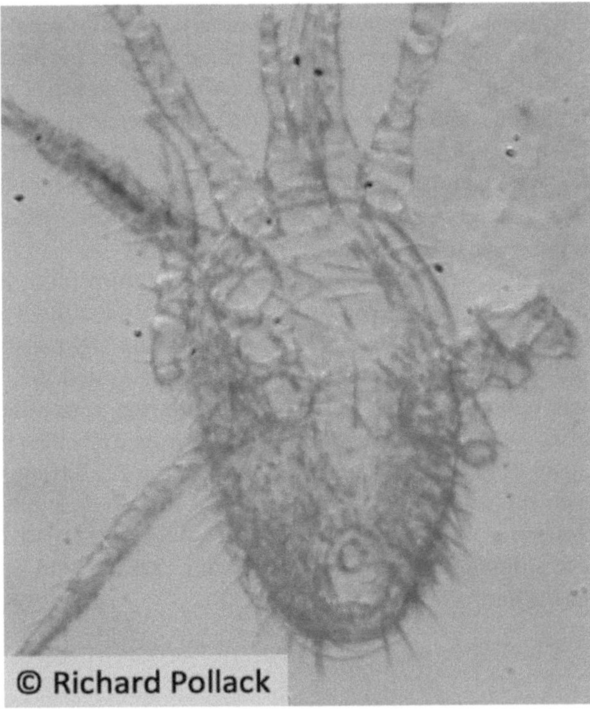

Fig. 4 Histologically cleared northern fowl mite

be in direct contact with a wall and immediately adjacent to an operable window. The exterior ledge was adorned with numerous active and abandoned bird nests. Northern fowl mites blood feed almost exclusively on their avian hosts. When the birds die, fledge, or otherwise depart, their mites increasingly wander in search of new hosts. Many thousands of these mites had readily wandered through the intact window screening and were then observed covering the walls and particularly the chair favored by the patient. Whereas some of the mites do bite people, the mites almost invariably soon die. The victims, however, suffer considerable dermatitis, discomfort, and often exceptional social stigma from the interactions. Once the bird nests were removed and the ledge cleaned, this patient's itching sensations almost immediately abated.

4   Seeing Things

A 60-year-old male complained for more than 5 years of unrelenting itching and irritated skin. His history of alcohol and illicit drug use and his visions of "seeing stars" were reasons that many medical caregivers tended to attribute his complaints mainly to his addictions. A rudimentary eye exam, mainly for acuity testing, revealed the presence of minor corneal opacities. Based upon a presumptive diagnosis of scabies, this patient was prescribed and treated multiple times with a scabicide but without any resolution of symptoms. He was then diagnosed as suffering from "delusions of parasitosis (DP)" (delusional infestation (DI)).

This case was presented in a clinical forum for discussion. Those attending suggested additional studies to probe for various kinds of viral, bacterial, and fungal infections, and they recommended appropriate therapies for each. The last name of the patient, unobscured on one projected slide at the conference, was reminiscent of a name I had commonly encountered in an area of southern Sudan. The patient confirmed that he resided and worked for about three decades in that region and that his home village was close to a river.

I suggested that—should the patient return for a follow-up eye appointment—he be further evaluated using a slit lamp, and with an added step to assume a head-down posture for several minutes to stir up any particulates that may have settled within the fluids of the eye. I was later notified that this was accomplished, and many motile worms were seen shimmering within the patient's eyes. These were almost certainly microfilariae (the motile embryos) of *Onchocerca volvulus*, the agent of "river blindness," a vector borne condition that frequently results in significant ocular and dermal disease in parts of sub-Saharan Africa and South America. Whether the microfilariae were causing or contributing to this patient's eye shimmering or to his skin itching remains unresolved, as the patient was lost to further follow-up.

5 Discussion

The preceding descriptions were provided to illustrate that, on occasion, patients who manifest with complaints and behaviors strikingly consistent with a DI may, indeed, be burdened with demonstrable biting, infesting, or irritating organisms.

Although many patients are referred with suggestions of a "DP" or "DI" component, it is wisest to presume nothing and to carefully review each case without prejudice. Whenever practical, inquire as to when the patient first noticed the problem, whether others in the patient's home or workplace have had similar complaints, and explore details of any travel history, animal contact, pest management efforts, vocational and avocational pursuits, and medication use.

Samples taken in the medical provider's office and maintained in a secure chain of custody apart from the patient are usually of greatest value. Those collected by the patient, however, should not be ignored, as they can sometimes be revealing. Repeated sampling and sample processing from the patient or their environment are often necessary to find the "needle in the haystack," though it can be critical to set time boundaries and limitations on the number and kind of samples accepted for processing. Be mindful that the absence of evidence is not the same as evidence of absence.

Several of the most complex and difficult cases have been resolved through a cooperative team effort, incorporating input from the patient, their family members, the health care providers, a public health entomologist or parasitologist, pest management professionals, and others.

Sorting the wheat from the chaff, so to speak, is necessary and requires more than just relevant expertise. Resolving many such cases also demands considerable time, patience, compassion, an open mind, an appreciation of alternative explanations, and the fortitude to not be swayed by the obvious.

6 Summary

Patients who manifest with complaints and behaviors consistent with a DI may, indeed, occasionally be burdened with demonstrable biting, infesting, or irritating organisms. Several cases are presented as examples to highlight the challenges to the patient and their medical providers. Insights are offered to help guide practitioners in considering diagnostic and management options.

Richard J. Pollack is a public health zoologist who has served Harvard University as a researcher and environmental public health officer. He is also the principal scientist at IdentifyUS LLC, an independent consulting firm that provides insights and guidance to assist identifying and resolving diverse cases of infestation. He has evaluated tens of thousands of specimens of genuine and imagined pests and parasites during his more than four decades in this field. While he's seen more than his share of patients suffering from delusional infestation (DI), he's also had a few particularly satisfying Eureka! moments when the presumably delusional patient was found, in fact, to be truly infested. His chapter focuses on when delusional patients are infested, an exception to the rule.

The Role of the Clinical Parasitology Laboratory in Delusional Infestations

Bobbi S. Pritt and Blaine A. Mathison

1 Introduction

The clinical parasitology laboratory plays an essential role in the evaluation of patients with suspected delusional infestation (DI). First and foremost, the laboratory personnel provide expert examination of patient specimens for evidence of actual arthropod infestation or other parasitic infection. This is a fundamental step, as many parasitic infections may have manifestations that overlap with DI. Similarly, there are many non-parasitic arthropods and objects that may be mistaken for parasites by patients (i.e., pseudoparasitism), and the laboratory can play an important role in excluding true parasitism in these cases. If careful examination of representative specimens fails to reveal a parasite, then laboratory experts can partner with the primary patient care team to document the laboratory results in a clear and meaningful way and provide recommendations for additional testing.

Despite the importance of laboratory testing in the evaluation of suspected DI, there may be limited local laboratory expertise for detection and identification of uncommon parasites. Medical parasitology is an inherently complex and

---

All images in this chapter are by Bobbi Pritt, colleagues at the Mayo Clinic, the CDC, or are in the public domain.

B. S. Pritt (✉)
Division of Clinical Microbiology, Mayo Clinic, Rochester, MN, USA
e-mail: pritt.bobbi@mayo.edu

B. A. Mathison
Institute for Clinical and Experimental Pathology, ARUP Laboratories, Salt Lake City, UT, USA

Department of Pathology, University of Utah, Salt Lake City, UT, USA
e-mail: blaine.mathison@aruplab.com

G. E. Ridge (ed.), *The Physician's Guide to Delusional Infestation*,
https://doi.org/10.1007/978-3-031-47032-5_12

challenging branch of laboratory medicine, with a vast number of parasites causing human infection, complex lifecycles, and diverse morphologic forms [1]. For these reasons, parasitology is best approached as an applied zoology, in which a practical and theoretical knowledge of biology, taxonomy, epidemiology, and anatomy is required for laboratory evaluation [2]. Unfortunately, this level of training is not usually available in medical laboratory science, pathology residency, and clinical microbiology fellowship training programs, particularly in resource-rich settings where there are a limited number of endemic parasites. As a result, there is a great deal of variability in the training received in parasite detection and identification by testing personnel, laboratory directors, and other laboratory leaders. Arthropod identification, in particular, receives little attention. Furthermore, the general rarity of many parasites in many settings limits the amount of exposure the testing personnel get to actual clinical specimens unless they are at a large referral center. Many testing personnel only see certain parasites (e.g., *Plasmodium* species, intestinal helminths other than *Enterobius vermicularis*) in proficiency testing specimens (e.g., those received through commercial providers), and these may be limited in quality and availability. This makes it challenging to gain and retain morphologic skills in parasite identification.

When expertise is not available locally, there are multiple external options for assistance. The US Centers for Disease Control and Prevention Division of Parasitic Diseases and Malaria (https://www.cdc.gov/dpdx/dxassistance.html) provides a valuable telediagnostic service for parasite identification and also accepts tissues and other specimens for testing. For identification of arthropods, an entomologist at the state public health laboratory, local university, or extension office may also be of assistance. The primary patient care team should work with local laboratory specialists to determine the best options for external consultant and arrange for transportation of specimens if appropriate.

To aid the patient-facing clinician in the evaluation of patients with potential DI, the following sections of this chapter and Table 1 review the primary parasitic infections that can mimic DI. Herein we focus on the helminth infections, primarily those with cutaneous involvement as DI typically has a cutaneous component [3]. For infestations caused by arthropods, the reader is referred to other chapters in this work, including Chap. 15. Non-parasitic organisms that are commonly submitted to the clinical parasitology laboratory are also discussed, along with their key identifying features. The chapter concludes with a discussion of how the primary patient care team and laboratory experts can work together to optimize the workup of patients with suspected DI and provides illustrative examples of true parasitism, pseudoparasitism, and DI.

Table 1 Parasitic diseases that may present in a similar fashion to delusional parasitosis

Disease (causal agents)	Geographic distribution	Relevant clinical manifestations	Diagnostic methods
Arthropod Agents			
Scabies (*Sarcoptes scabiei*)[a]	Worldwide	Classical variant: Pruritic papular rash, especially webbing between the fingers, folds of joints, penis, breast, shoulder blades; crusted ("Norwegian") variant: hyperkeratotic plaques over large areas of skin	Observation of mites, their eggs, and feces (scybala) in skin scrapings and biopsy specimens
Demodicosis (*Demodex folliculorum, D. brevis*)[a]	Worldwide	Generally asymptomatic, but mites have been found in association with rosacea, folliculitis, blepharitis, seborrheic dermatitis (unknown significance)	Observation of mites in skin scrapings and biopsy specimens
Zoonotic biting mites[a] (*Ornithonyssus, Laelaps, Dermanyssus, Pyemotes, Cheyletiella, Liponyssoides, Ophionyssus*, others)	Worldwide	Short-lived pruritic papular lesions, nonspecific dermatitis	Clinical presentation in association with the presence of mites and epidemiologic factors, such as known contact with infected natural hosts and/or habitat (e.g., bird's nest)
Hard ticks (*Ixodes, Amblyomma, Dermacentor, Hyalomma, Rhipicephalus*, others)	Worldwide (individual species may be geographically restricted)	Generally asymptomatic, possible discoloration or rash at the bite site; in rare cases paralysis and alpha-gal syndrome	Gross examination of tick; biopsy specimens of engorged ticks (e.g., those mistaken as thrombosed skin tags)
Soft ticks (*Ornithodoros, Carios*, others)[a]	Worldwide, but more common in warn tropical and subtropical climates	Generally asymptomatic; possible discoloration or rash at the bite site	Gross examination of tick, although soft ticks are rarely seen since they do not attach for prolonged periods of time (unlike hard ticks)

(continued)

Table 1 (continued)

Disease (causal agents)	Geographic distribution	Relevant clinical manifestations	Diagnostic methods
Bed bugs (*Cimex lectularius, C. hemipterus*, others)[a]	Worldwide	Rash characterized by erythematous macules/papules, often in a line or zigzag pattern; occasionally bullseye-form rash	Gross examination of bugs in conjunction with clinical presentation
Pediculosis (*Pediculus humanus humanus, P. h. capitis*)[a]	Worldwide	Pruritus due to allergic reactions to louse saliva, with or without rash	Observation of lice and nits on hair (head louse) or clothing (body louse)
Pthiriasis (*Pthirus pubis*)[a]	Worldwide	Pruritus due to allergic reactions to louse saliva, with or without erythematous rash or blue-gray macules (macula cerulea); usually around the groin and external genitalia	Observation of lice and their eggs (nits)
Furuncular myiasis (*Dermatobia, Cordylobia, Cuterebra, Wohlfahrtia*, others)[a]	*Dermatobia* (Latin America, Caribbean), *Cordylobia* (sub-Saharan Africa), *Cuterebra* (North America), *Wohlfahrtia* (Northern Hemisphere)	Non-migratory boils (furuncles)	Gross examination of extracted fly larvae; observation of fly larvae in skin biopsy specimens
Migratory (creeping dermal) myiasis (*Gasterophilus, Hypoderma*, others)[a]	Worldwide	Furuncle initially, evolving to migratory tracks and swelling	Gross examination of extracted fly larvae; observation of fly larvae in biopsy/debridement specimens
Wound (aural, nasal, cerebral) myiasis (*Lucilia, Phormia, Wohlfahrtia, Chrysomya bezziana, Cochliomyia hominivorax*, others)[a]	Worldwide	Wound containing larvae may be extensive and involve the mouth/gums, nasal cavity, ear, and rarely, brain; secondary bacterial infection common	Gross examination of extracted fly larvae; observation of fly larvae in biopsy/debridement specimens

Table 1 (continued)

Disease (causal agents)	Geographic distribution	Relevant clinical manifestations	Diagnostic methods
Fleas (*Ctenocephalides, Pulex*, others)[a]	Worldwide	Pruritic macules/papules	Clinical presentation in association with the presence of fleas and epidemiologic factors, such as known contact with infected natural hosts
Helminth Agents			
Cutaneous larva migrans (*Ancylostoma, Uncinaria, Bunostomum*, others)	Worldwide	Pruritic and erythematous serpentine tracks in the upper dermis	Clinical presentation
Strongyloidiasis/ground itch/larva currens (*Strongyloides stercoralis, S. fuelleborni*)	Worldwide, but more common in tropics and subtropics with poor sanitation	Pruritic and erythematous serpentine tracks in the upper dermis	Clinical presentation, detection of larvae in fecal and respiratory specimens
Gnathostomiasis (*Gnathostoma*)	Latin America, Southeast Asia, Japan, India, sub-Saharan Africa, Madagascar (distribution species-dependent)	Nodular migratory panniculitis; pruritic and erythematous serpentine tracks in the upper dermis	Serology (immunoblot); observation of larval worms in biopsy specimens; gross examination of extracted larvae
Lagochilascariasis (*Lagochilascaris minor*)	Central and South America	Large, painful skin abscesses, often on the neck	Observation of adult worms in biopsy specimens; extraction of adult worms from lesions, the ear, or mouth
Dioctophymiasis (*Dioctophyme renale*)	Worldwide	Migratory, subcutaneous nodules	Observation of larval worms in biopsy specimens
Onchocerciasis (*Onchocerca volvulus*)	Sub-Saharan Africa, Yemen, Latin America	Soft, palpable, discrete nodules up to 5 cm diameter; dermatitis that is itchy, papular, and lichenoid; occasionally depigmented (leopard skin), scaly and atrophic (lizard skin), or hyperkeratotic (elephant skin); hanging groin	Observation of microfilariae in skin snips; observation of adults and microfilariae in skin biopsy/nodulectomy; serology

(continued)

Table 1 (continued)

Disease (causal agents)	Geographic distribution	Relevant clinical manifestations	Diagnostic methods
Dirofilariasis (*Dirofilaria*)	North, Central, and South America, Europe, Asia, Africa (distribution species-dependent)	Tender, subcutaneous nodules; may be fixed or migratory; ectopic migration of adults to the eye	Observation of worms in biopsy specimens; extraction of adult worms from the eye
Zoonotic filarial (*Acanthocheilonema, Brenlia, Brugia, Dunnifilaria, Molinema, Onchocerca, Pelecitus*)	Worldwide (distribution species-dependent)	Tender, subcutaneous nodules; ocular infections	Observation of worms in biopsy specimens or extraction of worms from the eyes and skin
Dracunculiasis (*Dracunculus medinensis, Dracunculus* spp.)	Chad, Angola, Ethiopia, Mali, South Sudan	Painful blister that ruptures to create an ulcer, usually on lower extremities	Extraction of adult female worm from the ulcer
Schistosomiasis/Bilharziasis (*Schistosoma*)	Latin America, Africa, Middle East, Southeast Asia (distribution species-dependent)	Initial infection by cercariae: small, pruritic, maculopapular rash; genitourinary schistosomiasis: genital dermal nodules, ulcers, vaginal bleeding	Detection of eggs in wet mounts of stool; detection of eggs and adults in biopsy specimens; serology
Cercarial dermatitis (*Anserobilharzia, Austrobilharzia, Bilharziella, Bivetellobilharzia, Gigantobilharzia, Heterobilharzia, Ornithobilharzia, Trichobilharzia*, others)	Worldwide	Pruritus and erythema of the skin, followed by eruption of pruritic papules and vesicles	Clinical presentation in conjunction with an appropriate epidemiologic history (e.g., recently swimming in fresh or brackish water)
Sparganosis (*Spirometra*)	Worldwide	Subcutaneous nodules, usually painless, often migratory	Detection of larval tapeworm in biopsy specimens or examination of extracted tapeworm

[a] See related chapters

2　Helminthic Infections that May Mimic Delusional Infestations

Cutaneous Larva Migrans

Cutaneous larva migrans (CLM), also referred to as creeping eruption, is a zoonotic nematode infection caused by hookworm species that normally infect non-human mammals. The most common agents are *Ancylostoma braziliense*, *A. caninum*, and

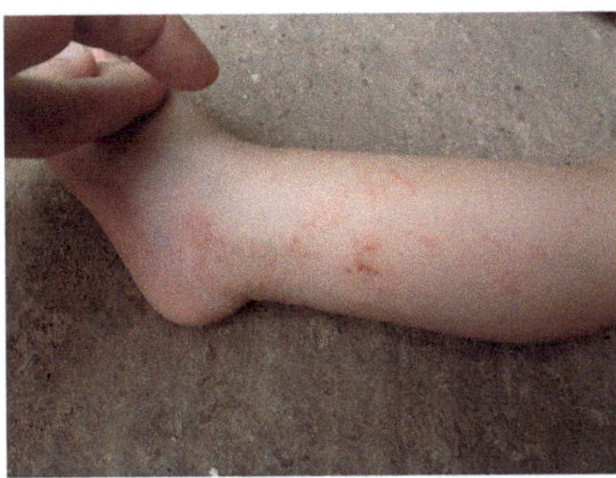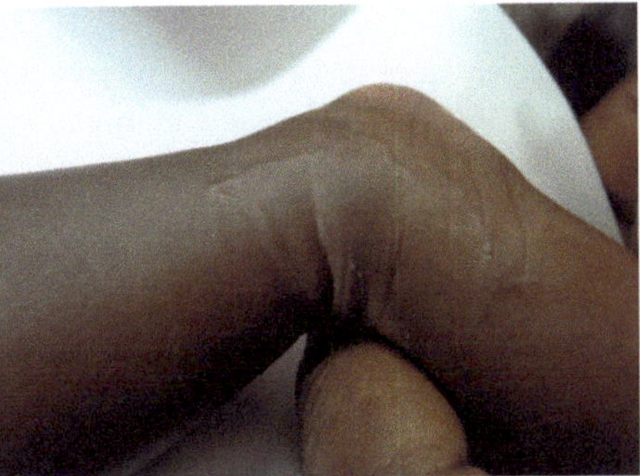

Fig. 1 Cutaneous larva migrans in light- and dark-skinned individuals. Note the raised tracks from the migrating larvae. Images courtesy of Dr. Richard Bradbury

Uncinaria stenocephala, which are intestinal parasites of dogs and cats, and *Bunostomum phlebotomum*, an intestinal parasite of cattle [4]. Human infection can occur worldwide, but is most common in warm, humid climates that support the free-living part of the parasites' life cycle. Infection occurs when people walk barefoot or in sandals on contaminated soil. Infectious L3 (filariform) larvae directly penetrate the skin. The larvae are unable to penetrate the skin basal membrane and further develop in the human host and eventually die at that site; some larvae may become arrested in deeper tissues after migration. Infection is most commonly seen in patients with frequent exposure to the environment or the natural hosts of the parasites, including construction workers, dog breeders, farmers, agriculturalists, and similar [4].

Clinically, patients with CLM present with small pruritic erythematous papules or vesicles that form serpentine and creeping paths in the upper layers of the epidermis (Fig. 1). The larvae advance at a rate of approximately 1–3 cm a day and usually die by 5–6 weeks after infection [5]. Although the disease is self-limiting, prolonged symptoms and complications such as impetigo, allergic reactions, and secondary bacterial infections often make treatment desirable. The recommended treatment is single dose 400-mg oral albendazole or single dose 12-mg oral ivermectin [6].

Diagnosis of CLM is made based on clinical findings in conjunction with epidemiological information [7]. Biopsies are typically not collected for CLM, and there are no serologic assays available. Stool examination for hookworm eggs is not indicated since the species responsible for CLM do not develop patent intestinal infections in the human host.

Strongyloidiasis (Ground Itch and Larva Currens)

Strongyloidiasis is a global nematode infection caused by *Strongyloides stercoralis* and *S. fuelleborni*. In the normal life cycle of *S. stercoralis*, female (only) nematodes reside in the intestinal epithelium of the human host. They produce eggs parthenogenetically (without fertilization from the male) which hatch in the lumen of the small intestine, releasing L1 (first instar or rhabditiform) larvae. The L1 larvae are shed in stool and take one of two pathways in soil. They either become infectious L3 (filariform) larvae and directly penetrate the skin of a new patient, usually someone walking barefoot or in sandals on contaminated soil, or they go on to become adult free-living nematodes. In this free-living stage, there are both males and females, which mate and produce eggs that hatch L1 larvae. The L1 larvae resulting from sexual reproduction become L3 larvae and must penetrate the skin of a host or die; the environmental cycle is not continuous. Once L3 larvae penetrate the skin, they migrate randomly through tissue to the small intestine where they become females; most larvae migrate through the lungs before reaching the small intestine, but the lung stage is not required. Some L1 larvae will become L3 larvae while still in the lumen of the small intestine; these L3 larvae will infect the intestinal epithelium causing autoinfection. This can result in chronic strongyloidiasis that can persist subclinically for decades. The disease can be especially problematic in immunocompromised patients, especially those undergoing solid organ transplantation, as L3 larvae can disseminate to other parts of the body, such as the lungs, kidneys, skin, and brain, causing severe hyperinfection and seeding intestinal bacteria in the process [8]. The life cycle of *S. fuelleborni* has some similarities to that of *S. stercoralis,* but it is passed in stool primarily as eggs rather than larvae and is not thought to have an autoinfective stage.

There are two skin conditions associated with strongyloidiasis that may be included in the clinical differential for DI: an urticarial dermatitis referred to as "ground itch" and *larva currens*. Ground itch occurs when L3 larvae penetrate the skin, often resulting in intense pruritis, congestion, and edema at the site of entry. It most commonly occurs on the feet but may occur anywhere on the body when there is contact with contaminated soil [9]. It can last up to 3 weeks [10]; however, strongyloidiasis is usually not in the clinical differential at this point, unless there are remarkable epidemiologic links. Larva currens usually occurs during autoinfection when L3 larvae penetrate the skin in the perianal area. Clinically, larva currens manifests as raised, pink to red, pruritic, serpentine tracks that can progress around 5–15 cm/h [11]. It has a similar appearance to that of CLM, but is usually located in the perineum, thighs, and buttocks, and advances much more rapidly [12]. Unlike ground itch, larva currens is usually pathognomonic for strongyloidiasis. Treatment of strongyloidiasis is ivermectin, 200 µg/kg orally, for 1–2 days.

Diagnosis of strongyloidiasis is challenging due to the low sensitivity of many of the most commonly available methods. While stool examination for L1 larvae is still considered the gold standard, the shedding of larvae can be low and intermittent and upwards of seven serial stool specimens should be examined [8]. Serologic assays are available, but sensitivity is ≤80% and many cross-react with other nematode infections. In patients with a high pre-test probability, EIA assays may be more sensitive than IFA assays [8]. The most sensitive assay for diagnosing strongyloidiasis is the agar plate culture (APC) method; however, this is not commonly used in the clinical laboratory, as it is time consuming and presents a potential health risk to laboratory staff handing unfixed stools and culture plates, as L3 larvae may penetrate skin. APC is performed by placing fresh stool on an agar plate, such as blood agar or any one of several specialty agars for the method. If L1 larvae are present in the stool, they will become L3 larvae and migrate across the plate, dragging bacteria with them that will show up as visual tracks [13]. Larvae on the plate should be examined microscopically to confirm the identification, as other nematodes such as hookworms and *Trichostrongylus* may produce similar tracks.

Gnathostomiasis

Gnathostomiasis is a zoonotic nematode infection caused by spiruroid nematodes in the genus *Gnathostoma*. Several species have been implicated in human infection, including *G. binucleatum* (Latin America), *G. doloresi* (East and Southeast Asia, Japan), *G. hispidum* (Southeast Asia, Europe, Australia), *G. malaysiae* (Southeast Asia), *G. nipponicum* (Japan, Korea, China), and *G. spinigerum* (East and Southeast Asia, India, Japan, southern Africa, Madagascar) [1]. Natural definitive hosts are carnivorous mammals; humans typically become infected after ingestion of an infectious L3 larva in an intermediate or paratenic host such as fish and frogs. *Gnathostoma* species cannot develop to maturity in the human host, and the parasite typically manifests as cutaneous larva migrans or visceral larva migrans (larva migrans profundus) involving advanced L3 larvae. In rare instances, larvae may develop to sexually immature adults, but patent infections in humans have not been documented.

The most common clinical manifestation of CLM caused by *Gnathostoma* is a nodular migratory panniculitis. Common sites are the limbs, face, back, abdomen, and breasts [14]. Migration through the dermis and subcutaneous tissues results in painful pruritic and erythematous edematous swellings [14, 15]. Superficial migration can result in skin abscesses, and spontaneous expulsion of the larva from the skin has been described [14]. Migratory lesions on the head and face could result in ocular or CNS infection, the latter of which can be life-threatening [14]. Eosinophilia is common during initial worm migration [14]. The primary treatment method is surgical removal of the worm. Albendazole (400 mg twice daily for 21 days) is treatment of choice, with ivermectin (0.2 mg/kg once daily for two consecutive days) as an alternative regimen [14].

Diagnosis of *Gnathostoma* CLM is performed by surgical extraction and morphologic examination of the larva, or detection and identification of the larva in skin biopsies processed for histology. Serologic assays for gnathostomiasis have been developed but may not be readily available outside of endemic areas. The most widely used serologic assay is an EIA that detects IgG targeting the L3 larva, but it has a relatively poor sensitivity and there may be cross reactivity with other nematodes [14, 15].

Lagochilascariasis

Lagochilascariasis is a rare and enigmatic zoonotic infection caused by the ascarid nematode *Lagochilascaris minor*. The parasite is naturally occurring in Latin America, from Mexico to South America, with the majority of human cases being documented in Brazil [16]. The natural hosts are believed to be wild felids and experimental models using cats and mice suggest a two-host life cycle. The route of human infection is unknown but is believed to be from eating raw or undercooked meat, especially rodents such as Guinea pigs, cavies, and agoutis. Autoinfection might be possible [16].

The most common clinical presentation of infection with *L. minor* is the presence of a subcutaneous abscess on the neck, middle ear, tonsils, mastoid, and sinuses. Patients may describe the sensation of something moving in the throat. Lesions on the neck often start as a small, painless tumor without fistula. As the disease progresses, the lesion may grow to a size of 5–12 cm, becoming increasingly painful, with or without fistula, and with drainage of a purulent exudate that may contain larval worms and eggs [16]. The spontaneous discharge of adult worms from the nose, throat, and ears has been documented [17]. Treatment for lagochilascariasis is not well described, but chemotherapeutic treatment usually involves multiple drugs, initiated with thiabendazole, followed by diethylcarbamazine (DEC) or mebendazole, and then levamisole [16].

Diagnosis of lagochilascariasis is achieved by the detection and identification of worms in biopsy specimens or the morphological examination of worms expelled from abscesses [17].

Dioctophymiasis

Dioctophymiasis is a rare zoonotic disease caused by *Dioctophyme renale*, commonly known as the "giant kidney worm." The parasite probably occurs worldwide, and the natural definitive hosts are primarily mustelid carnivores such as minks, raccoons, otters, and martens. Domestic dogs can serve as reservoir hosts. Human infection is very rare and is believed to be acquired following the consumption of raw or undercooked paratenic hosts, such as fish and amphibians [18].

As the common name suggests, infection usually involves the kidney, and most human infections are diagnosed by the finding of eggs in wet mounts of urine or the expulsion of adult worms in urine. Cutaneous involvement is very rare and not well described in the literature, but usually involves ectopic migration of L3 larvae. Clinically, it presents as a migratory skin nodule, usually without pain, itching, or discoloration [18]. Lesions can occur anywhere, such as in the abdominal area or on the eyelids. Treatment for cutaneous infections is surgical removal of the worm.

Diagnosis of cutaneous dioctophymiasis is made by the detection and morphologic identification of L3 larvae in biopsy specimens of skin nodules processed for histology [17, 18].

Onchocerciasis

Onchocerciasis is caused by filarial nematodes in the genus *Onchocerca*. Traditional human onchocerciasis, also known as river blindness, is caused by *O. volvulus* and is endemic to sub-Saharan Africa, Yemen, and Central and South America. There are also several zoonotic species that have been periodically reported from humans; these species and their distributions and natural hosts include *O. cervicalis* (worldwide; horses and other equids), *O. dewittei japonica* (Japan; pigs), *O. gutturosa* (Old World; cattle), *O. jakutensis* (Eurasia; deer), *O. lupi* (North America, Eurasia, North Africa; canids and felids) [1]. Onchocerciasis is a vector-borne disease; *O. volvulus* is transmitted by black flies in the genus *Simulium,* while the zoonotic species are transmitted by black flies or biting midges, depending on the species. Adult *Onchocerca* species reside in subcutaneous connective tissues of the definitive host. Gravid females release microfilariae (early first-stage larvae) into the skin. The microfilariae are picked up by an appropriate vector and develop into infectious L3 larvae. The L3 larvae migrate to the proboscis of the vector and are transmitted to the next host when the vector takes a blood meal. On occasion, microfilariae of *O. volvulus* will infect the eye; the host's inflammatory reaction to the larvae initiates corneal opacities or punctate keratitis that can further develop into corneal scarring and sclerosing keratitis (Fig. 2). This leads to "river blindness" which is an irreversible outcome [19].

There are different clinical presentations involving the skin with *Onchocerca* infections. The classic presentation of *O. volvulus* infection is formation of subcutaneous nodules (onchocercomata). The onchocercomata contain coiled adult

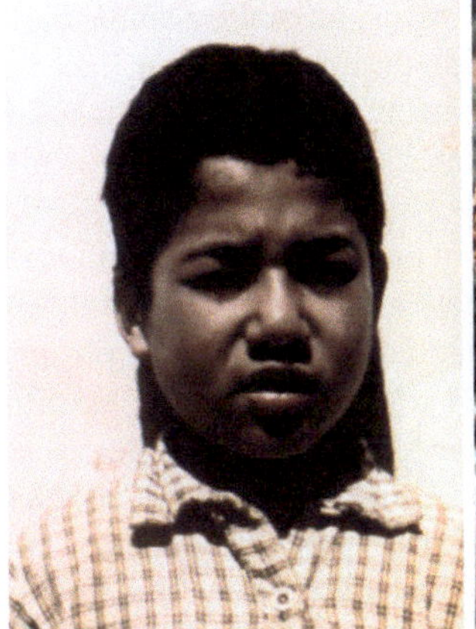

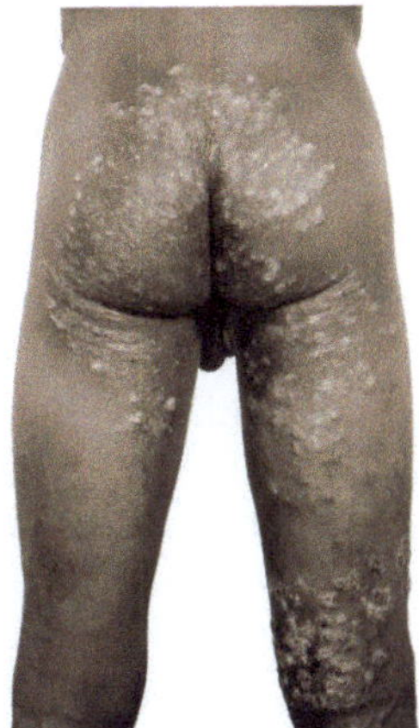

Fig. 2 Various presentations of onchocerciasis. The left panel shows a young Guatemalan boy with early ocular symptoms manifested by photophobia. Note the prominent swelling in his right forehead, likely representing an onchocercoma. The center panel shows a West African man with blindness and skin changes (e.g., depigmentation), while the right panel shows numerous cutaneous lesions in a Nigerian man with onchocerciasis. Images courtesy of the CDC Public Health Image Library

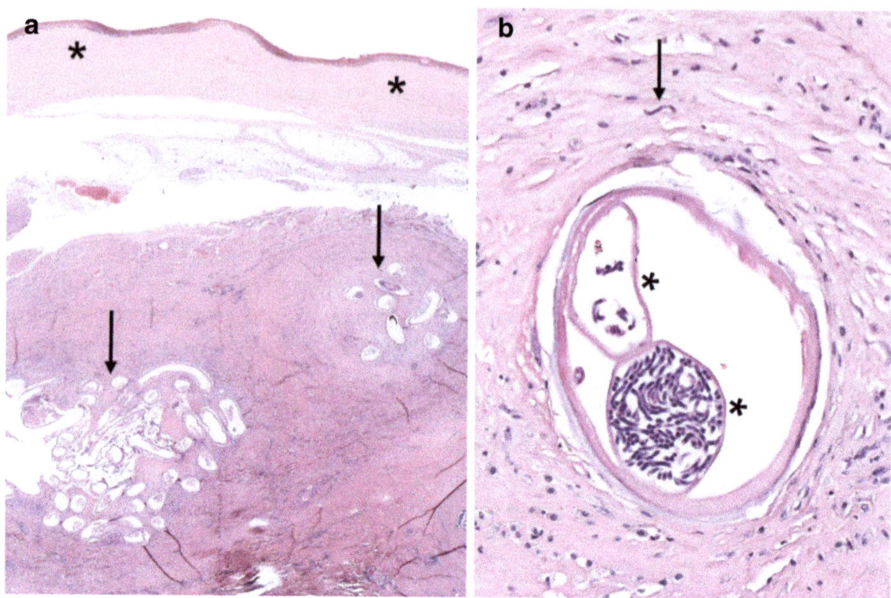

Fig. 3 Onchocercoma. At low magnification (**a**), clusters of adult *Onchocerca volvulus* (arrows) can be seen in cross and longitudinal sections. The overlying epidermis (asterisks) is intact. Higher magnification (**b**) reveals larvae within the double uterine tubes (asterisks), with free larvae (arrow) in the surrounding host tissue. The migrating larvae are the primary source of morbidity in onchocerciasis

nematodes (Figs. 2 and 3) and can occur anywhere on the body, although they are often associated with bony areas. As the disease progresses, an acute papular rash develops. This rash will progress into chronic papular dermatitis which may be associated with lichenification, papule formation, and atrophy. Lesions can take on a variety of morphotypes and may be spottily depigmented (so-called leopard skin), scaly and atrophic ("lizard skin"), or thickened and hyperkeratotic ("elephant skin") (Fig. 2). There is a particular morphotype seen in Yemen and the savanna regions of Africa called "sowda" in which patients will present with severe chronic dermatitis with dark black hyperpigmentation. Patients with chronic disease may also present with lymphedema of the grown (hanging groin) [20, 21]. Migration of the microfilariae from the onchocercomata results in itching of the skin [19]. In human cases of *O. lupi*, an emerging parasitosis in the Southwestern United States, skin nodules have been most commonly associated with the cervical spine and scalp, but also the forearm and superior rectus muscle [21]. The primary treatment for *O. volvulus* is ivermectin, which kills the microfilariae but not the adult worms. It therefore should be administered every 6 months for the life span of the adult worms (10–15 years) or for as long as the patient has physical evidence of infection [22]. DEC, which is frequently prescribed for patients with other filarial infections, such as loiasis (*Loa loa*) and lymphatic filariasis (*Wuchereria bancrofti*, *Brugia* spp.) can worsen ocular disease with *O. volvulus* and is therefore contraindicated in onchocerciasis. For this reason, patients with loiasis or lymphatic filariasis in areas that are also endemic for onchocerciasis have a negative test for *O. volvulus* infection before DEC is administered [22]. Doxycycline which targets *Wolbachia* (symbiotic bacteria in the nematode) has been used to treat onchocerciasis; however, it may not be feasible in endemic areas due to the frequency and duration of the drug required (200 mg/day for 4–6 weeks) [19].

The gold standard for the diagnosis of *O. volvulus* infection is the skin snip method which detects the microfilaria stage. Biopsies of subcutaneous nodules (onchocercomas) will reveal adult male and female worms, as well as microfilariae within the uterine tubes and free in the surrounding tissue (Fig. 3). EIA assays targeting *O. volvulus*-derived antigens, such as Ov16, Ov20, have been developed, but they may not be readily available in many diagnostic laboratories, especially outside of endemic areas. The Mazzotti patch test, which stimulates an inflammatory response with DEC, is used in endemic areas [19]. To date, cases of *O. lupi* and other zoonotic *Onchocerca* have been diagnosed by observation of adult worms in biopsy specimens [21].

Dirofilariasis

Dirofilariasis is a zoonotic filarial disease caused by members of the genus *Dirofilaria*. Like other filarial infections, dirofilariasis is a vector-borne disease and cycles between a mammalian definitive host and a mosquito or black fly vector in nature. There are two broad clinical presentations with human dirofilariasis: pulmonary dirofilariasis caused by *D. immitis* (dog heartworm) and *D. spectans*, and cutaneous dirofilariasis caused by several species in the subgenus *Nochtiella*. Herein we will focus on cutaneous dirofilariasis. Five species of *Nochtiella* have been reported causing human infection; these species, their geographic distributions, and natural hosts include *D. repens* (Europe, Asia, Africa; carnivores, primarily canids and felids), *D. striata* (Western Hemisphere; wild felids), *D. subdermata* (North America; porcupines), *D. tenuis* (eastern North America; raccoons), and *D. ursi* (northern North America, Russia, Japan; bears) [1]. In addition, a currently undescribed species referred to as *Dirofilaria* Genotype Hong Kong has been reported from dogs and humans in Southeast Asia [23], and a traveler returning to Austria from India [24].

The most common clinical presentation of cutaneous dirofilariasis is the presence of a small, painless subcutaneous nodule. Nodules can occur anywhere on the body, but commonly occur on the extremities, eyelids, breasts, and genitalia. Over time, nodules may become larger and inflamed, but without systemic symptoms. Nodules may be migratory or stationary. Peripheral eosinophilia is usually mild. In some cases, especially with *D. tenuis* and *D. repens*, adult worms will migrate to the subconjunctiva [25]. Treatment for dirofilariasis is surgical removal of the worm.

Diagnosis of cutaneous dirofilariasis is made by detection and identification of sub-adult and adult worms in biopsy specimens (Fig. 4). Patent infections in humans

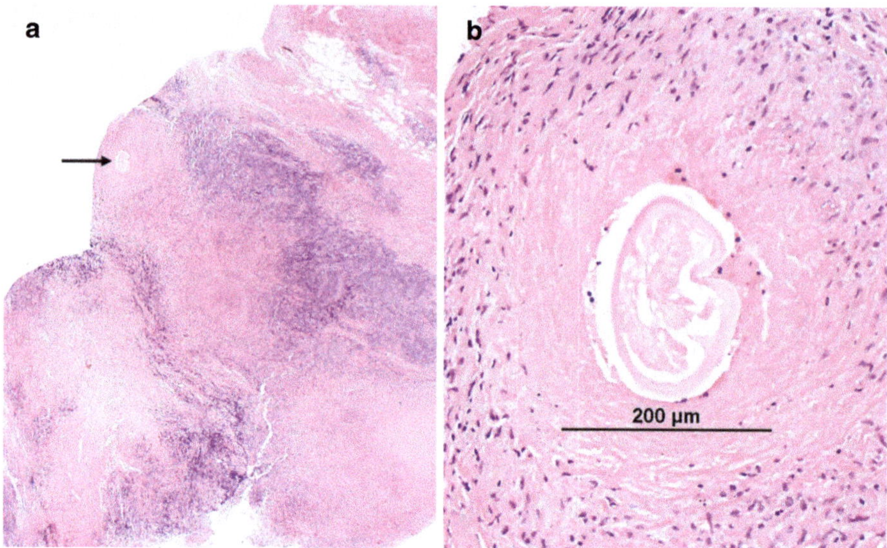

Fig. 4 Extrapulmonary dirofilariasis presenting as an extraocular mass. At low magnification (**a**), a cross section of a nematode can be seen within soft tissue with extensive associated inflammation and necrosis. (**b**) Higher magnification reveals a degenerating worm with thick cuticle and features consistent with *Dirofilaria* sp.

are extremely rare, so examination of blood films for microfilariae is not recommended. In cases of ocular infection, diagnosis is made by gross examination of extracted adult worms [25]. In the absence of epidemiologic data, loiasis should be in the differential for eye involvement and onchocerciasis for cutaneous disease.

3   Miscellaneous Zoonotic Filarial Nematodes

In addition to zoonotic *Onchocerca* and *Dirofilaria*, which were covered above, there are several other genera of zoonotic filarial nematodes that have been rarely reported from humans. *Acanthocheilonema* is a genus of filarial nematodes that parasitizes dogs and other canids worldwide. Several cases involving ocular infection have been reported in humans; *A. reconditum* has been implicated in some infections but in most cases the worms have not been characterized at the species level [26]. *Breinlia annulipapillata* is a parasite of marsupials in Australia that has been reported causing ocular filariasis in a human patient in Australia [27]. There have been multiple reports of zoonotic *Brugia* throughout the Americas, from the USA, Brazil, Colombia, and Peru. Most of the infections involved the lymph nodes of the groin, neck, head, arm, breast, and torso. The vast majority of these *Brugia* isolates have not been identified at the species level, but some of the North American cases have been tentatively attributed to *B. lepori* or *B. beaveri* [25]. There is a single report of a human case of *Dunnifilaria* causing intraocular infection in a patient in Malaysia that has been preliminarily attributed to *D. ramachandrani*, a parasite of the long-tailed giant rat in Malaysia [28]. *Meningonema peruzzii*, a monkey parasite in Africa, has been attributed to CNS infections in human patients in Africa [25]. *Molinema* are parasites of rodents in the Americas. Ocular cases in North America have been tentatively attributed to *M. arbuta* or *M. sprenti*; these cases were originally attributed to *Dipetalonema* [25]. *Pelecitus* is a genus of filarial nematodes that occurs nearly worldwide and has a broad host range. In 2011, intraocular infection with an unknown *Pelecitus* species was reported from a patient in Brazil [29]. Diagnosis of these zoonotic filarial infections is achieved by either detection of the worms in biopsy specimens or extraction of the worms from the eye. For most cases, removal of the worm is curative.

4 Dracunculiasis

Dracunculiasis, also known as Guinea worm disease (GWD), is caused by *Dracunculus medinensis*. Historically, this debilitating parasitic disease was widespread in 21 countries in Africa and Asia, but thanks to elimination efforts, GWD was only reported from focal areas in Angola, Chad, Ethiopia, Mali, and South Sudan in 2022 [30]. Additional cases in Cameroon were thought to be imported from Chad. Most human cases in recent years have been documented in Chad, where transmission is being driven primarily by domestic dogs that have become a reservoir host for the parasite [31]. The COVID-19 pandemic, political insecurity, and civil unrest are additional challenges to eradication. A second, possibly undescribed, species of *Dracunculus* has recently been reported from a human patient in Vietnam [32]. Human infection of *D. medinensis* is caused by the ingestion of infected copepods (microscopic freshwater crustaceans) harboring infective L3 larvae in contaminated drinking water or possibly by the ingestion of L3 larvae in infected fish or amphibians serving as paratenic hosts.

Clinical manifestations are usually associated with gravid female worms migrating through connective tissues and their emergence from the skin. Prior to emergence, a burning, painful blister forms on the skin, usually on the feet and lower extremities. After a few days, the blister bursts, resulting in formation of a shallow ulcer that contains L1 larvae, PMNs, macrophages, and eosinophils. As larvae are released, the worm starts to dry up, with the entire worm coming out of the lesion in a few weeks. The lesion heals quickly, but the track of the worm may become superinfected with bacteria which can result in incapacitation, septicemia, and tetanus [33]. Treatment of GWD involves physical removal of the female worm by gently winding the worm out on a stick (Fig. 5). This process must be done slowly and carefully, at the rate of about 1–3 cm of worm a day, otherwise the worm might rupture resulting in intense inflammation. During the process, the lesion should be kept clean and after removal of the worm, topical antibiotics should be applied to

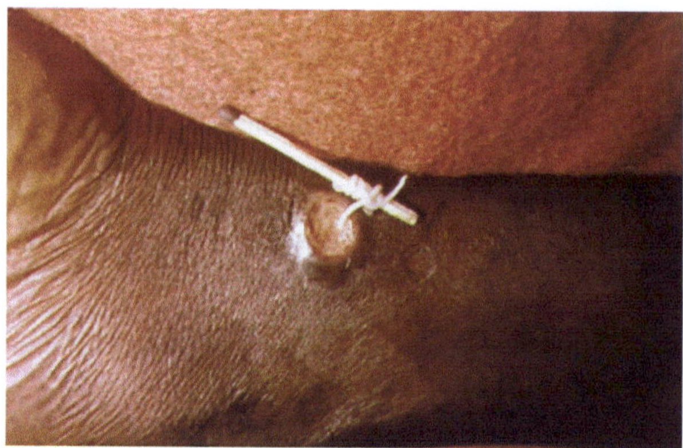

Fig. 5 Adult female *Dracunculus medinensis* in the process of removal by slow extraction and winding around a matchstick. Image courtesy of the US Centers for Disease Control and Prevention Public Health Image Library

prevent secondary bacterial infection [33]. There are no chemotherapeutic or surgical recommendations for treating GWD.

Diagnosis of GWD is performed by gross examination of the extracted worm [33]. Spargana, the larval form of zoonotic tapeworms in the genus *Spirometra* (see also below), may also present this way in areas endemic for GWD and should be in the morphologic differential [34]. EIA assays targeting Ig4 have been developed for the diagnosis of GWD but may not be readily available outside of endemic areas [33].

5   Schistosomiasis (Bilharziasis)

There are two broad clinical conditions caused by schistosome flukes that may mimic DI: cutaneous schistosomiasis, caused by members of the genus *Schistosoma* that cause patent infections in humans, and cercarial dermatitis, an inflammatory skin condition caused by zoonotic schistosomes. In this section of the chapter, we will discuss infections caused by *Schistosoma*; cercarial dermatitis is covered in the next section.

Human schistosomiasis is caused by six species in the genus *Schistosoma*: *S. mansoni* (sub-Saharan Africa, Arabian Peninsula, South America, Caribbean), *S. haematobium* (Africa, Middle East), *S. intercalatum* (Democratic Republic of the Congo), *S. guineensis* (West Africa), *S. japonicum* (China, Philippines, Indonesia), and *S. mekongi* (Mekong River Valley of Cambodia and Laos) [1]. In addition, human infections caused by a *S. mattheei* x *S. haematobium* hybrid have been reported from South Africa [35], and human infections caused by *S. haematobium* x *S. bovis* hybrids have been reported from Corsica [36]. Adult *Schistosoma* resides in the veins servicing various organs, the venous plexus of the urinary bladder for *S. haematobium,* and the mesenteric venules of the intestinal tract for the other species. Eggs are shed in feces or urine into water where the first larval stage (miracidium) infects a snail. From the snail emerges the infectious larval stage called a cercaria (Fig. 6). Humans become infected when swimming or bathing in water

Fig. 6 Cercaria of *Schistosoma mansoni.* This is the infective form for humans

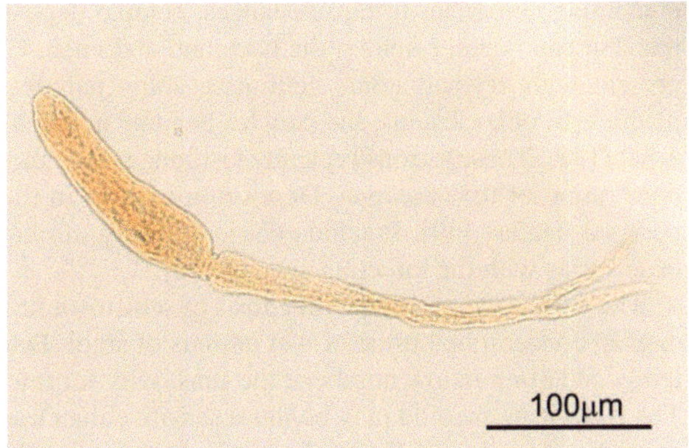

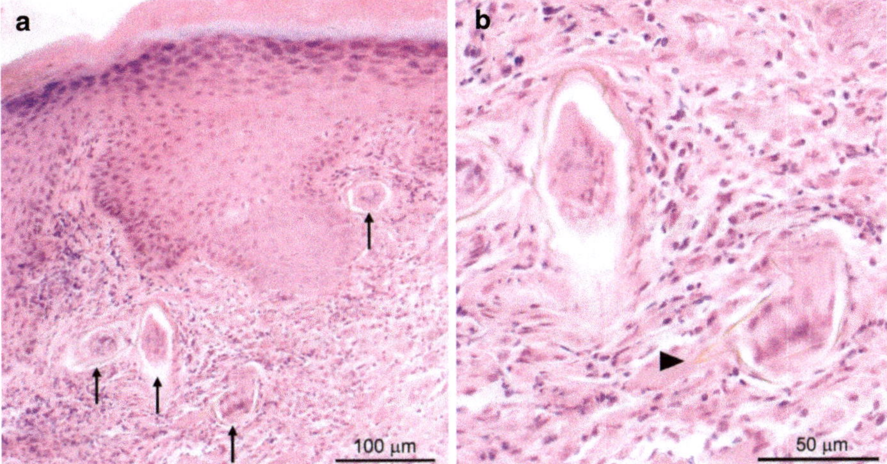

Fig. 7 *Schistosoma mansoni* eggs in skin. Lower magnification (**a**) shows four eggs within the dermis (arrows). The overlying epidermis is hyperkeratotic. Higher magnification (**b**) reveals the gold-brown shell of the eggs and internal nucleated miracidium. A large lateral spine is also visible (arrowhead), allowing identification to the species level

containing cercariae. The cercariae penetrate the skin, enter the bloodstream where they eventually mature into adult worms, pair up, mate, and lay eggs [37]. The eggs become lodged in nearby tissues, and many will work their way through the tissue, aided by the host inflammatory response, to be shed in feces or urine.

Schistosomiasis is one of the leading causes of morbidity and mortality among parasitic infections, usually caused by the inflammatory reaction to eggs lodged in tissues (Fig. 7) [37]. Schistosomiasis is likely to be in the clinical differential for DI during the first stage of infection when cercariae are penetrating the skin. The most common presentation is a local hypersensitivity reaction that manifests as small, itchy, maculopapular lesions. This usually occurs shortly after exposure to water containing cercariae. In rare instances, ectopic deposition of *Schistosoma* eggs in the skin can occur, often on the back and abdomen. Clinically, these ectopic cases present with erythematous, confluent, shiny papules, often forming zosteriform plaques. In older lesions, the papules become nodules with a rough granulomatous aspect [38]. These lesions typically last longer than those that form upon initial skin penetration of the cercariae. Deposition of eggs in the perianal area may result in perianal fistulae [39]. Praziquantel is the drug of choice for schistosomiasis. The dose varies with the infecting species [40].

The gold standard for the diagnosis of schistosomiasis is the detection of parasite eggs in concentrated urine or wet mounts of stool. Because eggs are shed intermittently and often in low numbers, the sensitivity for microscopic examination is low. The Kato-Katz method may be more sensitive than traditional concentration procedures for detection of *Schistosoma* spp. eggs in feces, but requires a fresh, unfixed specimen and is rarely performed outside of endemic areas. Eggs and adult worms may also be observed in biopsy specimens. Serologic assays are available, but their sensitivity and specificity are dependent on the antigen preparations used for the particular assay and the test procedure. Assays that use crude antigen may have a greater chance of cross reaction with other trematode or nematode infections [8].

6 Cercarial Dermatitis

Cercarial dermatitis is an inflammatory reaction caused by the penetration of zoonotic schistosomes into the skin. The condition has been reported worldwide. Some of the genera associated with cercarial dermatitis include *Anserobilharzia, Austrobilharzia, Bilharziella, Bivitellobilharzia, Gigantobilharzia, Heterobilharzia, Ornithobilharzia,* and *Trichobilharzia.* The natural definitive hosts for these schistosomes are birds, primarily those that live in aquatic environments, such as shore birds, wading birds, and waterfowl, and mammals, such as carnivores and marsupials [1].

Cercarial dermatitis begins with the penetration of human skin by the cercariae of zoonotic schistosomes. Primary contact may be asymptomatic or present with mild skin reactions, such as itching or reddening of the skin. After 12–48 h, transient macules, maculopapules, or vesicular papules form (Fig. 8). In some instances, papules may not form for nearly a week after exposure. Repeated infections can cause more severe reactions followed by diffuse edema and the formation of erythematous papules or papulovesicles. Penetration by a large number of cercariae may result in generalized symptoms such as lymphadenopathy, nausea, diarrhea, and fever. Scratching of the papules may result in secondary bacterial infections. Papules usually disappear in 4–10 days, but hyperpigmentation of the skin may remain for weeks [41]. Treatment of cercarial dermatitis is typically limited to soothing agents to combat symptoms. In heavy infections, antihistamines or corticosteroids could be considered [41]. Topical antibiotics or antiseptics may also be indicated in cases of secondary bacterial infection [42].

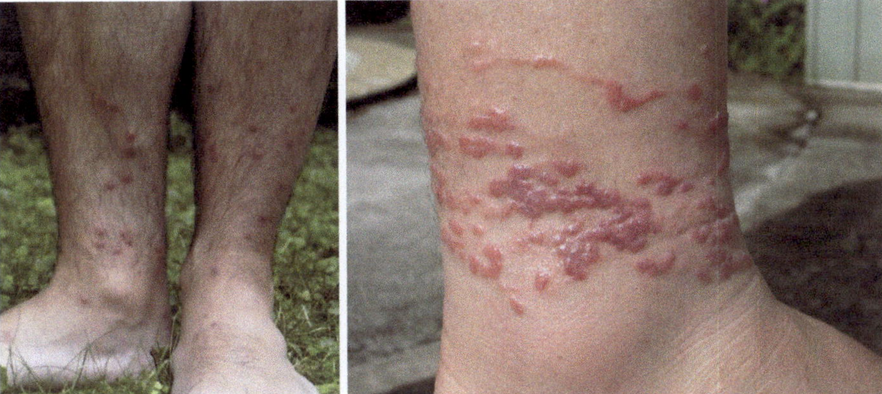

Fig. 8 An eruption of cercarial dermatitis of the lower legs after wading in the shallows of a lake demonstrated moderate (left) and heavy (right) infection. Left-hand image available at Wikimedia Commons (https://commons.wikimedia.org/wiki/File:Cercarial_dermatitis_lower_legs.jpg) under the GNU Free Documentation License. Image by Cornellier. Right hand image courtesy of Dr. Richard Bradbury, PhD

Diagnosis of cercarial dermatitis is usually made by the observation of characteristic clinical findings in conjunction with epidemiologic data, such as recent contact with fresh or brackish water.

7 Sparganosis

Sparganosis is a zoonotic cestode (tapeworm) infection caused by the larvae of members of the genus *Spirometra*. The epidemiology of species reported from humans can be challenging, as many of the identifications made by molecular studies may be based on incorrect data in public databases [43, 44]. Species implicated in human infection include *S. decipiens* (North and South America, Southeast Asia), *S. erinaceieuropaei* (Europe, Asia, Australia), *S. mansoni* (Europe, Africa, Asia, Oceania, South America), *S. mansonoides* (North America), and *S. ranarum* (Africa, Southeast Asia) [1]. However, it has been suggested that *S. erinaceieuropaei* and *S. decipiens* from Asia and *S. ranarum* from Africa and Asia might represent misidentifications of *S. mansoni* [43, 44]. The natural definitive hosts of *Spirometra* are carnivores, such as dogs and cats. Humans become infected after ingesting the proceroid larval stage of the cestode after drinking contaminated water containing the copepod first intermediate host or ingestion of the plerocercoid larval stage (sparganum) in raw or undercooked fish, amphibians, or reptiles that serve as second intermediate hosts. It has also been suggested that humans can acquire the sparganum from the use of frogs, snakes, or other intermediate hosts as a poultice applied to wounds or the surface of the eye [45]. After ingestion, the spargana undergo visceral migration and end up in various organs, including the skin and soft tissues, brain,

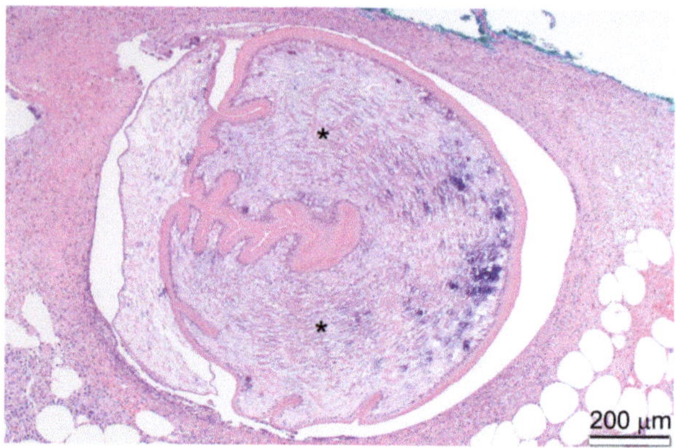

Fig. 9 Section of a sparganum (asterisks) within subcutaneous tissue with an associated acute inflammatory response

eyes, and lungs [45]. The most severe infections involve infection of the eyes and brain.

Cutaneous sparganosis usually presents as a slow-growing subcutaneous nodule, with or without pruritus, that is usually painless, but may be painful, and in some cases present with a hemorrhagic mass. Nodules may be migratory. Nodules can appear anywhere on the body, but are most common on the lower extremities, trunk, breast, and scrotum [46, 47]. In some cases, spargana may spontaneously emerge from sores or breaks in the skin [34]. Treatment for cutaneous sparganosis is surgical removal of the worm.

Diagnosis of cutaneous sparganosis is achieved by morphologic examination of an extracted sparganum or identification of spargana is biopsy specimens processed for histology (Fig. 9). When suspect spargana are observed in biopsy specimens, other larval tapeworms such as *Taenia solium* (cysticercosis) and *Echinococcus* should be considered.

8   Pseudoparasitism and Non-parasitic Organisms That May Be Submitted to the Clinical Parasitology Laboratory

There are a variety of organisms which are non-parasitic for the human host that are commonly submitted to clinical laboratories for identification. The definitive identification of these organisms (i.e., to the genus and species level) is often beyond the scope of most clinical microbiologists and parasitologists; however, it is important for those whose task is to identify parasites to recognize features that allow for true parasitism to be ruled out. Among some of the more common things submitted to laboratories are free-living fly larvae, earthworms, and horsehair worms [48]. Many of these organisms have aspects of their life cycle which make them likely to be found in houses, bathtubs, sinks, toilets, or latrines, which may give the casual observer the misconception that they were shed in feces, urine, vomitus, etc.

9 Free-Living Fly Larvae

Several fly larvae that breed in biofilms and organic waste and debris might be submitted to diagnostic labs after their finding in toilets, latrines, sinks, bathtubs, etc. Among the most common groups seen in the diagnostic laboratory are the drain flies (family Psychodidae) and rat-tailed maggots (family Syrphidae) [48]. A proper diagnosis can be hampered by the perpetuation in the literature of these flies causing myiasis, namely intestinal or urogenital myiasis, especially in the absence of physical evidence linking the flies to any pathologic findings [48].

Larvae of drain or sewer flies (Psychodidae) are probably the commonly submitted free-living fly larvae to the clinical laboratory. Members of this family are found worldwide. Their larvae (Fig. 10) develop in moist or shallow-water microhabitats, including plumbing drains in kitchens, bathrooms, and locker rooms. Despite being implicated in cases of myiasis, their mouthparts preclude them from being able to use human or other animal tissue as a nutritive source and their breathing apparatus is designed for capturing atmospheric air, something that would be quite challenging in the human alimentary canal or urogenital tract [49, 50].

Rat-tailed maggots are the larvae of hover flies. They get their common name from the possession of a long respiratory siphon that can be several times the length of the body (Fig. 11). This long siphon allows them to breathe while remaining submersed in water or other organic substrates, such as manure and sewage [51–53]. As with the psychodid flies, their presence in such places as toilets and latrines could lead to the false impression they were passed in stool or urine.

Fig. 10 Psychodidae larva. The presence of setae (hair-like projects) allows for easy differentiation of this insect larva from myiasis-causing fly larvae. Other features include a sclerotized head capsule (right end) and spiracular openings modified as extended tubes enclosed by a sclerotized cone (left end)

Fig. 11 Rat-tailed maggot. Note the characteristic long respiratory siphon at the posterior end of the larva. Available at Wikimedia Commons (https://commons.wikimedia.org/wiki/File:Rattail-maggot.jpg), public domain. Image by Brian Jones

10 Horsehair Worms

Horsehair worms (Nematomorpha) are a group of animals whose larvae are specialized parasites of arthropods such as insects. Adults are free-living in fresh or marine water, but the hosts are often terrestrial arthropods [54]. Because the horsehair worms must return to the water to complete their life cycle, some manipulate their insect host to drown itself, so the adult can emerge in the water [55]. As such, adult horsehair worms might be seen in standing water, such as toilets, bathtubs, or sinks if the host, for example, an anthropophilic cricket, grasshopper (Fig. 12) or cockroach, was to be compelled to jump into the water. They can be differentiated by human parasites by their very long, slender body.

Fig. 12 Horsehair worm emerging from a grasshopper. Note the characteristic long slender shape. Available through Wikimedia Commons (https://commons. wikimedia.org/wiki/ File:Horsehair_Worm_ (14629048952).jpg) through a Creative Commons Attribution-ShareAlike 2.0 Generic license. Image by Alastair Rae

11 Earthworms

Earthworms are ubiquitous free-living animals in soil, compost, and freshwater [56]. They are saprophytic, and there is no evidence that they cause pathological manifestations in humans [48]. On occasion, earthworms are submitted to the laboratory, usually when they are found in bathtubs or toilets. It is generally believed that their route to those sites is through cracks in pipes that are in direct contact with soil where the earthworms live.

12 Identification and Reporting of Non-Parasitic Organisms

In addition to the examples mentioned above, almost any anthropophilic arthropod that comes into one's home has the potential to being submitted to a clinical or diagnostic laboratory for identification. Some of these include, but are not limited to, stink bugs, booklice and barklice, ants, silverfish, and adults and larvae of carpet beetles. Why some patients feel compelled to bring these into their care providers is usually beyond the knowledge of the laboratory testing personnel. Many people are probably unfamiliar with the creatures they share their surroundings with and may have genuine concerns about whether the insect or arthropod is harmful. Others might be suffering from psychological disorders, such as Munchausen syndrome, whereby the patients are looking for care and attention. In cases of DI, specimens are often contaminated with material from the environment, and these can invariably include arthropods or parts thereof. While definitive identification of these organisms is often beyond the scope and ability of most diagnostic microbiologists or parasitologists, it is important for laboratory workers to at least be able to rule-out true human parasites.

Generally, it is discouraged from putting an identification, definitive or even general, on an organism that is not a true human parasite, so as not inadvertently create a link between the organism and any symptoms the patient may be experiencing. Instead, general verbiage for such cases includes "arthropod/insect not of medical importance," "organism does not resemble a known human parasite," "no parasites found," etc., is recommended. Ultimately, the degree of identification provided on the final laboratory report is under the purview of the laboratory director. If a health care provider wants a more definitive identification, even if for no other reason than to put the patient's mind at ease, specimens can be sent to specialists at an agricultural extension service, public health agencies (e.g., state public health entomologist), or a local university with an active entomology department and reference insect collection [48, 57].

13 Optimizing Collaboration Between the Primary Patient Care Team and Laboratory Experts

DI presents significant challenges for the patient, provider, and laboratorian alike. To ensure optimal laboratory evaluation for DI and true parasitism, the primary patient care team should contact the laboratory director responsible for parasitology testing as early into the evaluation as possible to discuss specimen collection, examination, report, and the differential diagnosis. Patients with DI frequently bring in multiple specimens for examination, and these usually consist of assorted non-parasitic materials (e.g., fibers, plant material, fragments of arthropods) in common household containers (e.g., small cardboard matchboxes; a.k.a., the "matchbook sign," plastic bags, glass canning jars, medicine bottles). Specimens are commonly accompanied by hand-written or typed notes from the patients with long descriptions of the "parasites" they are finding and the list of the symptoms they are experiencing. This large number of specimens can easily overwhelm the laboratory if not regulated. Thus, it is useful to ask the patient to select the specimens (e.g., 3–5) that they feel are most representative of what they have been observing. The patient can also be provided specimen containers with an appropriate fixative for collecting additional specimens. Providing specimen containers may help to assure the patient that his or her concerns are being taken seriously, while also ensuring that subsequent specimens are properly preserved for optimal laboratory examination [3]. Negative findings from one or more properly collected specimens from multiple specimen collections reduce the possibility of a true parasitic infection and may be indicated in certain circumstances [3].

If a true parasitic infection is suspected, then the laboratory director can provide information about available testing options and the types of specimens required for each test. In addition to parasitic infection, evaluation for other infectious and non-infectious conditions, such as HIV infection, syphilis, leprosy, tuberculosis, malignancies, endocrine disorders, neurologic conditions, nutritional deficiency, and ingestion of certain drugs or toxins, may be indicated [3]. The reader is referred to other chapters in this book for more information.

Finally, the laboratory director can discuss how laboratory findings will be reported and may consider if additional information in the report would be useful to the patient and primary clinician. One author (BSP) has received requests to provide a description of the material submitted, (e.g., no parasite present; object submitted most closely resembles plant material), which was felt to be helpful by the primary care team in documenting the findings.

14 Examples of Delusional Infestations, Pseudoparasitism, and True Parasitism

Case #1: Delusional Infestation

Case Presentation: A 48-year-old woman presented to a family medicine physician with a chief complaint of "worms coming out of my skin, eyes, and genitals." She also noted that parasitic insects had infested her house and were present in her heating ducts and tap water. She had been seen by multiple other family and internal medicine physicians over the past 6 months without relief and was seeking another opinion. She had brought a container of the objects she had retrieved from her body and house which contained what she thought were "parasitic larvae and pupae." Physical examination revealed numerous erythematous, ulcerated, and crusted lesions in various stages of healing on her extremities, trunk, neck, and face. Her previous medical history was unremarkable, and she was not currently taking any prescription or over-the-counter medications. She did not smoke or use any recreational drugs and had not traveled outside of the United States. She worked as a medical receptionist, was in a monogamous relationship with her husband, and did not have any pets. The physician ordered a number of laboratory tests and sent the specimen container to the clinical parasitology laboratory for evaluation.

Laboratory Findings: The specimen consisted of heterogenous organic and inorganic material resembling food particles, synthetic fibers, hair, plastic parts, and possible skin fragments/scabs (Fig. 13). The final laboratory report stated, "No parasite present."

Additional Investigation: After receiving the result of the parasitic evaluation, the referring physician contacted the laboratory director to seek guidance for further laboratory testing. He relayed that other tests were within normal limits, including a complete blood count, serum electrolytes, blood urea nitrogen, serum creatinine, and serum glucose. He planned to order additional testing to evaluate the patient for other possible medical conditions and inquired about the utility of submitting additional specimens for parasitic examination. The laboratory director suggested that one or two additional specimens could be submitted for parasitic evaluation and instructed the physician on how they should best be collected.

Final Interpretation: The rest of the evaluation did not show any evidence of parasitic infection or other condition that would account for the patient's symptoms. The physician ultimately concluded that the patient was suffering from delusional

Fig. 13 Specimen submitted as "parasitic larvae and pupae" that were collected from the patient's body and house

infestation based on the clinical history, physical exam findings, and laboratory results.

Clinical Outcome: The patient refused to accept the diagnosis but was willing to continue to see the physician for follow-up visits.

Comments: Delusional infestation is an extremely challenging condition to manage. The most important job of the clinical laboratory is to carefully examine the submitted specimens for actual parasites so that a true parasitic infection is not missed. If not, parasites are seen, the laboratory experts can partner with the primary clinical team to recommend additional testing, if indicated, and to provide guidance on how specimens should be collected and transported to the laboratory to ensure optimal specimen quality.

Case #2: Pseudoparasitism

Case Presentation: A 24-year-old woman presented to her primary care provider with concerns of parasitic infection after finding numerous live worm-like objects on a tampon she had recently removed. The objects were noted to be moving by the physician, who promptly submitted them to the laboratory for examination.

Laboratory Findings: Numerous phorid fly larva and pupae of varying sizes were identified within cotton fibers of a bloody tampon (Fig. 14). Pupal casings were also seen, indicating that at least one adult fly had hatched. Although a definitive identification was not obtained, the variability of larval size and shape was consistent with the presence of more than one genus.

Additional Investigation: Upon speaking with the patient's physician, the laboratory director learned that the flies were not seen in the body, but instead were found by the patient on a tampon that had been discarded in a waste basket several days beforehand.

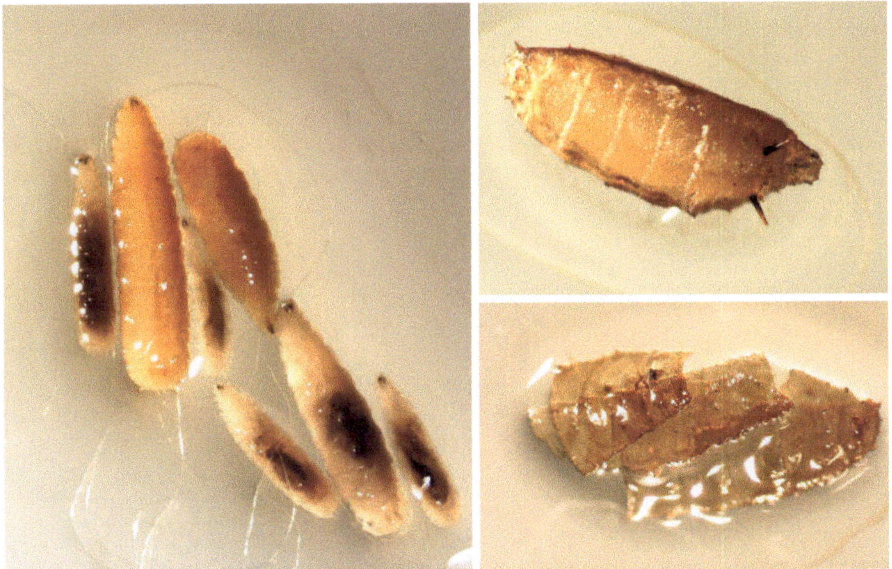

Fig. 14 Larvae (left), pupa (upper right), and pupal cases (lower right) obtained from a bloody tampon

Final Interpretation: The suspected presence of numerous phorid larval species, as well as pupae and pupal cases suggested that this was a case of environmental contamination (i.e., pseudoparasitism) rather than true parasitism. This was supported by the clinical information provided by the patient's physician.

Phorids (Diptera: Phoridae) are a family of than 250 genera commonly known as humpback or scuttle flies [58]. Many species are detritivores (i.e., consuming decaying organic matter) and are known to infest household trash receptacles. They have also been reported as rare causes of facultative wound myiasis [59].

Had this been a case of true parasitism/myiasis, the larvae would have been expected to be in the same stage (assuming a single inoculation event), and pupae and pupal casings would not have been present. Definitive diagnosis of vaginal/cavitary myiasis would have required definitive documentation of larvae in situ by the health care provider.

Clinical Outcome: The patient accepted the diagnosis and expressed relief with the explanation of the findings.

Comments: Cases of pseudoparasitism are commonly seen in the clinical laboratory. In most instances, the patient does not suffer from true delusional parasitosis (i.e., a psychiatric condition where people have an unshakeable belief that they are parasitized by bugs or worms) and can be persuaded to accept the diagnosis pseudoparasitism following expert laboratory evaluation. These are generally rewarding cases in that the diagnosis is accepted by the patient and alleviates the patient's concerns of parasitism.

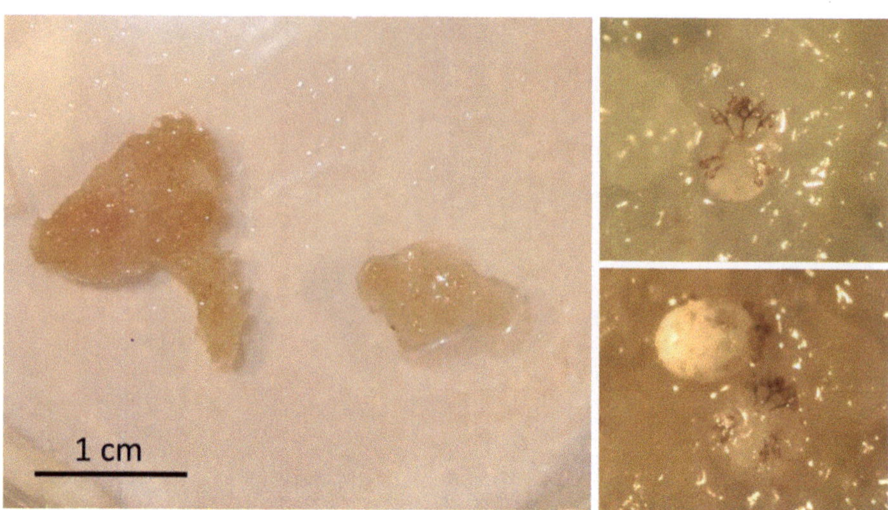

Fig. 15 Skin "scabs" (left) submitted from a patient with a chronic rash that did not resolve with topical steroid administration. Adult mites of *Sarcoptes scabiei* were identified by use of a dissecting microscope (righthand panels)

Case #3: True Parasitism Mimicking Delusional Infestation

Case Presentation: An 80-year-old man was noted to have an itchy rash that was not responding to topical steroid cream. The patient had dementia and limited mobility and resided in a skilled nursing facility. He had complained to his caregivers of "crawling sensations" on his abdomen and extremities for the weeks prior to medical investigation. Large regions of crusted, flaky skin were noted on physical exam, and several large crusts were submitted to the microbiology laboratory for examination (Fig. 15).

Laboratory Findings: Three tan-brown skin crusts measuring 0.3–1.5 cm in greatest dimension were received in the clinical parasitology laboratory (Fig. 15). Examination with a dissecting/stereo microscope revealed motile adult and nymphal *Sarcoptes scabiei* mites embedded within the skin. The identification was confirmed by examining scrapings obtained from the specimen by light microscopy. Abundant eggs and fecal pellets (scybala) were also identified. These findings are suggestion of a diagnosis of crusted scabies.

Additional Investigation: Given that crusted scabies is highly contagious, the parasitology laboratory director immediately contacted the skilled nursing facility where the patient resided to alert them of the diagnosis. The medical staff confirmed that the clinical presentation was consistent with the laboratory identification and indicated that several other patients had similar symptoms.

Final Interpretation: The clinical and laboratory features are consistent with crusted scabies (Norwegian scabies), a heavy infestation with the itch mite, *Sarcoptes scabiei* [60]. This form of scabies is most commonly seen in immuno-compromised and institutionalized patients and is characterized by the formation of thick crusts and vesicles over the body. Unlike the classical variant of scabies, itching may be minimal or absent. Crusted scabies is treated with both topical acaricides (e.g., 5% permethrin cream) and oral agents (ivermectin).

Clinical Outcome: The patient and other infected individuals at the facility received prompt treatment.

Comments: Outbreaks of crusted scabies have occurred among patients, staff, and visitors in institutional settings including long-term care facilities and hospitals. Debilitated individuals such as these patients are at increased risk of acquiring the crusted version of disease. Additionally, the prior use of steroid cream may have been a risk factor for the crusted variant as it produced localized immune suppression in this patient.

References

1. Mathison BA, Sapp SGH. An annotated checklist of the eukaryotic parasites of humans, exclusive of fungi and algae. Zookeys. 2021;1069:1–313.
2. Bradbury RS, Sapp SGH, Potters I, Mathison BA, Frean J, Mewara A, Sheorey H, Tamarozzi F, Couturier MR, Chiodini P, Pritt B. Where have all the diagnostic morphological parasitologists gone? J Clin Microbiol. 2022;60:e0098622.
3. Suh KN, Keystone JS. Delusional infestation (delusional parasitosis). In: Ryan ET, Hill DR, Solomon T, Aronson NE, Endy TP, editors. Hunter's tropical medicine and emerging infectious diseases. 10th ed. New York: Elsevier; 2020. p. 1132–6.
4. Stufano A, Foti C, Lovreglio P, Romita P, De Marco A, Lia RP, Otranto D, Iatta R. Occupational risk of cutaneous larva migrans: a case report and a systematic literature review. PLoS Negl Trop Dis. 2022;16:e0010330.
5. Caumes E, Danis M. From creeping eruption to hookworm-related cutaneous larva migrans. Lancet Infect Dis. 2004;4:659–60.
6. Caumes E. Treatment of cutaneous larva migrans. Clin Infect Dis. 2000;30:811–4.
7. Heukelbach J, Feldmeier H. Epidemiological and clinical characteristics of hookworm-related cutaneous larva migrans. Lancet Infect Dis. 2008;8:302–9.
8. Mathison BA, Pritt BS. Parasites of the gastrointestinal tract. In: Rezaei N, editor. Encyclopedia of infection and immunity, vol. 3. Oxford: Elsevier; 2022. p. 136–203.
9. Nutman TB. Human infection with Strongyloides stercoralis and other related Strongyloides species. Parasitology. 2017;144:263–73.
10. Meinking TL, Burkhart CN, Burkhart CG. Changing paradigms in parasitic infections: common dermatological helminthic infections and cutaneous myiasis. Clin Dermatol. 2003;21:407–16.
11. Arthur RP, Shelley WB. Larva currens; a distinctive variant of cutaneous larva migrans due to Strongyloides stercoralis. AMA Arch Derm. 1958;78:186–90.
12. Centers for Disease Control and Prevention. Parasites—Strongyloides; Resources for Health Professionals. 2022. https://www.cdc.gov/parasites/strongyloides/health_professionals/index.html.
13. Requena-Mendez A, Chiodini P, Bisoffi Z, Buonfrate D, Gotuzzo E, Munoz J. The laboratory diagnosis and follow up of strongyloidiasis: a systematic review. PLoS Negl Trop Dis. 2013;7:e2002.
14. Liu GH, Sun MM, Elsheikha HM, Fu YT, Sugiyama H, Ando K, Sohn WM, Zhu XQ, Yao C. Human gnathostomiasis: a neglected food-borne zoonosis. Parasit Vectors. 2020;13:616.
15. Herman JS, Chiodini PL. Gnathostomiasis, another emerging imported disease. Clin Microbiol Rev. 2009;22:484–92.

16. Campos DMB, Barbosa AP, Oliveira JA, Tavares GG, Cravo PVL, Ostermayer AL. Human lagochilascariasis-A rare helminthic disease. PLoS Negl Trop Dis. 2017;11:e0005510.
17. Orihel TC, Ash LR. Parasites in human tissues. Chicago: ASCP Press; 1994.
18. Tokiwa T, Ueda W, Takatsuka S, Okawa K, Onodera M, Ohta N, Akao N. The first genetically confirmed case of Dioctophyme renale (Nematoda: Dioctophymatida) in a patient with a subcutaneous nodule. Parasitol Int. 2014;63:143–7.
19. Brattig NW, Cheke RA, Garms R. Onchocerciasis (river blindness)—more than a century of research and control. Acta Trop. 2021;218:105677.
20. Mathison BA, Barnhill RL. Helminthic diseases. In: Barnhill's dermatopathology. New York, NY: McGraw-Hill; 2020. p. 696–708.
21. Cantey PT, Weeks J, Edwards M, Rao S, Ostovar GA, Dehority W, Alzona M, Swoboda S, Christiaens B, Ballan W, Hartley J, Terranella A, Weatherhead J, Dunn JJ, Marx DP, Hicks MJ, Rauch RA, Smith C, Dishop MK, Handler MH, Dudley RW, Chundu K, Hobohm D, Feiz-Erfan I, Hakes J, Berry RS, Stepensaski S, Greenfield B, Shroeder L, Bishop H, de Almeida M, Mathison B, Eberhard M. The Emergence of Zoonotic Onchocerca lupi Infection in the United States—a case-series. Clin Infect Dis. 2016;62:778–83.
22. Centers for Disease Control and Prevention. Parasites—Onchocerciasis (also known as River Blindness). 2019. https://www.cdc.gov/parasites/onchocerciasis/treatment.html.
23. To KK, Wong SS, Poon RW, Trendell-Smith NJ, Ngan AH, Lam JW, Tang TH, AhChong AK, Kan JC, Chan KH, Yuen KY. A novel Dirofilaria species causing human and canine infections in Hong Kong. J Clin Microbiol. 2012;50:3534–41.
24. Winkler S, Pollreisz A, Georgopoulos M, Bago-Horvath Z, Auer H, To KK, Krucken J, Poppert S, Walochnik J. Candidatus dirofilaria hongkongensis as causative agent of human ocular filariosis after travel to India. Emerg Infect Dis. 2017;23:1428–31.
25. Orihel TC, Eberhard ML. Zoonotic filariasis. Clin Microbiol Rev. 1998;11:366–81.
26. Otranto D, Eberhard ML. Zoonotic helminths affecting the human eye. Parasit Vectors. 2011;4:41.
27. Koehler AV, Robson JMB, Spratt DM, Hann J, Beveridge I, Walsh M, McDougall R, Bromley M, Hume A, Sheorey H, Gasser RB. Ocular Filariasis in Human Caused by Breinlia (Johnstonema) annulipapillata Nematode, Australia. Emerg Infect Dis. 2021;27:297–300.
28. Lam HH, Subrayan V, Khaw G. Intraocular filarial infection. Eye (Lond). 2010;24:1825–6.
29. Bain O, Otranto D, Diniz DG, dos Santos JN, de Oliveira NP, Frota de Almeida IN, Frota de Almeida RN, Frota de Almeida LN, Dantas-Torres F, de Almeida Sobrinho EF. Human intraocular filariasis caused by Pelecitus sp. nematode, Brazil. Emerg Infect Dis. 2011;17:867–9.
30. Hopkins DR, Weiss AJ, Yerian S, Sapp SGH, Cama VA. Progress Toward Global Eradication of Dracunculiasis—Worldwide, January 2021-June 2022. MMWR Morb Mortal Wkly Rep. 2022;71:1496–502.
31. Eberhard ML, Ruiz-Tiben E, Hopkins DR, Farrell C, Toe F, Weiss A, Withers PC, Jenks MH, Thiele EA, Cotton JA, Hance Z, Holroyd N, Cama VA, Tahir MA, Mounda T. The peculiar epidemiology of dracunculiasis in Chad. Am J Trop Med Hyg. 2014;90:61–70.
32. Thach PN, van Doorn HR, Bishop HS, Fox MS, Sapp SGH, Cama VA, Duyet LV. Human infection with an unknown species of Dracunculus in Vietnam. Int J Infect Dis. 2021;105:739–42.
33. Cairncross S, Muller R, Zagaria N. Dracunculiasis (Guinea worm disease) and the eradication initiative. Clin Microbiol Rev. 2002;15:223–46.
34. Eberhard ML, Thiele EA, Yembo GE, Yibi MS, Cama VA, Ruiz-Tiben E. Thirty-Seven Human Cases of Sparganosis from Ethiopia and South Sudan Caused by Spirometra Spp. Am J Trop Med Hyg. 2015;93:350–5.
35. Cnops L, Huyse T, Maniewski U, Soentjens P, Bottieau E, Van Esbroeck M, Clerinx J. Acute schistosomiasis with a schistosoma mattheei x schistosoma haematobium hybrid species in a cluster of 34 travelers infected in South Africa. Clin Infect Dis. 2021;72:1693–8.
36. Oleaga A, Rey O, Polack B, Grech-Angelini S, Quilichini Y, Perez-Sanchez R, Boireau P, Mulero S, Brunet A, Rognon A, Vallee I, Kincaid-Smith J, Allienne JF, Boissier

J. Epidemiological surveillance of schistosomiasis outbreak in Corsica (France): are animal reservoir hosts implicated in local transmission? PLoS Negl Trop Dis. 2019;13:e0007543.

37. Colley DG, Bustinduy AL, Secor WE, King CH. Human schistosomiasis. Lancet. 2014;383:2253–64.

38. Mota Lde S, Silva SF, Almeida FC, Mesquita Lde S, Teixeira RD, Soares AM. Ectopic cutaneous schistosomiasis—case report. An Bras Dermatol. 2014;89:646–8.

39. Mandili SY, Alkhotani N, Alem R, Filfilan WM. Case Report of Schistosomiasis Complicated with Perianal Fistula. IDCases. 2022;27:e01397.

40. Centers for Disease Control and Prevention. Parasites—Schistosomiasis. 2020. https://www.cdc.gov/parasites/schistosomiasis/health_professionals/index.html.

41. Horak P, Mikes L, Lichtenbergova L, Skala V, Soldanova M, Brant SV. Avian schistosomes and outbreaks of cercarial dermatitis. Clin Microbiol Rev. 2015;28:165–90.

42. Centers for Disease Control and Prevention. Parasites—Cercarial Dermatitis (also known as Swimmer's Itch). 2020. https://www.cdc.gov/parasites/swimmersitch/health_professionals/index.html.

43. Kuchta R, Kolodziej-Sobocinska M, Brabec J, Mlocicki D, Salamatin R, Scholz T. Sparganosis (Spirometra) in Europe in the Molecular Era. Clin Infect Dis. 2021;72:882–90.

44. Yamasaki H, Sanpool O, Rodpai R, Sadaow L, Laummaunwai P, Un M, Thanchomnang T, Laymanivong S, Aung WPP, Intapan PM, Maleewong W. Spirometra species from Asia: genetic diversity and taxonomic challenges. Parasitol Int. 2021;80:102181.

45. Wiwanitkit V. A review of human sparganosis in Thailand. Int J Infect Dis. 2005;9:312–6.

46. Kazemi A, Awosika O, Burgess C. Subcutaneous Sparganosis of the Breast. J Clin Aesthet Dermatol. 2018;11:26–7.

47. Liu W, Gong T, Chen S, Liu Q, Zhou H, He J, Wu Y, Li F, Liu Y. Epidemiology, diagnosis, and prevention of sparganosis in Asia. Animals (Basel). 2022;12:1578.

48. Mathison BA, Pritt BS. Laboratory identification of arthropod ectoparasites. Clin Microbiol Rev. 2014;27:48–67.

49. Williams FX. Biological studeis in Hawaiian water-loving insects, Part III Diptera or Flies—C, Tipulidae and Psychodidae. Proc Hawaiian Entomol Soc. 1944;11:313–38.

50. Vaillant F. Psychodidae-Psychodinae. In: Lindner E, editor. Die Fliegen der Palaearktischen Region Lieferung 287. Stuttgar: E. Schweizerbart'sche Verlagsbuchhandlung; 1971. p. 1–48.

51. University of Florida Institute of Food and Agricultural Sciences. Common name: drone fly, rat-tailed maggot. 2018. https://entnemdept.ufl.edu/creatures/livestock/rat-tailed_maggot.htm.

52. Townsend L. Rat-Tailed Maggots and Moth Flies. https://entomology.ca.uky.edu/ef500#:~:text=The%20elongate%20larvae%20live%20in,about%202%20to%203%20weeks.

53. Abass OA, Haggag AA, Badawy RM, Essa EE. Microbial Stress Resistance of Eristalis tenax Rat-Tailed Maggots. Egypt Acad J Biol Sci. 2020;13:275–82.

54. Hanelt B, Thomas F, Schmidt-Rhaesa A. Biology of the phylum nematomorpha. Adv Parasitol. 2005;59:243–305.

55. Thomas F, Schmidt-Rhaesa A, Martin G, Manu C, Durand P, Renaud F. Do hairworms (Nematomorpha) manipulate the water seeking behaviour of their terrestrial hosts? J Evol Biol. 2002;15:356–61.

56. Edwards CA, Bohlen PJ. Biology and ecology of earthworms. London, UK: Chapman & Hall; 1996.

57. Telford SR III, Mathison BA. Arthropods of medical importance. In: Carroll KC, Pfaller MA, Landry ML, McAdam AJ, Patel R, Richter SS, Warnock DW, editors. Manual of clinical microbiology. 12th ed. Washington, DC: ASM Press; 2019.

58. Merritt RW, Courtney GW, Keiper JB. Diptera: (Flies, Mosquitoes, Midges, Gnats). In: Resh V, Carde R, editors. Encyclopedia of insects. Cambridge, MA: Academic Press; 2009. p. 284–97.

59. Sherman RA. Wound myiasis in urban and suburban United States. Arch Intern Med. 2000;160:2004–14.

60. Centers for Disease Control and Prevention. Scabies. 2018. https://www.cdc.gov/dpdx/scabies/index.html.

A Matter of Time. Somatic Illness and Evolved Delusion. How Time Plays a Critical Role in Delusional Infestation

Gale E. Ridge

From the late 1900s to the present, Delusional Infestation (DI) has evolved from a rarely reported disorder to one that is quite common. Most cases result from under-lying undiagnosed medical conditions with an overlay of psychiatric investment which is protective against differential diagnosis. As time progresses, the psychiatric component evolves into an unshakable belief system that isolates patients from society and in some cases sends them into years of unnecessary suffering. Time is an important contributing element in the development of this disorder.

Delusional infestation (DI) has reality misidentified. Anxiety driven inaccurate parasite/pathogen descriptions by patients told to unsuspecting diagnosticians can be confounding. Many diagnosticians unwittingly are drawn into their patients' illusions because of a lack of knowledge and information about the disorder. DI is a multisystem condition. Apart from "meeting criteria for schizophrenia or another major mental illness," in most patients, it starts as an undiagnosed somatic disorder which evolves to include potent psychiatric commitment. This is Secondary DI. What patients experience is reality to them but delusion to us. The gulf between the two is what makes DI difficult to manage.

The German psychiatrist Jaspers called the disorder "Wahnarbeit." A literal translation is, "delusional work." It is when a patient tries to make sense of the world around them by interpreting memories, perceptions, and events that fit their delusional "themes." Patients reinterpret information and evidence that would not normally fit by making improbable associations and connections of them and tailoring these into their beliefs. These often highly intelligent articulate people select unlikely explanations over likely ones: house debris becomes "parasites," skin flakes become "eggs" and balls of lint, "cocoons."

G. E. Ridge (✉)
Department of Entomology, The Connecticut Agricultural Experiment Station,
New Haven, CT, USA
e-mail: gale.ridge@ct.gov

© The Author(s), under exclusive license to Springer Nature
Switzerland AG 2024
G. E. Ridge (ed.), *The Physician's Guide to Delusional Infestation*,
https://doi.org/10.1007/978-3-031-47032-5_13

In the late 1990s, I saw my first DI case. I had no prior experience or training with DI and there was nothing to corroborate the patient's claims since no arthropods were found. Patient numbers from thence forward increased. Over time their accounts, disorders, and sampling coalesced into a series of common themes. These themes include self-mutilation, repeated use of descriptors such as "biting, crawling, stinging, running sensations" along with particularly repeated use of the subject pronoun "they" and the object pronoun "them." Patients readily described bizarre biology as well as shared delusions, depression, stress, anger, lack of sleep, defensiveness while seeking affirmation, fear, loss of control, isolation, and breaking with their delusions but then retreating to them. Many knew they were perusing faulty thinking, but to own it would be unbearable. To admit "regret" can be daunting.

Following the 2007–2009 American recession, there was a considerable increase in patients seeking help. The type of patient changed. Before 2007, patients were principally older women who experienced menopause, hyper- and hypothyroidism, and/or use of legal or illegal drugs. Six to nine months after the beginning of the recession (December 2007), younger patients started to seek help. They had lost their jobs, businesses, suffered major social and financial dislocation or were forced to migrate for work abandoning settled situations with family and friends.

During the Connecticut fiscal year July 1, 2009–June 30, 2010, there was a 30% increase in DI cases. Since then, numbers have continued to rise (Table 1). In 2021, case numbers were exceptionally high. This was likely the result of the SARS-CoV-2 (COVID-19) pandemic. Many patients reported stress and isolation as contributing factors. In 2022, numbers dropped but resumed the established upward trend from years prior to the 2021 outlier. Observations show there is approximately a 6- to 9-month time lag between the beginning of negative societal events and DI patients starting to call for help.

Most patients I saw fall under secondary DI while primary DI in my experience is less common. Most are continuously anxious, worried, and tired. Often their opening words to me are "I'm desperate" and "I'm not mad/crazy." Most are correct! They are isolated, sleep deprived, they trust no one and depend on their own instincts and intuition which may have become faulty (Box 1). Winning patient trust takes time. It's not immediate and not easy [1]. Working with patients takes weeks to years of regular or intermittent interactions. My job is to exhaust all entomological options, then encourage and guide patients to their medical providers for help.

Follow-ups and interactions with physicians are regular. Physicians also need guidance since DI is not a routine condition. In my experience, patients need at least 18 months of physician follow-up care before the psychological and emotional components of the disorder are fully addressed.

Table 1 Delusional infestation (DI) case records between 1996 and 2022. The Connecticut Agricultural Experiment Station, Insect Information Office

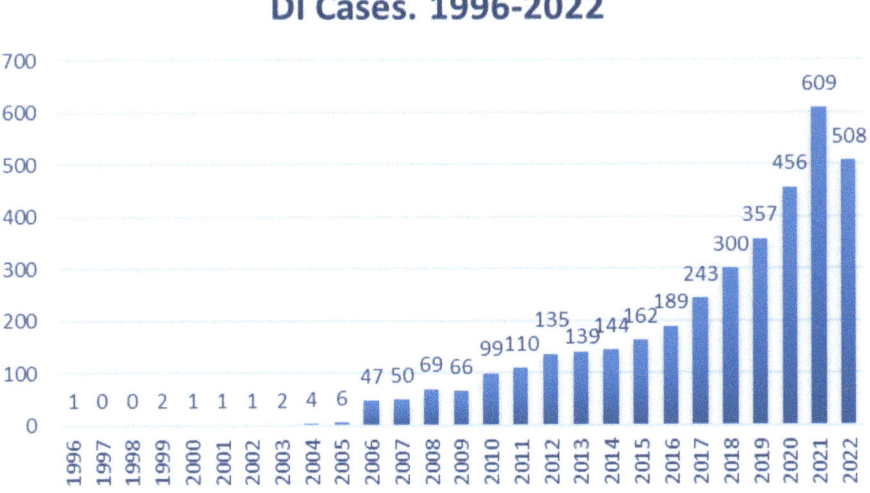

Box 1. Why Do DI Patients Seek Help and Then Don't Listen?
The problem with intuition vs reason without wisdom and how physicians and patients talk past each other.

Humans are fundamentally intuitive and not rational. At several levels, humans are innately ill-equipped to listen to reason. It's not that people don't work with reason, they do, but for many their arguments aim to support their own conclusions. Reason doesn't work like Lady Justice with her blindfold and scales who impartially weighs evidence and guides to wisdom. Reason for most people works as a lawyer or lobbyist by justifying our acts and judgments to others. Thus, reason is a viewpoint, evolved to help spin ideas, not help us learn. Reason is in fact subordinate to something deeper that lies in our underlying moral intuition whose conclusions reason serves and defends. This underlying moral intuition includes fairness, care, loyalty, liberty, sanctity, faith, and authority. These are moral pillars that carry weight and guide reason. They are the bedrock on which the unspoken tension of unreflective intuition and self-serving reason walk hand in hand. None of this can be fixed because it is normal. Only the power of experience, particular training, and ensuing wisdom may transcend these innate natural tendencies.

In the fields of science, a scientist's moral intuition must be left at the door. They cannot insert their own interpretation into research. It would be called speculation which leads to bias. Scientific thinking is curiosity driven by asking questions, testing hypotheses, observations, results, and logical conclusions based purely on generated data. Reason is managed by the intellect, not feelings. It is engendered from years of training and discipline. Outside the laboratory, scientists may discuss, interpret, argue, and entertain irrational intuitive thinking over evening drinks, but in the laboratory, it's "down to business."

And so, when a DI patient approaches a physician or expert for help both parties often talk past each other following completely different communication agendas. This is because there is a lack of understanding on both sides about how intuition and reason are working for each party. It results in the emotional intuition of the patient flaming out, while the physician or expert failing to understand why their patient isn't listening. In order to effectively communicate with these patients, physicians need to have the time, be understanding, sympathetic, and contemplative. As Plato once wrote:

"The greatest mistake in the treatment of disease is that there are physicians for the body and physicians for the soul, although the two cannot be separated."

Prophetic when working with DI patients. DI patients need emotional (psychiatric) support *and* somatic healing if they are ever to have any chance of becoming well.

Box 2. Core Cognitive Distortion

Core cognitive distortion is biased thinking that causes negative feedback loops which increase anxiety and depression. Some of these thought processes include:

Mental filtering: Selecting negatives and ignoring positives.

All-or-nothing, polarized thinking: Self-evaluation in the extremes often described as black and white thinking.

Overgeneralization: Based on a single incident, drawing a general conclusion about self-worth, ability, or performance.

Discounting the positives: "Mental illusion" where positive experience is transformed into negative experience.

Jumping to conclusions: Selecting a negative conclusion that is not supported by the facts. Putting yourself down.

Labeling: Biased overgeneralization of identity, e.g., "He's a drunk" or "I am stupid."

Emotional reasoning: Backwards thinking where emotions are falsely taken as evidence of truth. "I feel like a failure so I must be one."

Minimizing: Errors, fears, or imperfections are exaggerated while personal strengths and achievements minimized.

Mind reading: Assuming what others are thinking.

Personalization and circumstance blaming: A deflating tendency to interpret negative external events onto oneself in the form of blame. "Life has got it in for me" or "It's my fault."

Should and ought statements: Internalized negative expectations which may not be appropriate for the individual.

Catastrophic thinking: The assumption of the worst where ordinary worries escalate.

Fallacy of change: Expectation that others will change their behaviors to suit the thinker's needs and expectations.

Always being right: Arguing to prove you're right while discounting the facts and other people's feelings.

Confirmation bias: Selecting information to support a preconceived idea and/or belief.

Dunning-Kreuger effect: Tendency for uneducated/unskilled individuals to overestimate their abilities, while highly skilled/educated individuals underestimate their abilities.

1 DI Is a Common Disorder

Results from a paper written by Freudenmann et al. [2] which surveyed attendees at the 2009, 13th Congress of the European Society of Dermatology and Psychiatry (ESDaP) suggest that DI is common. In a more focused study, a Mayo Clinic survey of patient medical records from Olmsted County, Minnesota suggests 1.9/100,000/year suffer from various forms of DI [3]. Another paper described 147 DI patients who were seen at the Mayo Clinic over a 7-year period from 2001 to 2007 [4]. The patients represented several professions which included nurses, teachers, mechanics, hair stylers, office personnel, warehouse workers, engineers, caterers, auditors, professional drivers, construction workers, medical professionals, pharmacists, social workers, landscapers, and scientists [4]. Many had comorbid conditions which included hypertension, hypothyroidism, hyperlipidemia, pain, GERD, diabetes mellitus, asthma, anemia, fibromyalgia, hepatitis, IBS, peripheral neuropathy, fatigue, sleep apnea, cancer, degenerative joint disease, arthritis, or osteoarthritis. This suggests that DI is more common than currently believed.

Some patients experience transitory symptoms lasting a few days while others linger for years. The transitory clients may be many but are quickly cured by physicians who may not have been aware they were dealing with DI. These physicians would have identified particular medical conditions with attendant dermal or enteric symptoms, addressed them and promptly found a cure. In these cases, there would be no medical records or "doctor notes" referencing DI because a) physicians were ignorant of DI and b) even though there may have been verbal references of parasitism made by patients, it was ignored.

Throughout the interview process, I am acutely aware of working bimodally. This is where the psychological and somatic issues are combined in the thinking of the patient. They are not separate entities but one and directly talking a patient out of their concerns and behaviors will not work.

Thus, duality of the psychological and somatic illness working together has been described as a "Psychosomatic Disorder" [5]. Yet psychosomatic is not referenced in the APA DSM-5 or the ICD and is not formerly defined in standardized definitions or classifications. Psychosomatic disorders are when mental disorders are caused or exacerbated by underlying somatic disorders or illness. DI in many patients starts as an underlying undiagnosed medical condition that gains a psychiatric component over time. The two then feed off each other and become comingled. Additionally, all long-term patients exhibit some form of core cognitive distortion [1]. This is troublesome when patients seek help from physicians and don't listen because "they can't." (Box 2).

2 Clues

Patients hand over clues. One dominant trait with DI patients is that they often have "chattering minds." There's a determination to prove what they have is real and they are "not mad/crazy." Body language is often rigid, and narratives generated from feelings not facts and memories are selected for that purpose. Some patients develop habits of intense face reading waiting for a glance of negativity from the person they are talking with which can cause an immediate withdrawal into a defensive position. It is common for patients to invent novel parasite biologies and deeply invested clients often push back because of their desire for validation. Their need for validation has replaced the need for genuine information even though they say they are "desperate." It is important to note that many patients I see when redirected to other subjects other than the focus of their disorder often lighten up and conversations become healthy and normal. I use that to build on patient's strengths through their work and interests. It develops a relationship.

Cognitive dissonance, anger, frustration, physical mutilation, self-medication, lying, hyper-cleaning, social isolation, and relocation caused by a belief that their persecutors will be passed on to others, as well as fleeing hot zones such as perceived "infested homes" to vehicles or other safe refuges are common. Volume of samples is usually high. Often there is no thought of any danger in handling what is perceived as dangerous. I have seen patients happily reach into full bags of "infested samples" without any concern and present them, to quote Peter Lepping, "as trophies."

Paranoia and complex medical histories are prevalent in this population. Never once is depression mentioned by patients, but sometimes descriptions of stress are. Stress can affect several common skin conditions such as psoriasis, acne vulgaris, lichen simplex chronicus, seborrheic dermatitis, and atopic dermatitis. Often patients admit to underlying medical disorders managed by drugs. Others have no idea. All this confounds many medical professionals particularly when patients lament their woes and describe the "magical" biologies of their parasites. In more severe cases, allies are recruited such as nurses, caregivers, and/or family members and friends (Folie à deux, Folie à trois, Folie à familiar). They express paranoia while denying that they have mental health issues. These are haunted people.

I as an entomologist confirm or eliminate real arthropods. This requires fluency in arthropods of medical importance. Fortunately, there is a short list of medically significant arthropods in the Northeast United States where I work. These include head and body lice, bed bugs, several species of mites, biting flies, ticks, spiders, insects with urticating hairs, and thrips. In other parts of the world, the list is different.

3 Patients and Their Physicians

Many DI patients experience what they perceive as physician failure. This does not mean all practicing physicians miss DI cues, but some do. Avoidance and fear of physicians are common in this group. Numerous patients report negative experiences with physicians who exhibit lack of sensitivity, patience, and concern. For example, one patient's boyfriend was told by their physician, "She's nuts...... leave her!"

There is a tacit social code where patients feel subordinate and deferential to their physicians.

When they see their physician, they have hope for a cure. For whatever reason if their physician is perceived as failing them, these patients withdraw. Everything is real to them, and in their minds, they have been left to "discover" reasons why they are sick and develop a defensive attitude based on evolved cognitive dissonance. As they continue to struggle, the level of confirmation bias and investment of energy and resources increases. The amount of emotional weight they bear slowly overwhelms them and they sink into a self-inflicted "Swamp of despair and sadness." It can be very difficult to climb out, particularly when fix ideation and loyalty to initial erroneous information become hardwired into their minds.

One major clue a patient is in trouble is "migration behavior" better known as "doctor shopping" or "doctor hopping." Patients work their way through medical disciplines, other professional services, and close contacts. These include family, friends, family physicians, dermatologists, university extension personnel, hospital emergency departments, pest management professionals aka. pest control officers/ pest control operators, health departments, state and local social services, veterinarians, epidemiologists, dermatologists, psychodermatologists [6–12], and occasionally mental health professionals. But nothing works for them because their intuition constantly misleads them (Box 1).

It is common for patients to report positive scabies diagnoses with treatment by physicians who did not perform due diligence. Reasons for this range from physician indifference, a physician's fear of catching scabies, lack of medical entomological knowledge, guessing, over-reliance on medical textbooks, or the belief that a simple procedure to excise a mite from its burrow and slide mount it is time consuming and difficult. Diagnosis of scabies is easy and quick (see Chap. 13, Box 3 for scabies diagnosis instructions). A scabies diagnosis particularly with DI patients should never depend on only visual dermal surface features since many patients have already damaged their skin using chemical cleaners, pesticides, physical excoriation, and over-bathing. These injuries often mimic scabies infestations. A repeated theme from patients was that their physicians were willing to play the "placebo card" to placate their difficult time-consuming patients and get them "out the door." Examinations were cursory followed by scripts ordering permethrin creams and/or oral Ivermectin when no mites were found.

Poor judgment and glib comments made by physicians can result in catastrophic consequences for their patients. For example, one patient wrote:

> "At a visit around 3 years ago, the doctor told me there was nothing he could do and then leaned slowly forward in his chair and whispered in my ear, "perhaps you have Morgellons.""

He'd used the euphemism Morgellons - a subset of DI - as a suggested diagnosis and her trust of him sent her into years of subsequent suffering. His iatrogenic comment put an idea into her head, and she latched onto it.

4 Talking with Patients

I have noticed with DI patients that a multitude of experiences may build up until it becomes overwhelming which most people would not notice. An example of this is "black specks." A DI patient might see a small black speck in an unexpected location. Being guarded, they start to observe more. This builds until black specks are seen everywhere. For them the situation goes out of control. To find answers they search everywhere, because of the need to control the tangibles. With a perceived absence of control, patients resort to activities such as skin "picking" or repetitive behaviors seen as ritualistic cleaning. This provides a level of comfort for them.

Knowing how to ask questions can provide a torrent of information. I am also careful to respond to what patients say and to not react. I also make it understood that I am not responsible for a patient's behavior. An example of this is usage of cell phones in my presence. Many patients have their heads down swiping and tapping their phones. They are distracted in part because of insecurity, fear, anxiety, and a desire to prove a point by finding definitive pictures. They are distracted. To stop this, I either say I need to leave and deal with another matter and come back when "you are ready," or

> "I want your full attention, because what we are doing is important so please leave the room, finish what you need to do with the phone, and come back when you are ready."

What's in a Word?
The English language in certain respects is "simple" because many single words are asked to do a great deal of work. One of these is the easy to pronounce, one syllable word, "bite." This single word may be used to describe an entire spectrum of uncomfortable dermal sensations because there are no other choices in the English lexicon. With DI, the word is much like a large sledgehammer used to crack an egg. It's clumsy with a great deal of weight and when mis-directed can cause a great deal of damage both in information gathering and emotional cost. In many respects in the DI disorder, this word can tell all but say nothing. I see it as a cry for help, something is wrong and it's up to the physician to figure out what actually is going on. Treat the word as a red warning light on which to build an inquiry.

This puts full responsibility onto the patient, provides a boundary, and I don't have to fight my way through a phone's distractive chatter. Usually this results in the phone being put down.

To gain a comprehensive view of a patient's history, ask basic questions. Basic questions are important. They provide vital information and need to be asked in a particular way to avoid reinforcing a patient's belief system. The technique of reflecting or mirroring patient language back at them helps them hear their own words and can be helpful. Usually, it calms worries and concerns about what I might be thinking of them because I am repeating their words to them. I stay with what they have given me. This normalizes patient thoughts and builds trust.

As my time with patients increases and they begin to trust me, I start to insert alternative language. I'll start using "pinching sensation" over "bite, biting, or bitten" and encourage them to do the same. This redirects thinking to a less emotionally potent place and many respond. Storytelling is important with these patients. I tell stories about other patients I have seen who have had similar experiences. It makes them feel less alone. I also insert the concept of dealing with a mystery and "We" are detectives working together. By doing this I join with patients and open doors to considering other possible causes for their discomfort. Detectives look at everything to solve unsolved crimes. This concept is usually well accepted by patients.

With highly fixated patients once I have cleared them of parasites, I ask them to follow some activities. It helps them because they are actively doing something rather than being passive. It also plants seeds in their minds by directing them to alternative causes while avoiding direct conflict. Then I let it go and allow their incessant ruminations to do the rest of the work.

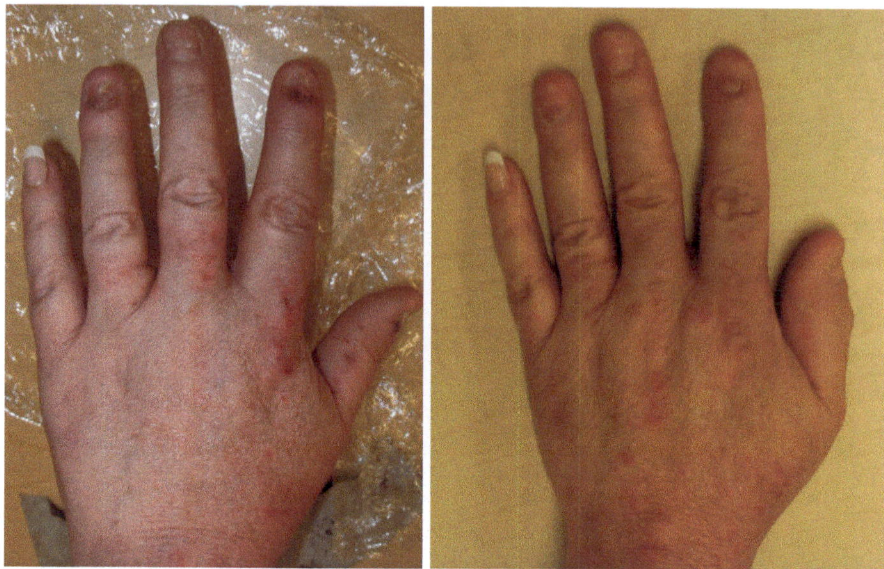

Fig. 1 **Left image.** Hand before a week of wearing protective cotton gloves at night and daytime cessation of picking. **Right image.** Hand after a week wearing protective cotton gloves at night and daytime cessation of picking

5 Picking Proof

If a patient is constantly picking at their skin I suggest to them that we follow a particular activity. Together, we take photographs of their lesions. Then I ask them to wear cotton gloves for one week at night and to not touch their injures during the day. A partner or family member may also be conscripted to assist. If patients follow my instructions, the injuries start to heal. I then have the patients return after a week and we take a second comparative set of photographs. Then the two sets of photographs are compared and discussed. By seeing the difference, I had a number of patients begin to think differently about their conditions (Fig. 1).

One of the hardest aspects of the condition for patients to deal with is to "own" their disorder. For many the difficulty of "unbelieving" a belief can almost be insurmountable. It can be done through normalizing their experience, but it takes a great deal of time. How they reached this place of DI isolation and angst took time and to roll back that time can take even longer. It is not to be underestimated.

6 A Matter of Time

Time is generally not considered a significant element in the practice of medicine other than for the identification of when a medical issue started or as a relevant part of a medical history investigation. DI is different. Time plays a significant role, particularly in the psychiatric development of the disorder. As time passes, the psychiatric component evolves from concern to a highly destructive form of emotional commitment and affirmation (Fig. 2). Additionally, patients can become addicted to their pain. In some patients, follow-up visits with neurologists reveal a lack of pain receptors in the skin while with others cortisone, endorphin, dopamine, or serotonin released by the body lessens the pain. Thus, assessing the length of time a patient has been in difficulties and the degree of their emotional investment is very important.

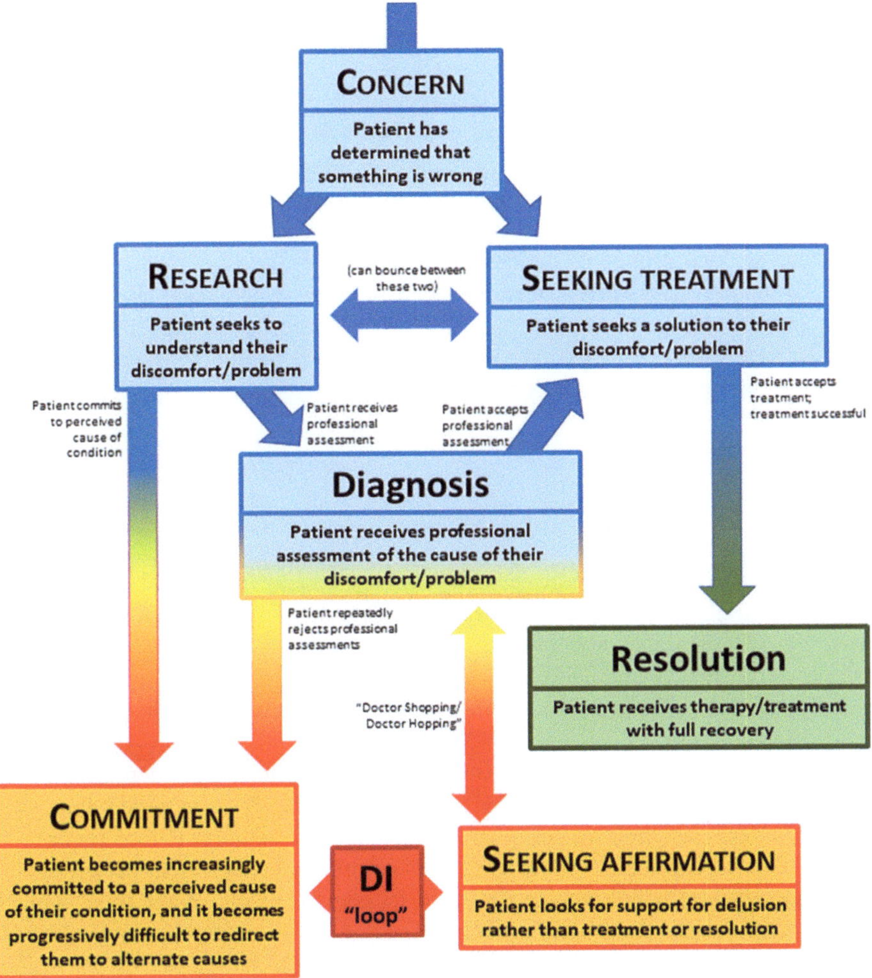

Fig. 2 Timeline of DI development from the patient's perspective. (Design by K. Dugas)

Romanov et al. [13] observed:

> … "longer duration of untreated psychosis is associated with poorer outcomes for patients with DI."

Lepping et al. [14] agreed. They wrote:

> …"suggest that longer duration of untreated psychosis in patients with DI is associated with significantly less favorable clinical outcomes."

In DI, often a patient's initial experience is seen as a novel worrying situation and adapting to accommodate this is a natural act of survival. The brain's circuitry has a built-in plasticity to learn new things, and new pathways of thought can be built. These evolve into habits, and the adaptive learning establishes to the point of excluding earlier experience [15].

In the early stages of DI, there is a separation of thinking associated with the disorder and other normal behaviors and responses. As a metaphor, DI might be seen as a cyst inside healthy tissue, isolated yet integrated. As time progresses, the cyst-like DI ideation metastasizes and the harder it becomes to treat. Early intervention is critical when caring for suspected DI patients.

There are two groups of patients, and both are responsive to the influence of time. Group one is less-invested patients because they experienced DI ideation for a short period of time, less than 6 months. The second group are more invested because their DI ideation is greater than 6 months. When patients reach and pass the 6-month tipping point, they become much harder to treat.

7 Less than 6 Months

Patients who experience symptoms of DI for less than six months exhibit worry and interest in what might be done to find a cure. They are concerned and seek information, they question, they ask for medical help, they search on the internet, and their conversations are mutual and evenly shared between practitioners or other professionals and themselves. Conversations are generally back and forth; trust is strong and optimism high.

If patients seek help before the 6-month tipping point, they often can be redirected. As time progresses, there is a shift in attitude. Trust in professional help erodes. Patients feel their concerns have not been met, they begin to look for answers that fit an evolving belief system as serious emotional commitment deepens. They begin to self-treat. Concerns give way to increased anxiety. After three months, there is an increase of self-harm, often accompanied by excessive cleaning of self and living space. Denial and self-isolation develop (Fig. 3).

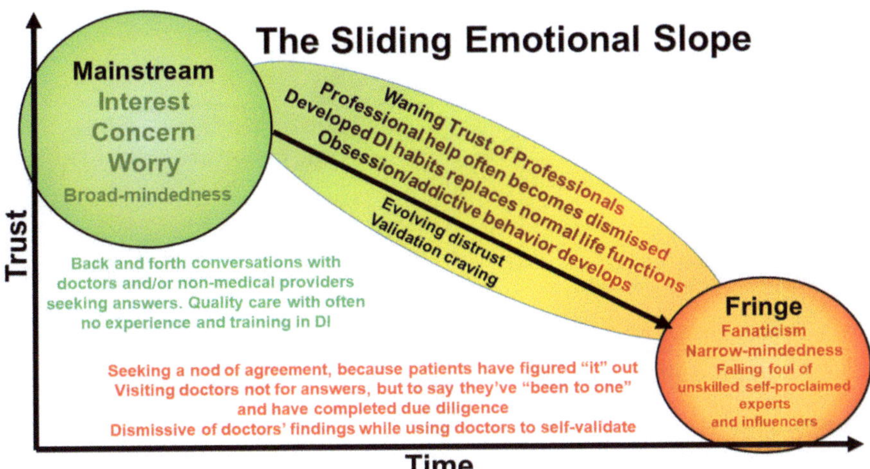

Fig. 3 Psychiatric investment time scale. As time progresses, patients become increasingly invested. (Design by K. Dugas)

8   Greater than 6 Months

Patients who experience symptoms of DI for more than six months exhibit increased psychological investment, questioning changes into a hunt for validation, and the search for professionals who they think can help them widens. Physicians are used to confirm beliefs over providing answers. Those professionals and physicians who do not provide desired answers are perceived as failed and are dismissed. Patients become increasingly pessimistic. Many exhibit intense focus on DI to the exclusion of everything. Trust is eroded. DI self-care habits and behaviors become prevalent, supported by selected misinformation. They descend into a make-believe world of misery, sunk cost fallacy, and confirmation bias. Conversations become monologues by the patients with denials of mental illness or madness. Alternate conversations on other subjects other than their personal issues are considerably diminished, if not erased. Often patterns of self-harm - painful or not - are well established. Skin picking/self-inflicted lesions or other forms of self-mutilation caused by underlying psychiatric issues prevail [16]. These patients become hard to reach (Figs. 3 and 4).

The longer a patient remains stuck in this state of mind, the more they will move toward total commitment and a form of "psychiatric addiction" more closely defined as "Disease model of DI addiction" (see glossary). Patients feel constantly under attack and this in part is governed by the "Lizard Brain" (see Box 3). Distrust pervades while validation is craved. Internet searches are focused and intense and many fall victim to online or in-person predators and influencers (Fig. 3).

The internet.   Internet algorithms are a serious problem for vulnerable DI patients. Major online search engines have engineered algorithms designed to maximize user engagement through steering material to them. An alternate universe is created, and

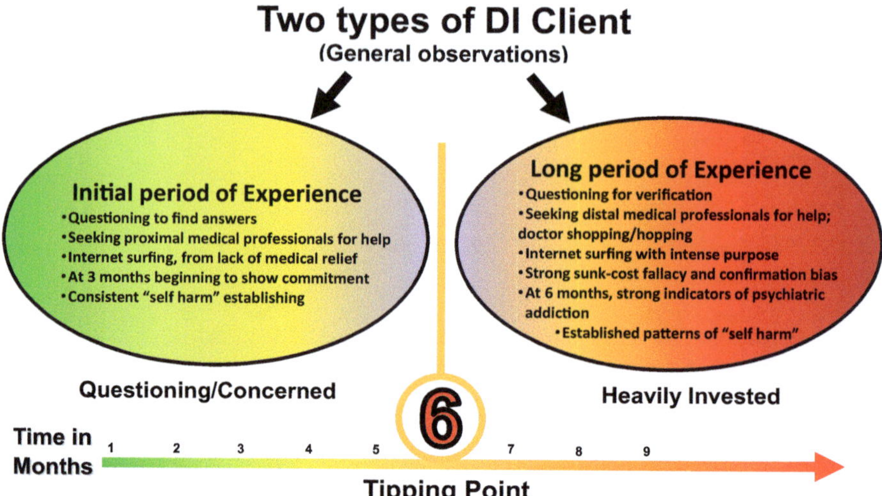

Fig. 4 The sliding emotional time slope from mainstream to fringe thinking. (Design by K. Dugas)

DI patients are drawn into it. Self-proclaimed "experts" and others with the disorder are often articulate and persuasive. "Misery loves company," and those practitioners who hawk it, know it. These individuals and organizations promote and cater to DI beliefs, and they aggressively defend their assertions irrespective of the facts. Unfortunately, sounding the alarm usually backfires. It can inadvertently promote these organizations and individuals by introducing more patients to them.

Management. Patience and understanding goes a long way. By the time many DI patients see me, they say I am "the end of the line" for them. Never debunk beliefs, avoid literal listening, and be prepared to get buffeted. Patients seek help but remain suspicious. Talk with patients not at them. I validate patients: what they perceive is real, even if what they believe is not.

Encourage patients to alter their internet search patterns and habits. Redirect them to web addresses that end in .edu or .gov, where much of the material is peer reviewed prior to publication. For the highly invested patients, tell them to stay away from the internet completely. Their judgment is compromised, and internet searches make a bad situation worse.

I encourage patients to pursue enjoyable activities and suggest if they are self-isolating to reconnect with family, friends, and colleagues. Isolation and lack of social connection and emotional bonding are common in DI. Many describe how they want to protect those they love from "being infested." It is noble but can do greater physical harm than smoking or obesity [17]. In the elderly, it can be lethal. One helpful technique is to encourage patients to go outside and look up. Looking skyward and walking to a high point to view the horizon is emotionally beneficial. By doing this, it also increases engagement in exercise.

Replace addictive thinking by changing word usage. Any fierce language should be softened through substitution. Using substitute words or short phrases while continuously and gently correcting patients when they slip into more potent language alters perspectives and shifts thinking.

Involve patients in their own inquiries and as it were, "walk" along with them in a combined "journey together." Where possible, include family, friends, colleagues, and other medical professionals in this process. Frame the collaboration as a "team working together."

Yet with all this, there comes a time when some patients are beyond reach. This is when it is time to disengage and say, "I cannot help you" (see the section, "Resigning from Care and Closing Obligation" in Part II).

9   The Future of DI Treatment

What should be done for the future of DI care? DI is not an uncommon disorder [18, 19]. Dr. Hinkle suggests that at any one time in the United States there may be thousands of active cases; most go unreported. DI is also undercounted because of public anxiety toward arthropods and/or other pathogens.

> **Box 3. The Lizard Brain/Limbic Cortex and the Disease Model of DI Addiction**
>
> The lizard brain is a set of structures (Amygdala, Hippocampus, Thalamus, Hypothalamus, Basal Ganglia, and Cingulate Gyrus) that sit above the spine and control upper brain function for emotion and memory. Most of the time this region remains silent, allowing other regions of the brain to perform normal daily duties. It is indifferent to what it perceives as normal. It becomes activated when a person is threatened or is in danger and quickly subverts upper brain functions that deal with appropriate social behavior in an act of perceived survival. In desperate situations, the lizard brain will do whatever it takes to survive. For example, any traumatic memories will be imprinted. If there are future events that match these memories, it will trigger a response and force patients to involuntarily perform acts of avoidance they would not otherwise do.
>
> The lizard brain is an important player in alcoholism and other addictions as well as DI. Ritualistic cleaning, tolerance of pain, and repetitive behavior are hallmarks of severely invested patients. The faulty mesolimbic/dopaminergic pathways in their brains alter psychological and subsequent behavioral processes which cause self-harm and non-chemical addictive behavior. It is described as a "Disease Model of DI Addiction." Patients perform compulsive acts of harm because they "can't help it." They see themselves constantly under attack and see their behaviors against their persecutors as protective and comforting.

Entertaining techniques of addressing DI by modeling after long standing successful programs such as Alcoholics Anonymous (AA) should be considered. AA's position is that alcoholism is a "physical compulsion, coupled with a mental obsession." Many aspects of DI fit well into these parameters. I suggest that AA or similar programs should be used by physicians as part of patient support in addition to regular medical care.

DI is a bifurcated disorder which is highly variable, it traverses many disciplines, it possesses common themes, and patients migrate from one professional to another. To address this, there needs to be a multi-disciplinary approach particularly in tracking and recording these patients. To paraphrase an African proverb:

"It takes a village to make you better."

If a physician suspects a patient has DI, inter-disciplinary coordination with other practitioners [20] and support from family, friends, and colleagues is a strategy that succeeds. This approach offers patients the best opportunity at returning to the quality of life they once enjoyed.

10   A Delusional Infestation Clinic

It has become clear that establishing a DI clinic in North America as seen in Europe where patients and those close to them can receive help is needed. The clinic would be populated by a staff of dermatologists, dieticians, epidemiologists, parasitologists, neurologists, psychodermatologists, counselors and psychiatrists etc., who would work together and not in isolation. For example, this allows a dermatologist who suspects a DI patient has neurological issues the opportunity to send them down the hall to see a colleague in the neurology department. The clinic might be part of a larger hospital where a community of physicians are already in place. Consultations might also be offered online for patients and families who can't travel.

By having a clinic not only would patients be with a community of physicians who are knowledgeable about DI, it would stop doctor shopping/hopping by patients and demoralizing searches by families, colleagues, and friends who are seeking help for their loved ones. Additionally, it would be an institution other professionals both inside and outside medicine might refer their patients or clients to. The clinic might also be connected to other medical groups in North America. The net result is that those struggling with DI will receive timely knowledgeable care by a group of connected physicians who can treat their patients efficiently and not lose them to the care they truly need.

11   A Condensed List of Suggestions for Working with DI Patients

No social media in the pocket!   For DI patients who are vulnerable to persuasion, internet access is a disaster. The most dangerous propaganda is information that is not recognized as dangerous, and most DI patients do not recognize these dangers.

Engineered algorithms contribute to DI delusion by funneling misleading information. These algorithms track a user's clicks and web visits, gather similar information, and steer it to their smartphones, tablets, or computers. Family, friends, and colleagues should consider internet algorithms deceptive and dangerous. The combination of anxiety, confirmation bias, sunk cost fallacy, and algorithms creates a potent scenario where patients rapidly become trapped in a particular form of negative thinking. Patients know something is wrong yet reject alternatives. They repeat behavior cycles becoming hamsters in their own wheels going nowhere. It's because they are constantly trying to rewrite their experiences and failing.

Ask how long the patient has experienced "believed parasitism." More than six months indicates deep investment that is resistant to corrective change. Less than six months means a less potent emotional investment and patients may be more easily persuaded to entertain other possible causes for their disorder.

Ask if patients have traveled. Patients who have traveled abroad can accidentally pick up human parasites from foreign sources. These may be from:

Endemic parasitic areas,
Drinking contaminated water,
Eating raw or unwashed foods.

Conversations. DI patients imagine and believe in things that are not there and are often very anxious and highly sensitive to perceived judgment and subterfuge. Be genuine but careful about what is said. Adjust language to be comfortable and reassuring because patients are primed not to hear you. Take time to actively listen. This will make it easier to have measured appropriate responses. Never display impatience or speak dismissively about patient experiences. Patients are in the heat of battle and feel attacked, so their interpretation of information will be entirely different from yours.

Time management. Always acknowledge patients concerns because what they are experiencing is real. Be welcoming, patient, yet set time limits with consultations. Open-ended meetings can turn into marathons. Suggested times, one hour for initial interviews, ½ hours for subsequent meetings. Include significant people in patients' lives as participants.

Compare patient conversations. By having conversations with patients and comparing the highly focused ideation of parasitism with daily activities such as gardening, favorite activities, pets, or the weather, physicians can assess whether patients have DI or are dealing with psychosis. Outside the ideation, DI patients have normal conversations. Most DI patients are not psychotic, they simply have a singular fixed idea (an obsession) supported by strong emotional attachment to misinformation.

Involve non-invested concerned family members and friends. Patients in most cases have family and friends who are affected to varying degrees by their ideation. Many want to help, but are rebuffed, feel ill-equipped or feel powerless. Others unwittingly enable, while others are dismissive. When possible, it is important to have people who are emotionally connected to patients present during interviews as observers and participants/helpers who are supportive. This reduces lying and denial by patients. They also can provide additional information patients may not be able to provide. In addition, it offers family or friends an opportunity to witness patient interactions with someone outside their relationships. It can be very enlightening for them. It empowers them with additional information they might use during patient care. These helpers can become potent collaborators, especially with patient behavior modifications at home. They can also become a patient's memory. Often patients don't recall instructions and/or advice due to their condition. Strong family and friendship bonds lead to an exponential increase of successful recoveries by DI patients. DI patients can't do it alone.

Managing sampling. Control sample volume to a handful of patient selected specimens. "Sample sign" as patients attempt to prove the existence of their persecutors is time consuming and often fruitless. By minimizing sample volume, it gives more time for other issues.

Involve patients in their own investigations. By involving patients in their own investigations, it evokes "partnerships" where providers and patients work together. It also stops the passive position many patients are put into where they do nothing while the physician does all the work. Allow patients to view their samples under high magnification, discuss what they see and talk about pest insects or other parasite morphologies, biologies, and behaviors in a friendly casual way. This promotes curiosity and imparts correct information. It takes the fear and anxiety many patients feel out of the process. When patients are involved, they are more likely to accept results.

Alcoholics Anonymous (AA) as a support model. Alcoholics Anonymous (AA) was founded in the 1930's as a voluntary supportive organization for recovering alcoholics. Based on mentoring and support by a community of sober alcoholics for less experienced sober alcoholics, it has produced a formidable success rate. AA is based on peer-to-peer relationships and community interactions. This inherited disease with subsequent addiction is managed by AA members without supportive drugs. AA supports the emotional well-being of its members and allows them to effectively control their cravings using peer-to-peer mentoring. The same principals could be applied to DI. Patients have an emotional addiction and craving rather than a chemical addiction and craving and for long-term cases treating emotional well-being using AA principles is as important as treating the somatic illnesses many DI patients suffer from. It is a very successful voluntary program which should be considered.

Follow ups with DI patients. To experience long-term success, it is critical that physicians and other professionals engage in follow-up care for at least 18 months. This stops recidivism regarding "Disease model of DI addiction" (see glossary). Long-term DI patients are emotionally addicted to their disorder and negative thinking, and it can take as long as 18 months to permanently reconfigure and restore neurological pathways in thinking that is healthy. DI patient brains need retraining, and this takes time. One benchmark concept is to consider a patient in recovery will need as long to recover as the time it took to become ill. For example, if a patient had DI for 2 years set aside 2 years for follow-up visits and check ins.

Diet. Make sure patients eat well. Many patients have poor diets and through depression, anxiety, and stress descend into a state of semi-starvation and malnutrition. Yet paradoxically many are over-weight due to poor high caloric diets and lack of exercise. The gut is the "second brain." If its function is compromised by medications, legal and/or illegal drugs, poor health, limited physical activity, a sedentary lifestyle, bad sleeping habits, and a poor diet, it can significantly influence patient health outcomes.

Conference calls. When possible, engage in conference calls with medical entomologists and other experts who have experience with DI and have patients present during these calls. Include those who are close to them, when possible, during these conference calls. A three or more-way conversation is highly beneficial to patients.

Humans are coded for pattern recognition [21]. This patterning in DI patients is potent. Having multiple people on a call with different disciplines and perspectives provides emotional support that may allow patients to reconsider alternatives. In time, it will change their pattern of thinking. These calls may give patients "space" to reevaluate, by seeing other perspectives. Mattson's abstract explains this:

> "Humans have long pondered the nature of their mind/brain and, particularly why its capacities for reasoning, communication and abstract thought are far superior to other species, including closely related anthropoids. This article considers superior pattern processing (SPP) as the fundamental basis of most, if not all, unique features of the human brain including intelligence, language, imagination, invention, and the belief in imaginary entities such as ghosts and gods. SPP involves the electrochemical, neuronal network-based, encoding, integration, and transfer to other individuals of perceived or mentally fabricated patterns. During human evolution, pattern processing capabilities became increasingly sophisticated as the result of expansion of the cerebral cortex, particularly the prefrontal cortex and regions involved in processing of images. Specific patterns, real or imagined, are reinforced by emotional experiences, indoctrination, and even psychedelic drugs. Impaired or dysregulated SPP is fundamental to cognitive and psychiatric disorders. A broader understanding of SPP mechanisms, and their roles in normal and abnormal function of the human brain, may enable the development of interventions that reduce irrational decisions and destructive behaviors [21]."

12   Two Cases Without Cures

DI is non-selective; it strikes anyone, anytime, and can be fatal. Two patients I worked with died. One was a medical researcher who died accidentally. Symptoms started immediately after his wife and son abandoned him. He self-prescribed medications and when I first met him, he was bathing in Lindane to rid himself of perceived parasites. His demeanor was hunched, head down, with no eye contact. His skin was bright pink. I saw him for several weeks until a call from a pest management professional collaborator told me he had died at home following a heart attack. Follow-up revealed severe depression from the abandonment. The second patient was a call center employee who had involved companions, colleagues, and managers (Folie à N) in the belief that the building they were working in was infested with parasites. 60 people were involved. She committed suicide and the "infestation" in the call center immediately stopped. She too, suffered from severe depression.

13   Three Cases with Cures

The first case was a businessman who had built a successful small American business. At the time of the 2008 American recession his small business had 23 employees, all of whom were personal friends. They had all contributed to the building of the business. Because of the recession, sales of his product suddenly stopped, and he faced the specter of imminent bankruptcy. For him, this meant the unthinkable: firing his friends and coworkers. He spiraled into depression and developed severe sensations of being bitten. He was referred to me. I eliminated arthropods as a cause and had a family meeting where 11 members of his close-knit family attended. Family members spoke of his depression, and their compassion made him break down in tears. Then a son suggested he ask his employees what they think he should do, and he did. His employees unanimously decided they would take severe pay cuts, keep working and the business going, and "ride-out" the recession together. No one was fired. His sensations of being bitten stopped and following counseling and with family support he returned to full health. I came away with the understanding that if there is strong family support, many patients can be cured. Case after case confirmed this.

The second case was a woman who suffered a reaction to a prescribed medication. Following her recovery, she wrote the following:

"Hello Dr. Ridge,

First, I want to thank you for attempting to solve my issue. I wanted to let you know that I have figured out what the nonexistent bugs problem was.

It turns out that I had hives from the cholesterol medication Rosuvastatin 20 mg. Which caused a number of side effects as well as causing pretty intense memory loss, confusion, and more. Which intern caused intermittent taking of the medication. That is why it took me so long to figure out what was happening. Some days I could barely think or function. I would forget the pills and then my mind would clear some. Then I would get back to taking the medication regularly and begin losing my memory, have more confusion and so on. This was a truly terrible experience. I am seriously reaching the point of being against pharmaceutical drugs in most cases. I hope this is useful to you for future bug biting situations as I seem to be collecting these pharmaceutical meds. that give me hives, etc."

The third case was in a large hospital laundry. A number of staff had started to complain of being bitten and I was asked to visit the facility. The laundry filled a very large basement area of the hospital. There was the main work area and a small corner room where linen was folded. When I entered this particular room, it was crackling dry. On one wall was a very large dehumidifier which was running 24 hours a day. I asked if the folding room had been built after the dehumidifier had been installed and the answer was, "Yes." What had happened was that the dehumidifier designed for a 4000 sq/ft floor space was confined in a "little-itty-bitty-space." The result was that it had severely dried out the room resulting in static electricity. Staff were being "zapped" not "bitten." Everything resolved once the dehumidifier was removed.

14 Summary

DI is non-selective, though people who live alone are at higher risk. It is a bifurcated disorder of psychological and somatic disease which is time sensitive. As patients linger, they often drift away from mainstream thinking, information gathering, and quality care to fringe fanaticism and the manipulation of well intentioned or maleficence of other long-time sufferers, unqualified care givers, or unscrupulous predators. Self-inflicted injury takes on a strong psychiatric component. Cure is most easily achieved through collaborations of physicians, experts, and close associates of patients during in person contacts resulting in restoration of a life once lived and enjoyed.

References

1. Beck AT. Thinking and depression: I. Idiosyncratic content and cognitive distortions. Arch Gen Psy. 1963;9(4):324–33.
2. Freudenmann RW, Kölle M, Schönfeldt-Lecuona C, Dieckmann S, Harth W, Lepping P. Delusional parasitosis and the matchbox sign revisited: the international perspective. J Compil Acta Derm-Venereol (letters to edit). 2010;90:517–9. https://doi.org/10.234 0/00015555-0909.
3. Bailey CH, Anderson LK, Lowe GC, Pittelkow MR, Bostwick JM, Davis MDP. A population-based study of the incidence of delusional infestation in Olmsted County, Minnesota, 1976-2010. Br J Derm. 2014;170(5):1130–5. https://doi.org/10.1111/bjd/12848.
4. Foster AA, Hylwa SA, Bury JE, Davis MDP, Pittelkow MR, Bostwick JM. Delusional infestations: clinical presentations in 147 patients seen at Mayo Clinic. J Am Acad Derm. 2012;67(4):1–10.
5. Bransfield RC, Friedman KJ. Differentiating psychosomatic, somatopsychic, multisystem illness and medical uncertainty (review). Healthcare. 2019;7(114):1–28.
6. Bewley A, Lepping P, Taylor RE, editors. Delusional infestation: psychodermatology in clinical practice (Lepping; Chapter 12). Springer; 2021. p. 441.
7. Nguyen CM, Danesh M, Beroukhim K, Sorenson E, Leon A, Koo J. Psychodermatology: a review. Pract Derm. 2015:49–54.

8. Freudenmann RW, Lepping P. Delusional infestation. Clin Microbiol Rev. 2009;22(4):690–732. https://doi.org/10.1128/CMR.00018-09.2.

9. Lepping P, Freudenmann RW. Chapter 10: Delusional infestation in childhood, adolescence, and adulthood. In: A clinical manual of child and adolescent psychocutaneous disorders. Pediatric Psychodermatology. 2012. pp 211–236. ISBN 978-3-11-027393-9.

10. Lepping P. Eine seltene Störung mit hohem Leidensdruck. CliniCum derma. 2017; 3/17.

11. Jaspers K. Allgemeine Psychopathologie. 9th ed. Heidelberg, Germany: Springer; 1973.

12. Jaspers K. General psychopathology. 4th ed. Manchester, UK: Manchester University Press; 1963.

13. Romanov DV, Lepping P, Bewley A, Huber M, Freudenmann RW, Lvov A, Squire SB, Noorthoorn EO. Longer duration of untreated psychosis is associated with poorer outcomes for patients with delusional infestation. Acta Derm Venereol. 2018;98:848–54.

14. Lepping P, Aboalkaz S, Squire SB, et al. Later age of onset and longer duration of untreated psychosis are associated with poor outcomes in delusional infestation. Acta Derm Venereol. 2020;100:adv00261. https://doi.org/10.2340/00015555-3625.

15. Park AJ, Harris AZ, Martyniuk KM, Chang CY, Abbas AI, Lowes DC, Kellendonk C, Gogos JA, Gordon JA. Reset of hippocampal-prefrontal circuitry facilitates learning. Nature. 2021;591:615–9.

16. Gieler U, Consolf SG, Tomas-Aragones L, et al. Self-inflicted lesions in dermatology: terminology and classification—a position paper from the European Society of Dermatology and Psychiatry (ESDaP). Acta Derm Venereol. 2013;93:4–12.

17. Holt-Lunstad J, Smith TB, Baker M, Harris T, Stephenson D. Loneliness, and social isolation as risk factors for mortality: a meta-analytic review. Pers Psych Sci. 2015;10(2):227–37. https://doi.org/10.1177/1745691614568352.

18. Trabert W. Epidemiology of delusional ectoparasitic infestation. Nervenarzt. 1991;62:165–1693.

19. Hylwa SA, Foster AA, Bury JE, et al. Delusional infestation is typically comorbid with other psychiatric diagnoses: review of 54 patients receiving psychiatric evaluation at Mayo Clinic. Psychosomatics. 2012;53(3):258–65.

20. Todd S, Squire SB, Bartlett R, Lepping P. Delusional infestation managed in a combined tropical medicine and psychiatry clinic. Trans R Soc Trop Med Hyg. 2018;00:1–6.

21. Mattson MP. Superior pattern processing is the essence of the evolved human brain. Front Neurosci. 2014;8:265. https://doi.org/10.3389/fnins.2014.00265.

Where the Emergency Department Fits In

Johnathan M. Sheele

Delusional Infestation (DI) or Ekbom syndrome is a fixed delusion where patients believe they are infested with parasites or inanimate objects [1–5]. Patients with DI will complain of abnormal cutaneous sensations attributed to their belief of having an infestation despite evidence to the contrary [6]. The Diagnostic and Statistical Manual of Mental Disorders (DSM-5) criteria for DI include that the duration of the delusion is ≥ 1 month, the patient does not have schizophrenia, manic, or depressive symptoms that have been brief relative to the delusion, functioning is not significantly impaired except for the delusion, and another condition or disorder does not better explain the symptoms [6]. Most people suffering from DI are older women, but there is a more equal sex distribution under age 50 years [5]. Objective data contradicting an infestation will not convince a person with DI that they do not have an infestation. They will not easily accept that they are suffering from an underlying psychiatric disorder [4]. Patients with DI can suffer socially, economically, physically, and mentally for years. Patients with DI may have already seen multiple medical specialists and now are presenting to the ED directly or indirectly related to their delusion. Patients suffering from DI can present diagnostic and disposition challenges for the emergency department (ED) clinician. There are very few published reports related to DI in the ED to help guide acute management [7–9]. Additionally, well-designed randomized trials related to DI are lacking.

The ED clinician needs to triage suspected or known DI patients into those previously worked up clinically for a secondary cause of their delusion from those who have had an incomplete investigation. Patients with known DI and who have no acute medical complaints can likely be discharged with minimal, if any, ED investigation. Patients with possible DI or who have been incompletely worked up may require a more thorough ED investigation to either exclude an actual infestation or

J. M. Sheele (✉)
Department of Emergency Medicine, Mayo Clinic, Jacksonville, FL, USA

© The Author(s), under exclusive license to Springer Nature
Switzerland AG 2024
G. E. Ridge (ed.), *The Physician's Guide to Delusional Infestation*,
https://doi.org/10.1007/978-3-031-47032-5_14

identify a treatable medical cause for their complaint [10] ED providers should avoid anchoring bias and complete a thorough history and physical exam.

ED clinicians should determine how long the patient's symptoms have been present, review previous diagnostic workups, determine the use of any self-prescribed and prescription treatments, identify pre-existing psychiatric diagnoses, determine if there is any concurrent drug and alcohol abuse, there are any known infectious exposures, and obtain a travel history [11, 12]. Acknowledge any recent neurological imaging studies, biopsies results, and whether the patient has been evaluated by dermatology, psychiatry, and infectious diseases. ED providers should review old medical records if they are available. ED clinicians should be aware of *folie à deux*, *folie trois*, and *folie à famille* (which are DI shared by two people, three people, or a family, respectively) as well as DI by proxy [4, 5, 8, 13–16].

Patients can suffer emergent conditions directly related to their delusions (e.g., self-harm, secondary infection from skin picking, etc.), intoxication and toxicity from DI treatments, and acute psychiatric conditions (e.g., suicidality) [4, 5, 17–19]. DI patients frequently present with the "specimen sign," where the patient brings in samples or specimens or they have videos or pictures of their infesting organism or object for the clinician to review [3, 5, 16, 20].

ED clinicians should perform a thorough psychosocial history [21]. DI patients may be guarded about answering psychiatric questions as they may not want to be perceived as suffering from a mental illness. The clinician should screen for homicidal or suicidal ideation, visual or auditory hallucinations, and self-harm behavior that can be acutely addressed in the ED [5, 21]. DI patients may perform maladaptive unsafe behaviors to eliminate or minimize their infestation [5]. Ensure that all dependents, such as children or the elderly, have a safe living environment and take appropriate actions as indicated.

Patients with DI may describe their infestation in great detail, but there will likely be biological implausibility [5]. The ED clinical should do a thorough physical exam, including examining the patient for any ectoparasites. Specifically, look for evidence of lice, mites, bed bugs, thrips, springtails, and fleas [5]. Lice and bed bugs can be visualized with the naked eye, and there are no specific diagnostic tests in the emergency department for scabies, lice, or bed bugs. Bedbugs and lice do not burrow into the skin or lay eggs under the skin. Scabies typically presents with a pruritic rash on specific body areas, such as skin creases. Scabies can be treated with antiparasitic drugs, but symptom resolution can take weeks. Fleas do not live on humans; their bites can cause maculopapular or papular urticaria [22]. Thrip bites can cause pink macules or papules and small puncta [22].

DI can be primary or secondary with a broad differential diagnosis [11, 23, 24]. The frequency of secondary medical reasons for DI in the ED is unknown but has been reported to be as high as 60% [25]. The ED clinician should consider stroke, brain masses, organic brain syndrome, autoimmune diseases, hypothyroidism, anemia, posterior reversible encephalopathy syndrome (PRES), cutaneous larval migrans, pruritis from systemic illness, substance abuse (e.g., cocaine, amphetamine, and marijuana, alcohol), infections, arthropods (e.g., avian mite-induced dermatitis, scabies, Grocer's itch (mites), bed bugs, lice, fleas, thrips, etc.), pet-induced dermatitis, fiberglass dermatitis, cholestasis, liver disease, vitamin deficiencies, renal failure, post infections like post-herpetic neuralgia, and psychiatric disorders (e.g., obsessive-compulsive disorder, skin picking, dermatitis artefacta, trichotillomania, hypochondriasis) [6, 7, 11, 26–32]. Some medications are associated with DI, including corticosteroids, some antibiotics (e.g., clarithromycin, ciprofloxacin, erythromycin), opiates (including opioid withdrawal), antivirals, dopamine agonists (e.g., for the treatment of Parkinson disease), methylphenidate, armodafinil, interferon alpha b2 plus ribavirin, modafinil, phenelzine, donepezil, bromide intoxication, and topiramate [3, 24, 33–39].

There are no formal guidelines related to the workup and treatment of DI in the ED, and the following are the author's opinions. No testing in the ED may be appropriate for a patient with suspected or known DI depending on the duration of the symptoms and the extent of previous evaluations. However, the ED workup could potentially include neuroimaging, serum electrolytes, complete blood counts, thyroid studies, liver studies, testing for human immunodeficiency virus (HIV), syphilis, pregnancy test (if applicable), and hepatitis testing, urine drug screens, autoimmune screening labs, vitamin B12 level (or B vitamin levels), and folate level [3, 7, 20, 24, 29, 32, 40–42]. Erythrocyte sedimentation rate and c-reactive protein can look for evidence of inflammation, although these tests are non-specific. If the ED clinician plans to start an antipsychotic, an electrocardiogram to examine the QT interval could be obtained, although it is not required [23] If indicated, Borrelia serologies, vasculitis screening, allergy testing, and autoimmune tests can all likely be deferred to outpatient providers [32].

The author believes that DI does not require initiation of treatment in the ED especially if the diagnosis has not been confirmed or there are questions related to outpatient follow-up. If the ED clinician strongly suspects DI and other medical causes have been ruled out, antipsychotic drugs such as pimozide (0.5–4 mg daily), risperidone (0.5–2 mg daily), aripiprazole (2–30 mg daily), haloperidol (0.5–5 mg daily), or olanzapine (2.5–12.5 mg daily) can be initiated [6, 11, 16, 23, 25, 26, 43–45]. Patients may be reluctant to take psychiatric medication to treat their perceived medical infestation, especially when they may already feel that the medical profession thinks they are "crazy" [25]. Letting the patient know that while antipsychotic medications are used to treat mental illness, they may also provide relief for their symptoms rather than curing them of an infestation [25]. However, patients should be counseled that the medication has been shown to help reduce symptoms over time and that they may require long-term therapy [5]. If starting a medication for DI, close outpatient follow-up is needed because stopping treatment is associated with the return of symptoms [46].

It may be beneficial to engage with the patient early in the ED encounter about what they hope to achieve from their visit. It may be helpful for the ED provider to acknowledge the patient's suffering [25]. However, it is unlikely to be helpful for the ED provider to tell someone with suspected DI that they suffer from a delusional disorder as patient rapport will likely suffer. Strategies for approaching the DI have been published for the outpatient provider but not specifically for the ED [6].

Goals of care in the ED for the DI patient can focus on improving the patient's symptoms rather than providing definitive diagnoses. The ED clinician should not feed into the delusion by lying to the patient and saying they believe that they have an infestation when the clinician suspects DI. Antiparasitic drug prescriptions should be avoided in suspected DI. When the clinician does not suspect a parasitic infection, prescribing an antiparasitic drug can reinforce the patient's delusion. The DI patient will be best served by long-term follow-up in a clinic with expertise in psychodermatology [47–50]. DI patients may be convinced to follow up in these specialty clinics by focusing the conversation on getting help from their suffering while further testing can be undertaken, as indicated, by other specialists outside of the ED. Primary care, infectious disease/travel medicine/tropical medicine physicians, psychiatry, and dermatology clinicians may or may not have the expertise or be comfortable managing DI [19, 43, 51]. Patients with DI may be unwilling to follow up with psychiatry because they are convinced that their symptoms are unrelated to mental illness. In the absence of a specialty psychodermatology clinic, optimal patient outcomes for DI may benefit, at least initially, from a multi-disciplinary approach [47, 49, 51].

DI in the ED may be frustrating for both the clinician and patient. However, the clinician needs to remember that DI patients have real suffering and are approached with genuine empathy. ED clinicians should avoid confrontation with the DI patient about the nature of their infestation because they are unlikely to change the mind of the delusional patient. The clinician should be honest and truthful about what can and cannot be done in the ED, and that medication may help control symptoms, but outpatient follow-up is critical. It is important for the ED clinician to exclude any treatable causes of their symptoms, assess the patient for a safe disposition, and address any emergent medical or psychosocial concerns.

References

1. Dewan P, Miller J, Musters C, Taylor RE, Bewley AP. Delusional infestation with unusual pathogens: a report of three cases. Clin Exp Dermatol. 2011;36(7):745–8. https://doi.org/10.1111/j.1365-2230.2011.04086.x.
2. Bewley AP, Lepping P, Freundenmann RW, Taylor R. Delusional parasitosis: time to call it delusional infestation. Br J Dermatol. 2010;163(1):1–2. https://doi.org/10.1111/j.1365-2133.2010.09841.x.
3. Heller MM, Wong JW, Lee ES, et al. Delusional infestations: clinical presentation, diagnosis, and treatment. Int J Dermatol. 2013;52(7):775–83. https://doi.org/10.1111/ijd.12067.
4. Hinkle NC. Ekbom syndrome: a delusional condition of "bugs in the skin". Curr Psychiatry Rep. 2011;13(3):178–86. https://doi.org/10.1007/s11920-011-0188-0.
5. Hinkle NC. Ekbom syndrome: the challenge of "invisible bug" infestations. Annu Rev Entomol. 2010;55:77–94. https://doi.org/10.1146/annurev.ento.54.110807.090514.
6. Moriarty N, Alam M, Kalus A, O'Connor K. Current understanding and approach to delusional infestation. Am J Med. 2019;132(12):1401–9. https://doi.org/10.1016/j.amjmed.2019.06.017.

7. Haas NL, Nicholson A, Haas MRC. Delusional parasitosis as presenting symptom of occipital lobe cerebrovascular accident. Am J Emerg Med. 2019;37(10):1990.e3–5. https://doi.org/10.1016/j.ajem.2019.158368.

8. Fisher JD. Emergency department presentation of 'delusional parasitosis by proxy'. Delusional parent, injured child. Am J Emerg Med. 2019;37(9):1806.e1–2. https://doi.org/10.1016/j.ajem.2019.05.058.

9. Boggild AK, Nicks BA, Yen L, et al. Delusional parasitosis: six-year experience with 23 consecutive cases at an academic medical center. Int J Infect Dis. 2010;14(4):e317–21. https://doi.org/10.1016/j.ijid.2009.05.018.

10. Desoubeaux G, Saada A, Bailly E, Guiguen C, Chandenier J. Ekbom's syndrome or real ectoparasitosis? An unexpected outcome of hidden springtails. Int J Dermatol. 2014;53(5):628–30. https://doi.org/10.1111/j.1365-4632.2012.05631.x.

11. Beach SR, Kroshinsky D, Kontos N. Case 37-2014. N Engl J Med. 2014;371(22):2115–23. https://doi.org/10.1056/nejmcpc1305989.

12. DeBonis K, Pierre JM. Psychosis, ivermectin toxicity, and "morgellons disease". Psychosomatics. 2011;52(3):295–6. https://doi.org/10.1016/j.psym.2011.01.006.

13. Sarkar S, Ghosal MK, Bandyopadhyay D, Ghosh SK, Guha P. Delusional parasitosis by proxy. Clin Exp Dermatol. 2009;34(7):487–8. https://doi.org/10.1111/j.1365-2230.2009.03555.x.

14. Giam A, Tung YL, Tibrewal P, Dhillon R, Bastiampillai T. Folie à deux and delusional parasitosis. Asian J Psychiatr. 2017;28:152–3. https://doi.org/10.1016/j.ajp.2017.04.027.

15. Yang EJ, Beck KM, Koo J. Folie à famille: a systematic review of shared delusional infestation. J Am Acad Dermatol. 2019;81(5):1211–5. https://doi.org/10.1016/j.jaad.2019.04.023.

16. Su P, Teo WL, Pan JY, Chan KL, Tey HL, Giam YC. Delusion of Parasitosis: a descriptive analysis of 88 patients at a tertiary skin Centre. Ann Acad Med Singap. 2018;47(7):266–8.

17. Meraj A, Din AU, Larsen L, Liskow BI. Self inflicted corneal abrasions due to delusional parasitosis. BMJ Case Rep. 2011;2011:4–7. https://doi.org/10.1136/bcr.04.2011.4106.

18. Hylwa SA, Ronkainen SD. Delusional infestation versus Morgellons disease. Clin Dermatol. 2018;36(6):714–8. https://doi.org/10.1016/j.clindermatol.2018.08.007.

19. Grant M. Between a rock and a hard place: balancing physician deception and patient self-harm in the management of delusional infestation. Am J Trop Med Hyg. 2020;102(1):3–4. https://doi.org/10.4269/ajtmh.19-0755.

20. Freudenmann RW, Lepping P, Huber M, et al. Delusional infestation, and the specimen sign: a European multicentre study in 148 consecutive cases. Br J Dermatol. 2012;167(2):247–51. https://doi.org/10.1111/j.1365-2133.2012.10995.x.

21. Hylwa SA, Foster AA, Bury JE, Davis MDP, Pittelkow MR, Bostwick JM. Delusional infestation is typically comorbid with other psychiatric diagnoses: review of 54 patients receiving psychiatric evaluation at Mayo Clinic. Psychosomatics. 2012;53(3):258–65. https://doi.org/10.1016/j.psym.2011.11.003.

22. Singh SMB. Insect bite reactions. Indian J Dermatol Venereol Leprol. 2013;79:151–64.

23. Brownstone ND, Koo J. Recent developments in psychodermatology and psychopharmacology for delusional patients. Cutis. 2021;107(1):5–6. https://doi.org/10.12788/cutis.0144.

24. Altunay IK, Ates B, Mercan S, Demirci GT, Kayaoglu S. Variable clinical presentations of secondary delusional infestation: an experience of six cases from a psychodermatology clinic. Int J Psychiatry Med. 2012;44(4):335–50. https://doi.org/10.2190/PM.44.4.d.

25. Lepping P, Huber M, Freudenmann RW. How to approach delusional infestation. BMJ. 2015;350(April):1–4. https://doi.org/10.1136/bmj.h1328.

26. Lepping P, Freudenmann RW. Delusional parasitosis: a new pathway for diagnosis and treatment. Clin Exp Dermatol. 2008;33(2):113–7. https://doi.org/10.1111/j.1365-2230.2007.02635.x.

27. Trigka K, Dousdampanis P, Fourtounas C. Delusional parasitosis: a rare cause of pruritus in hemodialysis patients. Int J Artif Organs. 2012;35(5):400–3. https://doi.org/10.5301/ijao.5000072.

28. Tsai SJ, Bin YC, Wang CW, et al. Delusional infestation in a patient with posterior reversible encephalopathy syndrome. Aust N Z J Psychiatry. 2016;50(12):1212–3. https://doi.org/10.1177/0004867416656259.
29. Ramirez-Bermudez J, Espinola-Nadurille M, Loza-Taylor N. Delusional parasitosis in neurological patients. Gen Hosp Psychiatry. 2010;32(3):294–9. https://doi.org/10.1016/j.genhosppsych.2009.10.006.
30. Campbell EH, Elston DM, Hawthorne JD, Beckert DR. And management of delusional parasitosis. J Am Acad Dermatol. 2019;80(5):1428–34. https://doi.org/10.1016/j.jaad.2018.12.012.
31. Maher S, Hallahan B, Flaherty G. Itching for a diagnosis—a travel medicine perspective on delusional infestation. Travel Med Infect Dis. 2017;18:70–2. https://doi.org/10.1016/j.tmaid.2017.05.011.
32. Vulink NC. Delusional infestation: state of the art. Acta Derm Venereol. 2016;96(4):58–63. https://doi.org/10.2340/00015555-2412.
33. Robaeys G, De Bie J, Van Ranst M, Buntinx F. An extremely rare case of delusional parasitosis in a chronic hepatitis C patient during pegylated interferon alpha-2b and ribavirin treatment. World J Gastroenterol. 2007;13(16):2379–80. https://doi.org/10.3748/wjg.v13.i16.2379.
34. Flann S, Shotbolt J, Kessel B, et al. Three cases of delusional parasitosis caused by dopamine agonists. Clin Exp Dermatol. 2010;35(7):740–2. https://doi.org/10.1111/j.1365-2230.2010.03810.x.
35. Marshall CL, Ellis C, Williams V, Taylor RE, Bewley AP. Iatrogenic delusional infestation: an observational study. Br J Dermatol. 2016;175(4):800–2. https://doi.org/10.1111/bjd.14558.
36. Howes CF, Sharp C. Delusional infestation in the treatment of ADHD with atomoxetine. BMJ Case Rep. 2018;2018:1–3. https://doi.org/10.1136/bcr-2018-226020.
37. Zhu TH, Werchan IA, Escamilla KV, Sebastian K, Hovinga CA, Reichenberg JS. Association between delusions of infestation and prescribed narcotic and stimulant use. J Psychiatr Pract. 2018;24(6):428–31. https://doi.org/10.1097/PRA.0000000000000338.
38. Mowla A, Asadipooya K. Delusional parasitosis following heroin withdrawal: a case report. Am J Addict. 2009;18(4):334–5. https://doi.org/10.1080/10550490902925888.
39. Ojeda-López C, Aguilar-Venegas LC, Tapia-Orozco M, Cervantes-Arriaga A, Rodríguez-Violante M. Delusional parasitosis as a treatment complication of Parkinson disease. Psychosomatics. 2015;56(6):696–9. https://doi.org/10.1016/j.psym.2015.07.004.
40. Mesquita J, Simoes S, Machado ÁSL. Ekbom's syndrome as the first manifestation of diabetes mellitus. J Neuropsychiatry Clin Neurosci. 2010;22(4):e36.
41. Duggal H, Singh I. Delusional parasitosis as a presenting feature of dementia. J Neuropsychiatry Clin Neurosci. 2010;22(1):11–2. https://doi.org/10.1176/appi.neuropsych.22.1.123-g.e11.
42. Foster AA, Hylwa SA, Bury JE, Davis MDP, Pittelkow MR, Bostwick JM. Delusional infestation: clinical presentation in 147 patients seen at Mayo Clinic. J Am Acad Dermatol. 2012;67(4):673.e1–673.e10. https://doi.org/10.1016/j.jaad.2011.12.012.
43. Freudenmann RW, Lepping P. Second-generation antipsychotics in primary and secondary delusional parasitosis: outcome and efficacy. J Clin Psychopharmacol. 2008;28(5):500–8. https://doi.org/10.1097/JCP.0b013e318185e774.
44. Mercan S, Altunay IK, Taskintuna N, Ogutcen O, Kayaoğlu S. Atypical antipsychotic drugs in the treatment of delusional parasitosis. Int J Psychiatry Med. 2007;37(1):29–37. https://doi.org/10.2190/M8M5-H1G2-1257-2017.
45. Kenchaiah BK, Kumar S, Tharyan P. Atypical anti-psychotics in delusional parasitosis: a retrospective case series of 20 patients. Int J Dermatol. 2010;49(1):95–100. https://doi.org/10.1111/j.1365-4632.2009.04312.x.
46. Wong S, Bewley A. Patients with delusional infestation (delusional parasitosis) often require prolonged treatment as recurrence of symptoms after cessation of treatment is common: an observational study. Br J Dermatol. 2011;165(4):893–6. https://doi.org/10.1111/j.1365-2133.2011.10426.x.
47. Todd S, Squire SB, Bartlett R, Lepping P. Delusional infestation managed in a combined tropical medicine and psychiatry clinic. Trans R Soc Trop Med Hyg. 2019;113(1):18–23. https://doi.org/10.1093/trstmh/try102.

48. Healy R, Taylor R, Dhoat S, Leschynska E, Bewley AP. Management of patients with delusional parasitosis in a joint dermatology/liaison psychiatry clinic. Br J Dermatol. 2009;161(1):197–9. https://doi.org/10.1111/j.1365-2133.2009.09183.x.
49. Altaf K, Mohandas P, Marshall C, Taylor R, Bewley A. Managing patients with delusional infestations in an integrated psychodermatology clinic is much more cost-effective than a general dermatology or primary care setting. Br J Dermatol. 2017;177(2):544–5. https://doi.org/10.1111/bjd.15088.
50. Goulding JMR, Harper N, Kennedy L, Martin KR. Cost-effectiveness in psychodermatology: a case series. Acta Derm Venereol. 2017;97(5):663–4. https://doi.org/10.2340/00015555-2620.
51. Wong YL, Affleck A, Stewart AM. Delusional infestation: perspectives from scottish dermatologists and a 10-year case series from a single Centre. Acta Derm Venereol. 2018;98(4):441–5. https://doi.org/10.2340/00015555-2875.

Brief Descriptions of Some Common Medically Significant Arthropods Related to Delusional Infestation

Gale E. Ridge, Lyle Buss, and Katherine Dugas

1 What Is an Infestation?

Having an insect, spider, or mite walking across a persons body whether feeding or not, does not mean there is an infestation. Often these are temporary situations which quickly resolve without significant intervention. Beach et al. define a "true infestation" as:

> "a parasite population is feeding, reproducing, and sustaining itself over time on a particular host [1]."

All arthropods (arachnids and insects) belong to the animal kingdom. They possess respiratory, nervous, circulatory, skeletal, muscular, digestive, reproductive, and sensory systems comparable to large animals, only they are very small. Arthropods are tiny animals. They feel pain, hunger, thirst, fear, experience curiosity, and possess the capacity to learn. They have specific and consistent biologies, behaviors, and needs, and live in specific habitats and ecologies. Most importantly, medically-significant arthropod biologies are well-documented and can be objectively observed.

G. E. Ridge (✉)
Department of Entomology, The Connecticut Agricultural Experiment Station,
New Haven, CT, USA
e-mail: gale.ridge@ct.gov

L. Buss
Department of Entomology & Nematology, University of Florida, Gainesville, FL, USA
e-mail: ljbuss@ufl.edu

K. Dugas
Agricultural Research Technician I, The Connecticut Agricultural Experiment Station,
New Haven, CT, USA
e-mail: katherine.dugas@ct.gov

© The Author(s), under exclusive license to Springer Nature Switzerland AG 2024
G. E. Ridge (ed.), *The Physician's Guide to Delusional Infestation*,
https://doi.org/10.1007/978-3-031-47032-5_15

DI patients often recount plausible descriptions of "infestations" and "parasites" and unless physicians are familiar with known human parasites, they may be misled into believing what their patients are saying is true. Patient's descriptions may misdirect differential diagnosis, good judgment, and timely treatment so should not be trusted.

It is important for physicians to possess some basic knowledge of medically significant parasites and arthropods' geographic distributions, biologies, and behaviors in the region of the world they work in. This is helpful for differential diagnosis by sorting out what is real or possible and what is flawed patient thinking.

This section sets its primary focus on medically significant arthropods found in temperate North America and should not be considered comprehensive. Discussions regarding potential parasite infestations must acknowledge the geographic location of the client and clinic, and so may include species not necessarily covered here; for example, human bot flies and chigoe fleas are prevalent medical concerns in subtropical or tropical regions of the world where they are endemic. However, it should be noted that geographic limitations are not typically considered by patients when conducting their own researches. Their primary searches are for names of parasites and the symptoms they cause. These patients often ignore or overlook geographical distribution of these parasites, either because of ignorance or convenience. It is important to ask patients about any recent travel histories.

Instructions using this section.

This section is designed for quick reference and is organized in systematic order, e.g., mites, ticks, spiders, and insects. Where appropriate, societal myths are addressed in boxes. Information considered critical to differential diagnoses are headed as "Important."

> **Box 1. How to manage the term "Bite" with DI patients.**
> The Merriam-Webster Medical Dictionary (57), describes a bite as a wound caused by occlusion, "the bringing of the opposing surfaces of the teeth of the two jaws into contact." Biting is the act that results in a wound. The term "bite" is also colloquially used to describe the act and/or result of arthropod blood-feeding, as well as a defensive/aggressive act such as stinging.
>
> In a medical sense, a horse fly could be loosely described as "biting"; it painfully lacerates the host's skin with scissor-like mouthparts and laps up blood from the wound. Strictly speaking, many hematophagous (blood-feeding) arthropods do not "bite,." For example, bed bugs and mosquitoes both have sophisticated piercing-sucking mouthparts that are carefully inserted into the hosts' skin for the purpose of blood feeding. This process does not cause laceration or external bleeding; it is normally well after these insects have left that any discomfort is experienced. Any resultant wound is self-inflicted from scratching.

The term "bite" can be problematic with DI patients. It has a psychologically triggering effect often founded on memories of previous negative encounters with arthropods. It is a catch-all term used to describe a selection of negative experiences patients have had with arthropods. Patients may incorrectly perceive and equate current ongoing sensations of pain, itching and discomfort as "biting" and therefore as active feeding or parasitism. This can reinforce incorrect beliefs and lead to an overwhelming urge to control or stop the "biting," heightening anxiety and potentially increasing discomfort if they attempt to self-treat or intervene.

If a patient continually fixates on this term, it is recommended to switch to less inflammatory language which is both more medically correct and less emotionally distressing. Encouraging patients to describe their discomfort in alternate terms will lower anxiety and reduce negative reinforcement.

Suggestions of alternate terminology to replace "bite:"

Prickling/pricking sensation	Pins and needles
Pinching	Itching Burning sensation Tickling sensation Running sensation

Box 2. Identifying an Arthropod Dermal Injury and Consequences.
It is *impossible* to confirm arthropod feeding described by patients as "bites" simply by examining lesions. DI patients are very vocal and can dominate conversations. Never allow patient descriptions to lead differential diagnosis if DI is suspected; consider all possible causes. An arthropod *must* be captured at the site of injury and identified by a trained expert to have conclusive evidence. *Never* speculate about arthropod or other parasite identification particularly with DI patients, who may latch on to or misinterpret the speculation as fact. Mistaken identification can amplify the delusion and make an already difficult situation worse. It is advisable for medical professionals to establish a relationship with an entomologist and/or parasitologist in case identification services are needed.

2   Mites (Class Arachnida, Subclass Acari)

Scabies

Name and Distribution

The human scabies mite, *Sarcoptes scabiei* var. *hominis*.

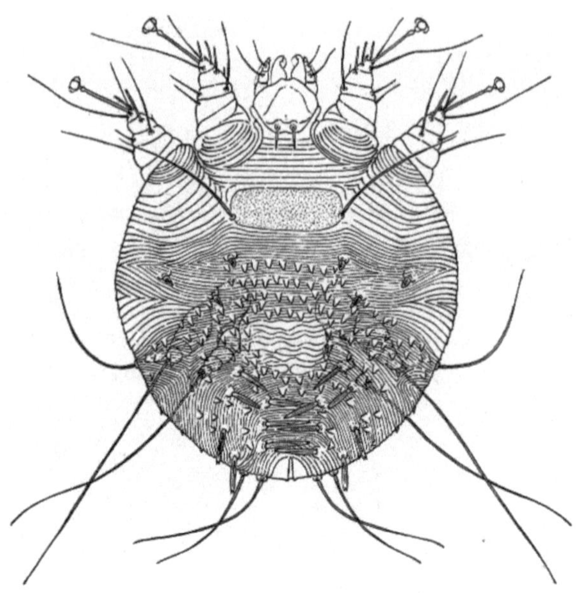

Fig. 1 Sarcoptes scabiei var. hominis. Image. National Pest Management Association, USA [2]

Human scabies mites are found worldwide, more commonly in tropical regions and areas with poverty and overcrowding. They are less common in the USA, but occasionally are a problem in nursing homes or older patients (Fig. 1).

Life Cycle and Behavior

Mites are transferred from one person to another by *prolonged* direct skin-to-skin contact. Casual brief touching such as a handshake may not provide enough time for mites to transfer. Rarely can they be spread via clothing or bedding. Mites can survive off a host for only 1 or 2 days under normal room conditions. Mated females excavate 1–2 cm (0.79 inch) long tunnels in the stratum corneum/epidermal junction and back fill the tunnels with eggs. Unmated females and males live in small, excavated pouches on the surface of the skin.

There are three immature stages: larva, protonymph, and tritonymph. They all burrow. Development from egg to adult is 8–15 days. There are brief periods when mites walk across the skin surface before burrowing into the skin. Life span is 1–2 months.

Medical Importance

The areas of the body that scabies mites most commonly infest include webbing between the fingers, wrists, elbows, axillae, buttocks, and genitals. Activity of mites can cause significant pruritis accompanied by a variable papular polymorphic rash and/or nodular lesions. These are reticuloendothelial system (RES) reactions to mite antigens. Scabies mites are incompetent vectors of disease-causing pathogens, but secondary infection can originate from scratching. Patients with no history of scabies may take up to 3 weeks to evoke an immune response. A rare, severe form of scabies is crusted (Norwegian) scabies, expressed as a thick crusted scaly skin and rash with mild itching. It can be associated with immunocompromised individuals.

Box 3. Myths: Scabies.
"Scabies comes from unclean conditions."
Mite biology is independent of personal hygiene. Scabies is transmitted through close contact with an infested person.
"Testing for scabies is difficult."
Confirmation of a scabies mite infestation using a skin scraping and light microscopy examination takes 4–5 min (see Box 4).

DI patients often present with scabies-like symptoms of itchy, irritated skin, particularly on their hands, because of excessive bathing and use of harsh chemicals. They often receive a misdiagnosis of scabies based on a visual examination and these symptoms. Box 4 contains information on more accurately diagnosing a true scabies mite infestation.

Box 4. Collecting Scabies Mites.*
It is imperative to perform a skin scraping to excise mites, eggs, and/or feces from burrows followed by compound light microscope examination for a scabies diagnosis. The entire process takes 4–5 minutes. Those with veterinary experience are likely very familiar with this procedure:

1. Put oil or a saline solution onto the affected area. Tunnels might also be found by coloring the area with India ink or gentian violet (tunnels come into profile because the colors penetrate the skin), with subsequent removal using rubbing alcohol.
2. Scrape along tunnels to remove mites or other material.
3. Mount collected mites or other material onto slide and examine. If mites are not found, this avoids unnecessary treatment of often damaged skin.

***For a thorough explanation on techniques for collecting, mounting, and identifying scabies mites, please go to Appendices A and B at the end of Chap. 9.**

> **Box 5. The Entomologists' Perspective on Conjecture and Scabies Diagnoses.**
> Entomologists in diagnostic labs frequently encounter clients claiming to be infested by parasites. This belief is further supported by previous physician diagnoses that were based on conjecture or cursory visual examinations. Giving patients suspect scabies diagnoses without isolating and confirming the presence of mites is problematic. It reinforces the patient's belief that they are truly infested by "bugs." When prescription medications fail (because mites are absent), it further translates to a belief that the medical community is either incompetent or the patient possesses exotic parasites "new to science."
>
> *Do not diagnose scabies and/or prescribe treatment without first physically isolating and identifying the mites.*

Bird Mites

Names and Distribution

There are principally three species of bird mites (families Macronyssidae and Dermanyssidae) that infest domestic and wild birds:

Northern fowl mite, *Ornithonyssus sylviarum* Canestrini and Fanzago.

Tropical fowl mite, *Ornithonyssus bursa* (Berlese) (Fig. 2).

Poultry red mite, *Dermanyssus gallinae* De Geer (Fig. 3).

Distribution of all three mites is worldwide.

The poultry red mite is mostly found in industrial poultry operations or family backyard/garden chicken coops.

> **Box 6. Myths: Bird Mites.**
> *"Bird mites feed on humans."*
>
> Bird mites have evolved to feed on birds, and it is impossible for them to feed on humans. Their mouthparts are too short to penetrate human skin. If they attempt to feed, it may feel like being pricked by pins.
>
> *"Bird mite infestations are contagious."*
>
> It isn't possible for bird mite infestations to "spread" to family members, their homes, or their vehicles. Even if a live mite was carried on a person's clothing, it would not be able to reproduce and would die.
>
> *"Bird mites are too small to see."*
>
> Bird mites are small (~ 1 mm) but they are visible to the naked eye. When bird mites are present, people usually see them crawling over smooth surfaces such as walls or window sills.
>
> *"Bird mite infestations can last months or years."*
>
> Bird mite infestations in homes are closely tied to wild bird activity. The mites will disappear within 2–3 weeks from the time the birds leave. Long-term "biting" problems on people are NOT caused by bird mites. If pet birds are present, have them checked by a veterinarian for mites.

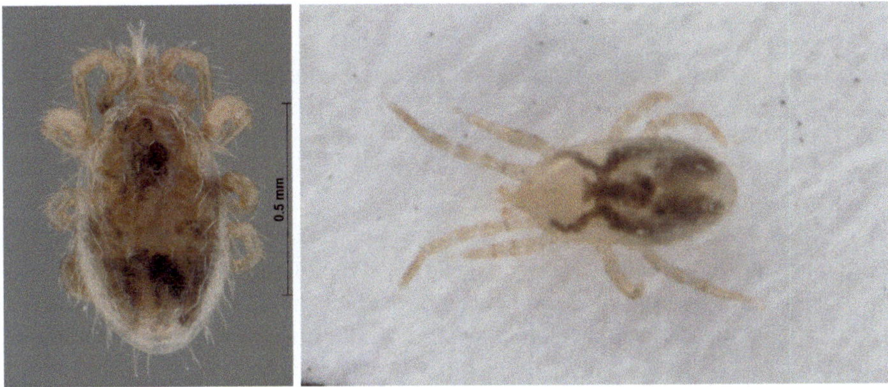

Fig. 2 Tropical fowl mite, *Ornithonyssus bursa*. Photos: Lyle Buss, Univ. of Florida

Fig. 3 Poultry red mite, *Dermanyssus gallinae.* Photo: Lyle Buss, Univ. of Florida

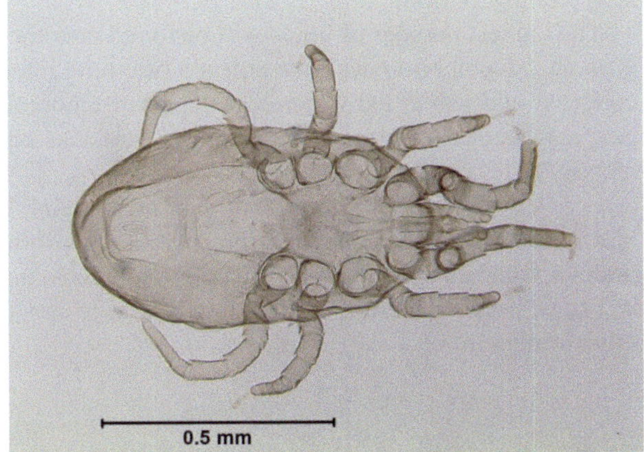

Life Cycle and Behavior

Birds are the hosts of these parasitic, hematophagous mites. They are pests of domestic poultry and wild birds, including pigeons, starlings, and house sparrows that nest around buildings.

Bird mites *cannot* feed on humans. Birds have very thin skin to keep them physically light for flight. It is very different from the thicker skin of land-bound humans and other animals. The short mouthparts of bird mites are adapted to feed on very thin skin, and are dysfunctional on any other skin type.

Northern fowl mites spend their entire lives on their bird hosts while tropical fowl mites and poultry red mites spend most of their time in the nests, periodically moving onto the hosts to feed. The three mite species have five life stages: egg, larva (6 legs), protonymph (8 legs), deutonymph, and adult [2]. All three species develop from egg to adult in ~7 days. When separated from their hosts, poultry red mites survive up to 8 months [3], while Northern and Tropical fowl mites survive up to 21 days.

> **Important!**
> Bird mites *cannot* live on humans and when separated from their natural hosts die from desiccation.

Medical Importance

Bird mites are very small (~ 1mm), and can be seen with the naked eye. People may encounter mites through physical contact with chickens or while in poultry facilities, but most unexpected encounters involve wild birds nesting in or on buildings. The problem usually starts after young birds have fledged and the nest is abandoned. Mites left behind begin wandering in search of a new bird host. In doing so, mites may crawl onto people and attempt to feed. Their mouthparts are too short to feed successfully on humans, but they can cause pruritic dermatosis through dermal contact, probing, or both.

The highest number of mites will be found near the source, which in most cases is an abandoned bird nest. To confirm a bird mite infestation, specimens should be collected and submitted to an acarologist or entomologist for identification. The pest management/control industry in general lacks equipment and trained staff to identify mites on site.

People searching for the cause of their mysterious "biting" will frequently blame bird mites, especially if they use the internet for information. Patients should avoid internet searches or limit their research to reputable university and government run websites, because there are numerous bird mite-specific sites that provide blatant misinformation.

Chigger Mites

Names and Distribution

Chigger mites are members of a complex family, the Trombiculidae. About 15 species feed on humans or their domestic animals. Common names include (Fig. 4):

"Harvest bugs, harvest mites, red bug, berry mite, scrub itch mite or chiggers [4]."

Chigger mites are found in the following habitats (Table 1).

Life Cycle and Behavior

Adult trombiculid mites are harmless to humans. They are predaceous, feeding on tiny arthropods on the ground. Females lay their eggs in the soil or on ground debris. The larvae that emerge are parasitic; this is the stage that can be a problem to humans. The larvae (chiggers) climb onto vegetation to ambush potential hosts that

Fig. 4 Trombicula (N) autumnalis. Image. National Pest Management Association, USA [2]

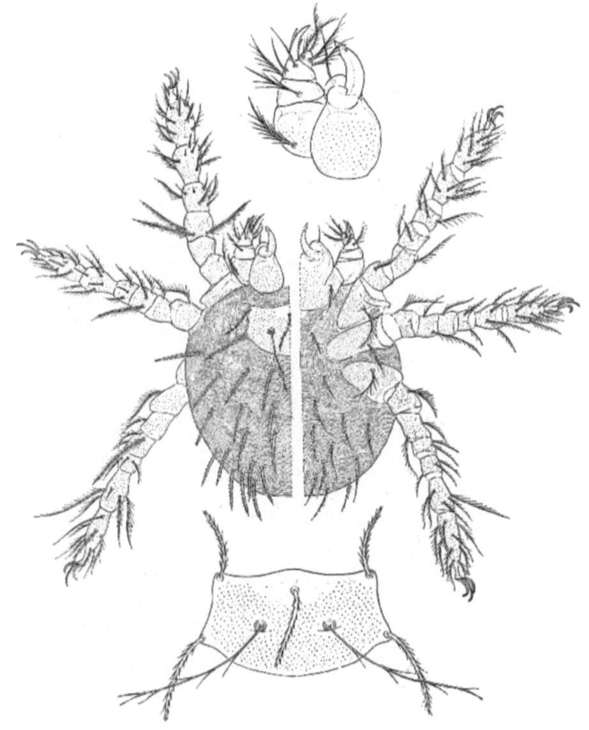

Table 1 Common Trombiculidae mites:

Common Trombiculidae mites	Habitats
Europe: *Trombicula* (*Neotrombicula*) *autumnalis*	Damp soils in open grasslands, especially near rabbit warrens
Australia, Southeast Asia, and Pacific islands: *Trombicula sarcina*	Savannah with smaller areas of grassland and scrub
Americas and Caribbean islands: *Trombicula* (*Eutrombicula*) *alfreddugesi*	Transition areas between forest and grasslands, on the edge of swamps
Eastern half of the United States and eastern Canada: *Eutrombicula splendens*	Prefers swamps, bogs, and low-lying areas with fallen trees and rotting stumps. Is dry area tolerant
South-Central United States: *Eutrombicula lipovskyi*	Prefers moist areas in and around streams and swamps with decaying logs and stumps

walk by. They feed on a variety of animals, including amphibians, reptiles, birds, and mammals. Humans are accidental hosts, not true hosts.

Important!
Ask patients about travel and/or whether they might have walked in habitats that support trombiculid mites.

It is important to be familiar with trombiculid species and their habitats for the geographic region you currently work in.

When these mites get onto humans, they often gravitate to areas of the body constricted by clothing such as the ankles, waist, and armpits. They insert their mouthparts into the skin and secrete digestive enzymes in their saliva. This breaks down skin cells and lymph, which the mites ingest. They are not hematophagous.

Medical Importance

Since humans are not a true host, chigger mites only survive on humans for up to 2 days. Their feeding triggers an immune response, leading to severe irritation and swelling which may include papules, hives, and/or a rash. By the time affected areas become intensely pruritic, the mites have fallen off. Dermal injury by the mites can be complex, both mechanically and enzymatically. Secondary bacterial infection from scratching is possible. Healing takes 1–2 weeks. Ask if patients have traveled to chigger mite endemic areas.

> **Box 7. Myths: Chigger Mites.**
> *"Chiggers/harvest scrub itch mites burrow into the skin."*
> Chigger mites do not burrow; they create a small depression in the skin using tube-like mouthparts called the stylostome.

Other Mites of Medical Significance, Including Harmless Mites

Scabies mites, bird mites, and chigger mites are frequently referenced by patients, but there are other species of mites to consider. Following is a list of species in order of prevalence that can cause dermatitis (Table 2), Figs. 5 and 6. Table 3 lists additional mite species found associated with buildings that are not parasitic.

Table 2 Common Mites

Mite common name	Mite scientific name	Source or host
Walking dandruff in cats	*Cheyletiella blakei*	Cats
Feline scabies (Notoedric mange)	*Notoedres cati*	Cats
Canine scabies (Sarcoptic mange)	*Sarcoptes scabiei var canis*	Dogs
Tropical rat mite	*Ornithonyssus bacoti*	Rats and other rodents
Straw/hay itch mite Oak leaf itch mite	*Pyemotes ventricosus* *Pyemotes herfsi*	Straw or Hay Oak galls
Rabbit fur mite	*Cheyletiella parasitivorax*	Rabbits
Follicle mites	*Demodex folliculorum,* *Demodex brevis*	Human skin. Can be asymptomatic or symptomatic

Fig. 5 Follicle mite,
Demodex folliculorum.
Photo: Jerry Butler, Univ.
of Florida

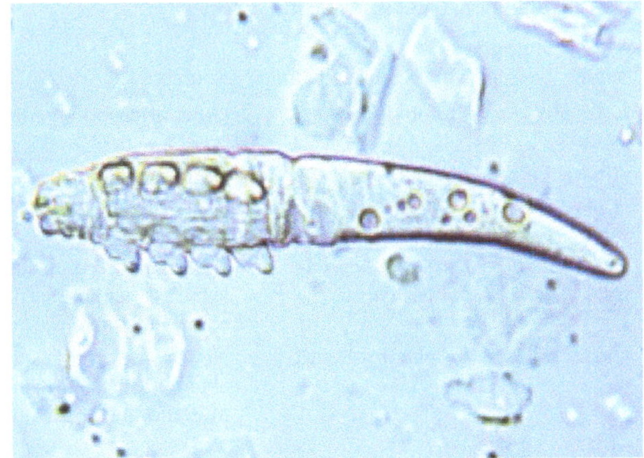

Fig. 6 Tropical rat mite,
Ornithonyssus bacoti on a
millimeter ruler. Visual
appearance is identical to a
bird mite. Photo: Lyle
Buss, Univ. of Florida

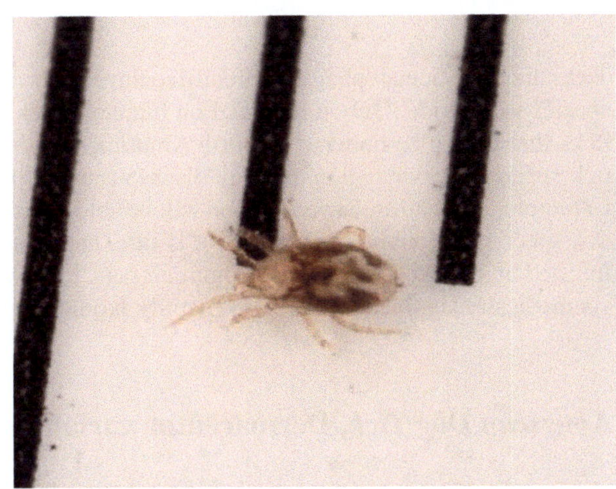

Note. In 1951, Jay Traver published a paper in the Proceedings of the Entomological Society of Washington describing a particular species of mite, *Dermatophagoides scheremetewskyi*, as a parasite of humans. She claimed that her scalp was infested by the mite [5]. This is incorrect! Later research revealed that Traver's mite was not a parasitic mite, but a species of common house dust mite. Jay Traver suffered from DI.

Box 8. Myths: "Paper Mites / Black Pepper Mites."
"Paper mites / Black pepper mites have infested my home/office."
   The paper mite myth arose from clerical workers trying to explain persistent itching sensations at their workplace. The black pepper mite myth evolved from homeowners who saw unexplained "black specks" in their homes. Neither of these alleged mites exist.

Table 3 Additional non-parasitic mite species

Mite common name	Mite scientific name	Source	Medical significance
Red velvet mites	Trombidiidae	Other arthropod predators	Harmless
American house dust mite	*Dermatophagoides farinae*	Buildings. Feed on shed skin cells and prefer moist warm locations	Contact dermatitis. Respiratory allergies involving asthma
European dust mite	*Dermatophagoides pteronyssinus*	Buildings. Feed on shed skin cells and prefer moist warm locations	Contact dermatitis. Respiratory allergies involving asthma
Clover mites	*Bryobia praetiosa*	New lawns or lawns that abut buildings allowing migration into structures	Harmless. May leave a red smudge when crushed

3 Ticks (Family Ixodidae)

Ticks are large hematophagous (blood-feeding) mites. There are approximately 900 species worldwide. Only a few feed on humans. Following are the more common ticks that parasitize humans in North America. It is important to be familiar with tick species in your region and any disease-causing pathogens they might vector. Entomologists who are in your region will be able to provide information about local tick species and public health risks. It is rare, but some DI patients describe being infested by ticks that have come from their pets. Following are brief descriptions of six medically significant hard ticks (Family Ixodidae) [6].

American Dog Tick, **Dermacentor variabilis** *(Say)*

The American dog tick (Fig. 7) is distributed east of the American Rocky Mountains and in limited areas on the American Pacific coast. This tick is occasionally called the wood tick.

Life Cycle and Behavior

American dog ticks overwinter in the soil, becoming active from mid-April to September. Adult females are active feeders, while males may briefly feed. There are three stages of development: larva (can survive 11 months without feeding), nymph (can survive 6 months without feeding), and adult (can survive 2 years without feeding). On a host, larvae feed for 2–14 days, nymphs feed for 1–3 days, and adults feed for 5–14 days. Females develop their egg clutches over 4–10 days and lay 4,000–6,500 eggs on the ground. Between 26 and 40 days later, the eggs hatch.

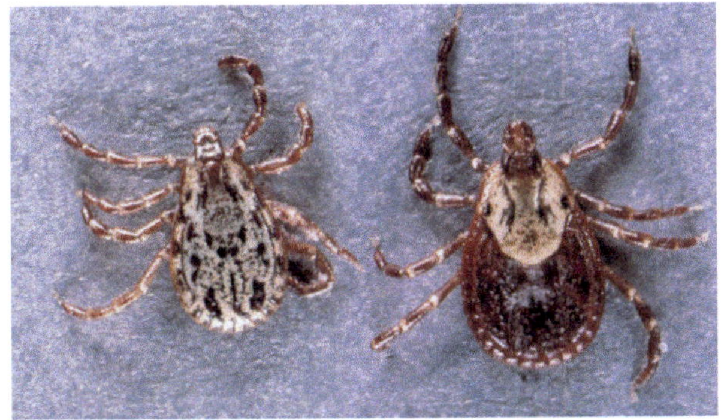

Fig. 7 American dog tick, *Dermacentor variabilis*. Male (left) & female (right). Photo: Jerry Butler, Univ. of Florida

These are passive questing ticks that wait in ambush on low vegetation to hook onto a host as it walks by. Dogs are common hosts, but the ticks will feed on any medium-sized animals.

Medical Importance

The American dog tick can vector the disease-causing pathogens for Tularemia (*Francisella tularensis*) and Rocky Mountain spotted fever (RMSF, *Rickettsia rickettsia*). Tularemia can cause distinctive S-shaped lesions in the webbing between the fingers. This can be mistaken for scabies. RMSF affects small peripheral blood vessels, causing rashes which develop 2–14 days following tick feeding. This is a seasonal pathogen and is present between April and September.

The Blacklegged/Deer Tick, Ixodes scapularis

The blacklegged or deer tick (Fig. 8) is found throughout eastern North America. The western blacklegged tick, *I. pacificus* (Cooley and Kohls) is found along the west coast, Utah, Arizona, and New Mexico (USA).

Life Cycle and Behavior

Blacklegged ticks have a 2-year life cycle. Adults are active September–May, although questing (host-seeking) activity pauses when air temperatures are below freezing and when snowpack prevents them from accessing hosts. Adult females lay up to 2,000 eggs in the late spring. Nymphs (larvae from the previous year) are active May–August. Newly hatched larvae are active August–September. Like the American dog tick, these are passive questing ticks that climb onto low vegetation

Fig. 8 Blacklegged tick,
Ixodes scapularis. Photo:
Lyle Buss, Univ. of Florida

and hook onto a host as it passes by. Preferred hosts for immatures are small- and medium-sized animals such as rodents and birds. Deer are the principal hosts for adults. All life stages can feed on humans.

> **Box 9. Myths: Lyme Disease.**
> *"Lack of a "bulls-eye" erythema migrans (EM) rash rules out Lyme disease."*
>   Not all cases of Lyme disease present with a rash.
>   *"All EM rashes have a "bulls-eye" appearance."*
>   This is more typical of older rashes and is only noted in less than half of cases.
>   *"Any dermal reaction to a tick bite indicates Lyme or other tick-borne disease."*
>   Common reactions to tick bites are pruritus and localized pain. EM rashes rarely itch and are usually not painful.

Medical Importance

Nymph and adult blacklegged tick feeding can vector the bacterial spirochaete that causes Lyme disease (*Borrelia burgdorferi*), the bacterial pathogen that causes human granulocytic ehrlichiosis (*Ehrlichia* spp.), and the malaria-like protozoan that causes human babesiosis (*Babesia microti*).

Cases of tick-borne disease increase between May and September in the northeast United States because of increased human outdoor activity in habitats with tick populations. The nymphs, which are active in the spring and summer, are smaller and less likely to be detected and removed prior to pathogen transmission.

Common symptoms associated with Lyme disease are flu-like: fatigue, muscle or joint pain, stiffness, fever, chills, and headache. Since none of these symptoms are specific clinical markers, Lyme disease could be overlooked when patients seek medical care in areas where Lyme disease is uncommon. It is important to check if a patient had recently visited a Lyme disease endemic area.

In 70–90% of Lyme disease infections, a "bulls-eye" erythema migrans (EM) rash will occur at the site of the tick bite 3–30 days following pathogen transmission. This is a clinical indicator of early infection, although the absence of a rash does not rule out Lyme disease. Note that any tick bite will cause a temporary dermal reaction [6]. Additionally, EM rashes may be cryptic depending on location and skin type.

The Lone Star Tick, Amblyomma americanum (L)

The lone star tick (Fig. 9) is found throughout the southeastern United States, where it is responsible for most of the human tick bites. Its range has been expanding northward and it is now increasingly found along coastal New England. This tick is named for the conspicuous white spot on the scutum of the adult female, not because it is from Texas, "the lone-star state." This is a common misperception.

> **Box 10. Myths: Ticks.**
> *"Ticks can fly, jump, and drop out of trees onto hosts."*
> Ticks' legs are adapted for walking, climbing, and clinging. They do not have wings or the capacity to jump. American dog ticks and blacklegged ticks climb onto low-growth vegetation (ankle to knee height) and hook onto passing hosts using the front pair of legs; lone star ticks actively seek and approach

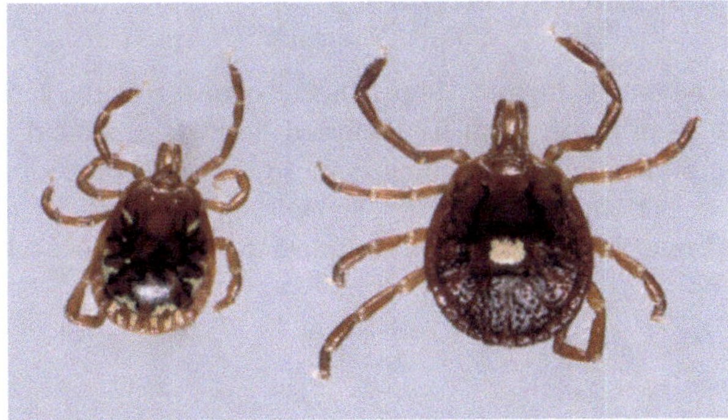

Fig. 9 Lone star tick, *Amblyomma americanum.* Photo: Jerry Butler, Univ. of Florida

hosts. Ticks can be transferred when a person brushes up against or touches clothing or animals with ticks on them. People may assume ticks are only found in tall grass and brushy areas, and therefore avoid these areas while walking, but ticks can be picked up in open bare locations as well.
"Anyone bitten by a lone star tick will develop a red meat allergy."
Only a small percentage of people bitten by lone star ticks have developed a red meat allergy (alpha-gal syndrome).

Life Cycle and Behavior

Lone star tick adults are active during spring, nymphs are active in April through mid-summer, and larvae in late summer into early fall/autumn. These ticks are more aggressive than the American dog and blacklegged ticks because they actively move toward a host rather than engaging in passive questing [6, 7]. They have a wide host range and feed on most mammals.

Medical Importance

The lone star tick is a vector for the pathogens that cause ehrlichiosis in both humans (*Ehrlichiosis chaffeensis*) and dogs (*E. ewingii*). They also vector the disease-causing pathogens for Tularemia (*Francisella tularensis*), Rocky Mountain Spotted Fever (RMSF, *Rickettsia rickettsia*), spotted fever rickettsiosis (*R. parkeri*), and Heartland virus disease(a Phlebovirus, Family Bunyaviridae). Although they do not transmit the causative agent of Lyme disease (*B. burgdorferi*), lone star ticks are linked with a Lyme-like illness called southern tick-associated rash illness (STARI), the causal agent of which is currently unknown. Bites from the lone star tick have also caused some individuals to develop an allergy to red meat, known as alpha-gal syndrome.

The Brown Dog/Kennel Tick, **Rhipicephalus sanguineus** *(Latreille)*

The brown dog tick (Fig. 10) is also known as the kennel tick because it is associated primarily with its preferred host, dogs. Found worldwide [8] though more prevalent in warmer climates, this tick infests outdoor and indoor areas where dogs are present, such as homes, kennels, animal hospitals, and pet bedding. In cold temperate climates, they can complete their entire life cycle inside buildings.

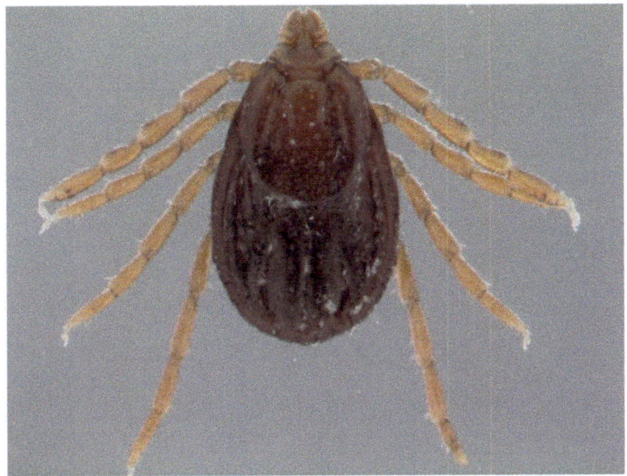

Fig. 10 Brown dog tick, *Rhipicephalus sanguineus*. Photo: Lyle Buss, Univ. of Florida

Life Cycle and Behavior

The brown dog tick can complete its life cycle in 2 months under favorable conditions. When infesting indoor locations, the ticks seek protected areas such as cracks and crevices. This is a three-host tick, meaning each stage abandons its host after taking a blood meal. Following a blood meal, immatures molt to the next stage and then seek another host. Adult females leave their host to lay eggs. Brown dog ticks that cannot find their principal host (dogs) may feed on alternate hosts such as cats, rodents, or birds.

Medical Importance

This tick can vector the disease-causing pathogens for canine ehrlichiosis (*Ehrlichia canis*) and canine babesiosis (*Babesia canis* and *B. gibsoni*). Since it readily moves from host to host, it can easily transmit pathogens among dogs, especially in infested public kennels, shelters, or animal hospitals.

Recent studies have demonstrated that at higher natural temperatures, this tick will more readily feed on humans and rabbits. With climate change perpetuating warmer and longer summers, the risk of human parasitism has increased [8, 9]. This has led to a greater risk for transmission of zoonotic pathogens such as *Rickettsia conorii* and *Rickettsia rickettsia* to humans.

The Asian Longhorned Tick, Haemaphysalis longicornis *Neumann*

The Asian longhorned tick (Fig. 11) is a new introduction to North America. It was first detected in 2017 on a sheep at a New Jersey farm. Native to eastern China, Russia, the Koreas, and Japan, it has recently become established along the eastern

Fig. 11 Asian longhorned tick, *Haemaphysalis longicornis* Neumann. Photo: K. Dugas, CAES

United States seaboard [7]. This tick is an important livestock pest in Australia and New Zealand.

Life Cycle and Behavior

The Asian longhorned tick has a similar biology to the blacklegged tick. Additionally, this tick can reproduce asexually (parthenogenisis); males have not been found in North America. Although its primary hosts are cattle, horses, sheep, and goats, it can also feed on native wildlife, pets, and humans.

Medical Importance

It is a competent vector of *Borrelia*, *Babesia*, *Anaplasma*, *Ehrlichia*, *Rickettsia*, thrombocytopenia syndrome virus (SFTSV), and Powassan virus in humans [7, 10].

The Gulf Coast Tick, Amblyomma maculatum *Koch*

The Gulf Coast tick (Fig. 12) was historically distributed along the southeastern United States and along the Gulf and Atlantic coasts. It has expanded its range northward along the eastern seaboard into New England. It was first detected in Connecticut late summer 2020 [7].

Fig. 12 Gulf Coast tick,
Amblyomma maculatum.
Photo: Jerry Butler, Univ.
of Florida

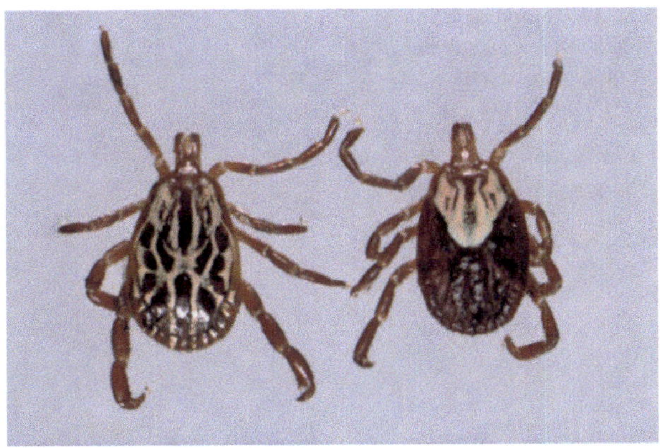

Life Cycle and Behavior

The Gulf Coast tick prefers natural grass prairies, coastal uplands, and grassy agricultural fields. It principally feeds on birds, wild mammals, and agricultural animals including cattle. Increasingly they are reported biting humans. They are closely related to the lone star tick. This tick is most active between April and September, but in southern regions of the United States, February to September. The adult male and female markings are similar to American dog ticks, but their mouthparts are elongated and they have red bodies with clean silver markings. This is a 3-host tick where a blood meal is required at each stage of development. The immature stages feed on small rodents and birds, while adults feed on cattle, bears, dogs, and coyotes.

Medical Importance

The Gulf Coast tick is a competent vector of the disease-causing pathogen American tick bite fever (*Rickettsia parkeri),* a Gram-negative intracellular bacterium that causes a form of mild spotted fever in humans. While feeding, the tick can cause tick paralysis, a progressive form of paralysis that starts with the feet and moves up the body. This can be lethal in small children and pets. If a child or pet develops unexplained paralysis, check to see if a tick might have attached at the base of the skull or neck.

4 Spiders (Class Arachnida, Order Araneae)

Few living creatures elicit anxiety more than spiders. Spiders are predators, feeding mostly on insects and other arthropods. They don't feed on humans but may bite in self-defense. Even though many spiders are capable of biting humans, they rarely do so, instead fleeing or retreating into a shelter when disturbed. Bites may

Fig. 13 Brown recluse spider, *Loxosceles reclusa*. Photo: Richard Vetter, UC Davis, California

occur when spiders are crushed against bare skin. Generally, the spiders capable of biting people (possessing mouthparts with the capacity to penetrate mammalian skin) are large and obvious. Spider bites are almost always a one-time occurrence. Patients that present with multiple mysterious "bites" on their bodies are not dealing with spiders.

Spiders are predators, and possess venom for hunting. The venom of most spiders causes only minor pain and swelling in vertebrates, but a few spiders are medically significant. In North America, this mainly includes the recluse and widow spiders.

Recluse Spiders (Family Sicariidae, Loxosceles spp.)

There are two recluse spiders of concern. The brown recluse spider (*Loxosceles reclusa*) (Fig. 13) is native to North America and the Mediterranean recluse spider (*L. rufescens*) has become cosmopolitan through worldwide travel and trade.

Name and Distribution

1. **Brown recluse spider, *Loxosceles reclusa* Gertsch & Mulaik.**
Found in the south-central portion of the United States and Mexico.
US states it is found in are:

> Arkansas (high populations), Missouri (high populations), southeastern Nebraska, Kansas, northern Louisiana, Oklahoma, western Tennessee, western Kentucky, southern Indiana, southern Illinois, and northeastern Texas.

2. **Mediterranean recluse spider, *Loxosceles rufescens* (Dufour).**

It originated in the Mediterranean area and western Asia and now can be found in other regions of the world. It is listed as one of the most invasive spider species in the world. In the United States, its distribution is spotty. It prefers caves, man-made tunnels, and damp basements or cellars and likes to feed on cockroaches and silverfish. The spider is nearly identical in appearance to the brown recluse.

Important!
Check that recluse spiders are endemic in the region you currently work. There have been numerous misdiagnoses of brown recluse spider envenomization in non-endemic regions. Bacterial infections and similar dermatological issues can mimic necrotic lesions caused by recluse spider venom.

Life Cycle and Behavior of the Brown Recluse

Brown recluse spiders are small (1 in./2.5 cm) and light brown, with a darker fiddle-shaped pattern on the dorsal surface behind the head (the cephalothorax). They have *six* eyes that are grouped in *three* pairs following a semi-circle (Fig. 14). Since few spiders share this character (most spiders possess eight eyes), eye arrangement can be used as a quick way to rule out brown recluse spiders. Do not solely rely on the "fiddle pattern" for identification. This pattern can be easily misinterpreted by nervous patients. Confirm identification of any spider with an arachnologist or entomologist before proceeding.

Brown recluse spiders can live up to 2 years. It takes 1 year for development from egg to adult. The spiders are drought and starvation-resistant, living for long periods without food or water. They build irregular webs within shelters such as in woodpiles, under tree bark, sheds, garages, cellars/basements, and other dry poorly-lit undisturbed locations. They are nocturnal and shy.

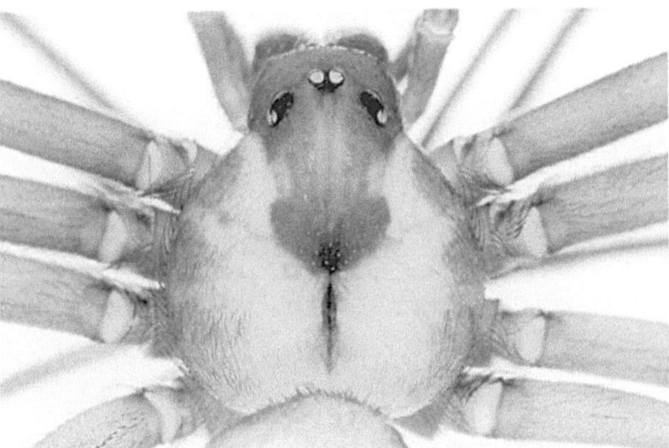

Fig. 14 Brown recluse spider eye pattern. Photo: Richard Vetter. UC Davis, California

Medical Importance

These small-fanged spiders are not aggressive and bite in self-defense when compressed or crushed against bare skin. For example, bites may occur when a sleeping person rolls over onto a spider or puts on clothing in which a spider is hiding. Envenomization injury is distinctive and can cause necrosis.

Richard Vetter (UC Davis, California, USA), an expert in brown recluse spiders described in his book, "The Brown Recluse Spider" [11], that most brown recluse spider bites are unremarkable and mild. They often heal on their own without any medical intervention. Serious cases involving severe skin necrosis are not the norm and account for up to 10% of all brown recluse bite cases. In his book, Vetter describes the appearance of the skin resulting from a brown recluse envenomization:

> "One of the first events in a brown recluse bite is the collapse of the capillaries at the bite site or destruction of the capillary tissues (destruction of endothelial cells lining the blood vessels) which prevents red blood cells from getting to the affected area. This loss of blood flow to the area (ischemia) is the reason the skin turns white. The skin may also be purplish from blood leaking into the surrounding tissue, causing bruising." [11]

To confirm any suspected brown recluse spider "bite," the patient must be living or recently moved from a region where recluse spiders are found, and the spider must also be collected and identified.

Assumptions for the cause of brown recluse "bites" are numerous. Physicians in non-endemic brown recluse regions should be very cautious when diagnosing idiopathic necrotic wounds. There are numerous other conditions which should be explored. See Box 11.

Box 11. Conditions That Might Prompt Misdiagnoses of Brown Recluse Spider-Caused Necrotic Lesions.

1. Chemical burn
2. Diabetic ulcer
3. Erythema multiforme
4. Fungal infection
5. Herpes simplex
6. Herpes zoster
7. Localized vasculitis
8. Lyme disease
9. Lymphomatoid papulosis
10. Poison ivy/oak dermatitis
11. Pyoderma gangrenosum
12. Squamous cell carcinoma
13. *Staphylococcus* or *Streptococcus* species infections
14. Syphilitic chancre [12].

Box 12. Myths: Brown Recluse Spiders.
"Brown recluse spiders readily bite people."

These spiders possess very small weak fangs that cannot penetrate fabric, only thin skin. In order to bite, the spider must be pressed against bare skin. They are shy and retiring.

Due to confirmation bias supported by rampant misinformation about the spider - particularly on the internet and the graphic photographs of purported brown recluse "bites" - social anxiety regarding these spiders is prevalent.

Widow Spiders (Family Theridiidae, **Latrodectus** *spp.)*

Name and Distribution.

These are the most important species of widow spiders in North America, along with their general distribution.

Northern black widow, *Latrodectus variolus.* Eastern half of the US.

Southern black widow, *L. mactans.* Eastern US up to southern New England.

Brown widow, *L. geometricus* (Fig. 15). Southern US, and many subtropical regions of the world.

Western black widow, *L. hesperus.* Western half of the US.

Fig. 15 Brown widow spider, *Latrodectus geometricus.* Photo: Lyle Buss, Univ. of Florida

Life Cycle and Behavior

Black widow spiders live for 1 year. They build loose disorganized three-dimensional webs which they sit to one side on. They have poor eyesight and use vibration to detect captured prey. Widow spiders are shy and reclusive, choosing quiet undisturbed locations such as stone walls, log piles, quiet garages, rubbish heaps, or outside toilets.

Medical Importance

Widow spiders are not aggressive, prefer to flee, and only bite in self-defense. Females bite while males do not. Bites usually happen when someone unknowingly compresses or crushes a spider against bare skin, usually by inserting a hand into a location where a spider is hiding. Widow spider venom contains a neurotoxin, but they can "dry bite," injecting no venom. A full envenomization can cause significant acute illness that requires hospitalization. It is highly unlikely that an ambulatory patient is dealing with an active envenomization. To be certain a patient has been bitten by a widow spider, it is important to collect the spider for identification.

> **Important!**
> Check if widow spiders are endemic to your region.

> **Box 13. Myths: Black Widow Spiders.**
> *"All black widow females eat their mates."*
> Most male black widows escape after mating. The common name "black widow" is based on biased observations of captive specimens, where males had limited opportunity to escape and were eaten.
> *"Black widow bites are frequently fatal."*
> Fatal bites from black widow spiders are very rare. Symptoms of a black widow bite often include localized pain at site of the bite, cramping and stiffening of abdominal muscles, nausea, sweating, difficulty breathing, restlessness, fever, and increased blood pressure. These symptoms may continue for several days. Medical care/hospitalization is recommended until symptoms resolve.

5　Insects (Class Insecta)

Bed Bugs (Order Hemiptera, Family Cimicidae)

Names and Distribution.

Two principal species of human-feeding bed bugs are the common bed bug (*Cimex lectularius* L.) (Fig. 16), found worldwide in temperate regions and the tropical bed bug (*C. hemipterus* (F)) (Fig. 17), confined to tropical and subtropical regions of the world.

Life Cycle and Behavior

Human-feeding bed bugs are temporary hematophagous (blood-feeding) ectoparasites. They do not live on the host. They have piercing-sucking mouthparts which are inserted in the host's skin to draw blood. Their saliva has anesthetic properties and so feeding is not felt. Bed bugs feed once every 8–10 days to effect development (nymphs) and reproduction (adults). Development from egg to adult is 4–6 weeks. They live near places where humans rest or sleep and synchronize feeding behaviors with their hosts' habits and activities. They are delicate, shy, and reclusive, preferring to hide in cracks and crevices.

Medical Importance

Bed bugs are incompetent vectors of disease-causing pathogens. Secondary infection is possible from scratching the feeding sites. Dermal responses per individual ranges from nothing to nodular prurigo. Systemic reactions requiring hospitalization are very rare. Skin irritation usually clears up within 48 hours. Bed

Fig. 16 Common bed bug, *Cimex lectularius*. Photo: Michael Thomas, CAES

Fig. 17 Tropical bed bug, *Cimex hemipterus*. Photo: Lyle Buss, Univ. of Florida

bugs garner significant psychological malaise and anxiety. Some individuals experience a psychological reaction to bed bugs similar to the five stages of grief, and inappropriate coping behaviors may occur.

When a patient presents or complains about mysterious "bites," bed bugs are frequently implicated. It is suggested that a professional exterminator perform an inspection of patient housing, or, if insects have been found, to have them identified by an entomologist or other expert. To those unfamiliar with insect taxonomy, a number of insect species can appear superficially similar to bed bugs. Small and newly-established populations of bed bugs can be difficult to detect, requiring an exterminator who has experience with these insects. Cases of mistaken or assumed identity result in unnecessary distress and hardship to a patient. If no evidence of a bed bug infestation is found, then other possible causes for reported "biting" sensations by a patient should be investigated.

> **Box 14. Myths: Bed Bugs.**
> *"Bed bugs feed in rows, "breakfast, lunch, and dinner."*
> This myth is perpetuated by misunderstood observations. Bed bugs are capillary feeders, prefer to feed once, and do not inherently feed in straight lines. Physical restrictions of access to the skin by clothing for multiple insects may lead to the appearance of linear feeding.
> *"Bed bugs hop, jump, and/or fly."*
> Bed bugs' legs are developed for walking, climbing, and clinging. While they are descended from flying ancestors, modern bed bugs have only vestigial "wing pads" rather than fully-developed wings. They are physically incapable of hopping, jumping, or flying.
> *"Bed bugs prefer filthy places."*
> This myth is perpetuated by human class culture. Bed bugs are incapable of comprehending concepts such as cleanliness, luxury, or class. Bed bug populations can be found both in luxury settings and abject squalor.
> *"Bed bugs live on people."*
> Bed bugs are temporary ectoparasites, meaning they spend most of their time off the host; they only interact directly with the host while feeding. Bed bugs are predominantly found in locations adjacent to where human hosts sleep or rest. Sedentary living situations may result in insects lingering on the host's body, but this is atypical behavior.

Human Lice

Name and Distribution

Sucking lice (Order Psocodea, Superfamily Anoplura) are hematophagous permanent ectoparasites of mammals, including humans [13].

There are three species that parasitize humans. These are as follows:

Human head louse, *Pediculus humanus capitis* De Geer (Fig. 18).
Human body louse, *Pediculus humanus humanus* Linnaeus and
Human pubic/crab louse, *Pthirus pubis* (Linnaeus).

Distribution of these three species of lice is worldwide.

Head and body lice are morphologically identical subspecies (morphotypes) separated by where they live on the human body.

Life Cycle and Behavior

Head lice cement their eggs to hair shafts *next* to the scalp where there is high humidity. Body lice (nicknamed "seam squirrels") lay their eggs along fabric seams of clothing the host is wearing, and crab lice lay their eggs on pubic and perianal

Fig. 18 Head lice, *Pediculus humanus capitis*. Photo Lyle Buss, Univ. of Florida

hair. All three species feed multiple times a day by inserting their piercing-sucking mouthparts into the skin to draw blood. Development from egg to adult is approximately 3 weeks. There are three nymphal stages of development before moulting to adults.

Medical Importance

Lice can cause localized skin irritation. With high numbers there can be dermal harm with thickening and darkening of the skin. Additionally, human lice can vector several human disease-causing pathogens. These are epidemic typhus (*Rickettsia prowazekii*), relapsing fever (*Borrelia recurrentis),* and the Trench fevers (*Bartonella quintana* and *Acinetobacter baumannii).* These pathogens are often reported in jails/prisons, refugee camps, and developing countries.

Louse activity can be confused with psoriasis, basal cell and squamous cell carcinoma, underlying psychiatric and/or undiagnosed medical conditions, tinea capitis, secondary bacterial infection, or neuropathy. As lice live directly on their human hosts, they are some of the first parasites that can be ruled out when a patient complains of "bites." When dealing with a DI patient, thoroughly check for lice.

Biting Midges and Flies, Bot Flies, and Mosquitoes (Order Diptera)

A variety of flies attack people to get a blood meal. Most females bite as they need a high protein source (blood) for egg production. Many males do not bite and survive by feeding on the nectar of flowers. Biting flies are not true parasites, as they only visit their hosts briefly. Bot flies are true parasites as their larvae develop within their hosts.

Names and Distribution

Horse flies, deer flies (Family Tabanidae),—found worldwide.

Stable fly, (*Stomoxys calcitrans* (Linnaeus), Family Muscidae)—found nearly worldwide.

Black flies, (Family Simuliidae)—found worldwide.

Biting midges, (*Culicoides* spp., *Leptoconops* spp., Family Ceratopogonidae)—found worldwide, although more abundant in coastal areas (Fig 19.)

Human bot fly, (*Dermatobia hominis* (Linnaeus Jr. in Pallas), Family Oestridae)—found in Central and South America from Mexico to Argentina excluding Chile.

Mosquitoes, (Family Culicidae)—found worldwide. Highest densities in the tropics.

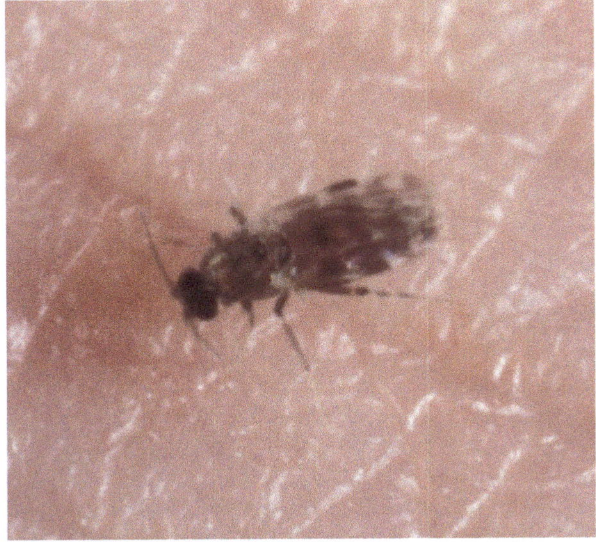

Fig. 19 Biting midge, *Culicoides barbosai.* Photo: Lyle Buss, Uni. of Florida

Biologies and Behaviors

Flies that lacerate the skin. Flies that bite do so because they are hematophagous (blood-feeding). These flies lacerate the skin with their mouthparts, and then lap up the blood. This is different from ants, bees, and wasps, which sting rather than bite in defense of a nest or self-protection.

Several types of hematophagous flies may bite humans. Most of them are large and obvious. Horse flies, deer flies, and stable flies are the size of a house fly or larger. In many areas proximal to natural bodies of water or wetlands, black flies can be abundant biters. They usually measure around 2–5 mm in length. Because they live outdoors and are not exceedingly small, large biting flies are not usually implicated as the causal agent by DI patients.

The smallest biting flies, commonly called no-see-ums, punkies, or biting midges, are 1–2 mm in length. Biting midges breed in a variety of habitats, but mostly semi-aquatic areas near bodies of water, salt marshes, and in soils high in organic matter. Only females bite. Where they are common, they can disrupt outdoor human and animal activities. Being small, people will often feel their painful bite but don't see them, hence the name "no-see-ums." Their small size also allows them to enter buildings through standard window screens.

Flies that pierce the skin. Mosquito mouthparts are adapted for piercing skin and drawing blood. This makes the mechanism for blood feeding different from biting flies. Much of what constitutes a mosquito "bite" is the host's immune reaction post-feeding and irritation due to scratching.

Flies than live in the skin. Human bot flies don't bite, but their larvae develop under the skin of vertebrates. Because of this, some people suspect them as the cause of their skin wounds. Human bot flies are endemic to New World tropical regions (Mexico, Central America, and most of South America); to become infested, a patient would have to live in or have visited one of these areas. Rodent and rabbit bot flies (*Cuterebra* spp.) are endemic to North America, but very rarely attack humans [14].

Medical Importance

In some regions of the world, biting midges transmit parasites to humans, but in the United States they are a nuisance pest. Most people feel the annoying bites, but some people experience stronger allergic responses. Some patients suffering from unidentified "biting "will sometimes blame these flies. However, they are not indoor insects and would not be responsible for continuous patient experience.

Patients who feel "bites" at home often collect specimens for identification. It is important to identify these specimens to make sure they are not the cause. Identification of biting midges is difficult and should be performed by an entomologist. In North America, the biting midges that attack people are in the genus *Culicoides*. However, there are many other genera of biting midges that feed only on animals. In Florida, for example, midges in the genera *Forcipomyia* and *Dasyhelia* are sometimes found in homes, but they do not bite people.

Sometimes people will use the colloquial terms "sand flies" or "sand fleas" to describe unexplained biting sensations, especially in coastal areas. However, "sand flies" more correctly refers to a group of flies in the family Psychodidae, and the human biters in this group are not found in the USA. "Sand fleas" usually refers to jumping crustaceans in the family Talitridae, which do not bite humans.

Box 15. Myths: Biting Flies.
"No-see-ums live in homes."
No-see-ums may occasionally enter homes through open windows, doors, or through screening, but they do not live or breed indoors.
"No-see-ums are invisible."
No-see-ums are very small, but still visible to the naked eye. They are much smaller than mosquitoes and are difficult to see on the skin.
"No-see-ums live in people."
No-see-ums bite humans or animals to obtain a blood meal, and then return to semi-aquatic habitats to lay eggs.

Fleas (Order Siphonaptera)

Name and Distribution

While there are approximately 2,500 known species of fleas worldwide, those significant for human health include rodent fleas (*Xenopsylla* spp.) and other domestic/urban wildlife fleas (Family Pulicidae) [15] (Fig. 20).

The fleas most likely to be encountered in a medical situation are associated with domestic animals and livestock. Common domestic animal fleas include the cat *Ctenocephalides felis* (Bouché) and dog *C. canis* (Curtis) fleas. Distribution of these groups and species is worldwide.By far, the most common flea around humans in North America is the cat flea. Humans are not a true host, but cat fleas are common on pet dogs and cats, as well as opossums, raccoons, and other wild animals. If patients feel bites in their homes, they should call a pest management professional to inspect for fleas, especially if dogs or cats are present.

The human flea, *Pulex irritans* Linnaeus, is considered cosmopolitan though it is uncommon in western industrialized nations.

Not all species of fleas are human "biters," and so a regional knowledge of flea species is suggested.

Fig. 20 Cat flea,
Ctenocephalides felis.
Photo. K. Dugas, CAES

Life Cycle and Behavior

Rodent and domestic/urban animal fleas have similar biologies. Development from egg to adult is from 3 to 6 weeks. Females lay non-sticky eggs on the fur of host animals. These roll off the pelt into the bedding below. Hairy white translucent worm-like larvae hatch and scavenge on debris and adult flea feces. Adult fleas do not fully digest their blood meals for the purpose of feeding their offspring. Flea larvae feed on the fecal material to pick up vitamin B 12. If they do not get the vitamin, they will not develop. Following the larval stage, larvae pupate in silken cocoons and emerge as adults.

> **Important!**
> Be familiar with the flea species endemic to the region you currently work in.

Medical Importance

Flea larvae are scavengers and do not bite. The adult fleas are parasitic and live on the host. In humans, flea bites occur only on lower legs. If there are lesions on the upper body, these might be from the carrying, hugging, or co-sleeping with an infested pet. Lesions from flea bites are usually round red spots which can become pruritic.

Tungiasis. Occasionally DI patients may self-diagnose with Tungiasis caused by the chigoe flea, *Tunga penetrans* (Linnaeus). This flea attacks people and is unusual in that the adult female burrows into the skin, usually the feet. They are found in subtropical and tropical regions of the world, including the West Indies, Mexico into South America, and Africa. They are not found in North America.

> **Box 16. Myths: Fleas.**
> *"Fleas can fly."*
>   Adult fleas are wingless, but can jump, giving the illusion of flight. Flea legs possess an elastic protein resilin pad that when compressed, acts like a spring. When jumping, the pad releases stored energy and catapults the flea into a tumbling arc. Cat fleas jump an average distance of 8 inches, dog and human fleas 12 inches, while rodent fleas jump 7 inches.

Springtails (Class Entognatha, Order Collembola)

Name and Distribution

Springtails (Order Collembola) (Fig. 21) are primitive, moisture-loving hexapods (see glossary), that live in leaf litter and soil. They are tiny and can easily wander into buildings under doors or around windows. This often occurs during heavy rains or severe droughts. Occasionally they are found in the soil of over-watered house-plants. They are reputed to be one of the most abundant of all microscopic animals worldwide numbering 100,000 per square meter in suitable conditions. They have a worldwide distribution.

Life Cycle and Behavior

Many springtail species are active upper soil horizon decomposers. They fragment decaying organic material such as leaf litter. They make nutrients available to many species of microbes and fungi and improve soil conditions for plants. Many species

Fig. 21. Springtails, order Collembola. Photo. Lyle Buss, Univ. of Florida

possess a furcula, a spine-like ventral spring which can propel them away from danger to safety, hence the name.

Medical Importance

Springtails are medically harmless. Why then, are springtails included in this guide? A paper written by Altschuler et al., published in 2004 [16] is often cited as "proof" by patients claiming springtails are infesting them (Box 17). These patients mistakenly believe springtails are the cause of their discomfort and seek erroneous affirmation online.

For a DI patient looking for the cause of their discomfort, the physical presence of numerous small and erratically-jumping springtails inside a building can cause significant distress. Springtails may move into buildings temporarily because of adverse weather events, but cannot survive indoors long-term. They will disappear when environmental conditions stabilize, or the building becomes too dry for them to survive.

> **Important!**
> Springtails are medically harmless and do not bite or parasitize humans.

> **Box 17. Myths: Springtails.**
> *"Springtails bite and parasitize humans."*
>
> Altschuler et al. created an urban myth when in 2004, they published an erroneous paper about the medical significance of springtails. The authors made a claim that springtails cause dermal "itching, stinging/biting and crawling sensations on or under their skin, which are often associated with excoriations, discoloration, scaling, tunneling or sores." They supported their claims with imagery of skin scrapings purportedly containing embedded springtails [16].
>
> In 2008, Christiansen and Bernard, both Collembolan experts, challenged the paper as erroneous and concluded the published images had been significantly edited. Their comments regarding a series of panels progressing from low to high contrast in the Altschuler et al. paper were:
>
> > "The mere labelling of a vaguely-recognizable blob, as in Fig. 2, does not validate the identification of the parts, which in this case is more similar to identifying animals by looking at clouds" [17]. (Also see Box 4: Pareidolia and Fictional Biologies)
>
> Lim et al. [18] supported the Christiansen & Bernard paper by affirming that Collembola are incapable of causing medical harm to humans. The Altschuler paper continues to be used by some organizations and DI patients to support their belief systems.

6 Non-arthropod Parasites: Nematodes

For more details on this subject, please read Part III Chapter 12, "The Role of the Clinical Parasitology Laboratory in Delusional Infestations," written by Bobbi S. Pritt and Blaine A. Mathison. They provide excellent information about non-arthropod parasites.

> **Box 18. Pareidolia and Fictional Biologies.**
> Humans inherently seek patterns. Pareidolia is a human tendency to see meaningful images in random ambiguous shapes such as seeing faces in clouds, landscapes, even burned toast. Rorschach's inkblot test is an example. This phenomenon is thought to be the consequence of the human brain attempting to rapidly recognize patterns and faces; it is effectively "filling in the blanks" of an incomplete image and erroneously interpreting the results.
>
> When confronted with poorly-understood sensations and when direct answers are not forthcoming, the brains of DI sufferers also attempt to "fill in the blanks." Their anxiety manifests as an erroneous explanation of their symptoms—a fictional parasite biology. Often entire life cycles will be intricately described by DI patients, which may or may not be inspired by real-world parasites or fictional ones from books, movies, or social media. These fictional parasites' life cycles will often easily and drastically change to "fit" any variation in symptoms, in direct contradiction to how a real-world biology operates. A DI patient's continual commitment to fictional biologies delays diagnosis, hinders consent to treatment, and ultimately prolongs physical and mental distress.
>
> Having a basic working knowledge or a reference of real-world parasites can help a care professional quickly recognize these signs and attempt to guide patients away from this pattern-seeking behavior (Fig. 22).

Fig. 22. Pareidolia. Late nineteenth century photograph. What appears to be a large face between the man and woman is in fact a small child wearing a bonnet sitting on the father's knee. Vegetation in the background between the child and the standing mother on the right resembles human hair

References

1. Beach SR, Kroshinsky D, Kontos N. Case 37-2014: a 35-year-old woman with suspected mite infestation. N Engl J Med. 2014;371:2115–23.
2. Baker EW, Evans TM, Gould DJ, Hull WB, Keegan HL. A manual of parasitic mites of medical or economic importance, a technical publication. New York: Nat Pest Con Ass Inc.; 1956. p. 169.
3. Chauve C. The poultry red mite *Dermanyssus gallinae* (De Geer 1778): current situation and future prospects for control. Vet Para. 1998;73:239–45.
4. Kettle DS. Medical and veterinary entomology. New York: Wiley and Sons; 1984. p. 658.
5. Traver JR. Unusual scalp dermatitis in humans caused by the mite, Dermatophagoides (Acarine: Epidermoptidae). Proc Ent Soc Wash. 1951;53:1–25.
6. Stafford KC. Tick management handbook. An integrated guide for homeowners, pest control operators, and public health officials for the prevention of tick-associated disease. New Haven: The Connecticut Agricultural Experiment Station; 2004. p. 1–66.
7. Stafford KC. Emerging and exotic ticks in Connecticut. CT Wildlife Mag. 2020:Sept.-Oct:16–7.
8. Dantas-Torres F. Biology and ecology of the brown dog tick, *Rhipicephalus sanguineus*. Para and Vect. 2010;3:26. (Open access)
9. Gray JS, Dautel H, Estrada-Peňa A, Kahl O, Lindgren E. Effects of climate change on ticks and tick-borne diseases in Europe. Interdis Persp Inf Dis. 2009;593232.
10. Pritt BS. Haemaphysalis longicornis Is in the United States and Biting Humans: Where Do We Go From Here? Inf Dis Soc of Amer. 2020;20. https://doi.org/10.1093/cid/ciz451.
11. Vetter RS. The brown recluse spider. Cornell Uni Press; 2015. p. 200.
12. Vetter RS. Medical myths. West J Med. 2000;173:357–8.
13. Light JE, Smith VS, Allen JM, Durden LA, Reed DL. Evolutionary history of mammalian sucking lice (Phthiraptera: Anoplura). BMC Evol Bio. 2010;10(292):1–15.

14. Stafford, KC, Ridge GE, Molaei G, Zarb C, Bevilacqua P. Rabbit Bot Fly Furnicular, Tracheopulmonary, and Human Bot Fly Infestations in Connecticut (Oesteridae: Cuterebrinae). J Med Ent. 2020;XX(X):1–7. https://doi.org/10.1093/jme/tjaa181.
15. World Health Organization (WHO). Public health significance of urban pests. Regional office for Europe. 2008. p. 569.
16. Altschuler DZ, Crutcher M, Dulceanu N, Cervantes BA, Terinte C, Sorkin LN. Collembola (springtails) (Arthropoda: Hexapoda:Entognatha) found in scrapings from individuals diagnosed with delusory parasitosis. J New York Ent Soc. 2004;112(1):87–95.
17. Christiansen KA, Bernard EC. Critique of the article "Collembola (Springtails) (Arthropoda: Hexapoda: Entognatha) found in scrapings from individuals diagnosed with delusory parasitosis". Ent News. 2008;119(5):537–40.
18. Lim CSH, Lim SL, Chew FT, Ong TC, Deharveng L. Collembola are unlikely to cause human dermatitis. J Insect Sci. 2009;9(3):1. https://doi.org/10.1673/031.009.0301.

Glossary

ACEi Angiotensin-converting enzyme inhibitor.

Acne vulgaris Blockage and/or inflammation of pilosebaceous units (sebaceous hair follicle glands).

Aetiology In medicine, it is the cause(s) or origin(s) of disease and/or disorder.

Algorithm A set of rules to solve a problem. Used by internet designers to learn users' habits and to gather and direct selected information to them.

Amygdala The emotion center of the brain.

Anchoring bias The favoring of initial information which influences final decisions.

Arachnid Terrestrial spiders and mites.

ARB Angiotensin II receptor blockers.

Atopic dermatitis A condition that makes the skin red and itchy.

Arthropod Very small animal with jointed feet, e.g., terrestrial insects and arachnids.

CCB Calcium channel blocker.

Chat rooms Popular locations on the internet where patients interact.

Comedo From the Latin word *"comedere,"* meaning "to eat up." Historically, comedo described parasitic worms. In current medical terminology, it is used to suggest the wormlike appearance of expressed material from the skin.

Comet sign Particular form of dermatitis caused by the European straw itch mite, *Pyemotes ventricosus*.

Confirmation bias Selecting information to support one's own belief system.

Core cognitive distortion Negative thought patterns that cause people to view reality inaccurately. Untreated, these can increase anxiety, depression, and poor judgement.

Dermatitis artefacta or factitious dermatitis A psychocutaneous disorder is when patients consciously inflict self-harm on skin, nails, hair, and/or mucosae for emotional release when under stress, attract attention, and/or evade responsibility. Patients typically deny self-induced harm.

© The Editor(s) (if applicable) and The Author(s), under exclusive license to Springer Nature Switzerland AG 2024
G. E. Ridge (ed.), *The Physician's Guide to Delusional Infestation*,
https://doi.org/10.1007/978-3-031-47032-5

Darier's disease Wartlike blemishes on the skin. Sites commonly seen on head, upper arms, upper torso, elbows, and knees.

Delusion Symptoms or signs involving content of thought. A belief that is demonstrably untrue or not shared by others, usually based on incorrect inference about external reality. The belief is firmly held with conviction and is not, or is only briefly, susceptible to modification by experience or evidence that contradicts it. The belief is not ordinarily accepted by other members or the person's culture or subculture, i.e., it is not an article of religious faith (WHO: ICD-11).

Delusional state A false fixed belief and immutable conviction despite all efforts by physicians to persuade otherwise.

Dermatitis In general terms, an irritation of the skin.

Dermatographism Skin writing. Skin reaction to light scratches as red raised lines or marks. These last usually up to 30 min and then disappear.

Disease model of DI addiction Faulty mesolimbic pathway/dopaminergic pathways in the brain that alter psychological and subsequent behavioral processes to cause self-harm and nonchemical addictive behavior.

Dysesthesia A generic term for dermal abnormal sensations which can include burning, tingling, stinging, crawling, and pruritus. In multiple sclerosis, it can be caused by neuropathic or neurogenic pain.

Eczema Red inflamed itchy patches on the skin.

Elbow angle sign Known as cubitus valgus where the forearm angle away from the body is greater than normal. This may suggest Turner or Noonan syndromes.

Emotional support human (ESH) A patient-selected trusted person who not only provides emotional support for the patient but also supports the physician in treatment and follow-up care.

Erythema Abnormal redness of the skin due to inflammation.

Eosinophils Higher than normal count of a certain type of white blood cell. May suggest parasitic infection, allergic reaction, or cancer.

FGA First-generation antipsychotics.

Folie à deux/trois Several people who share a delusion.

Folliculitis Inflammation of the hair follicles.

Formication An irritating sensation that mimics insects crawling on the skin.

GERD Gastroesophageal reflux disease.

Hexapods Subphylum Hexapoda. Greek for "six legs." This includes insects, Collembola, Protura, and Diplura.

HMG-CoA Reductase inhibitor-3-hydroxy-3-methyl-glutaryl-coenzyme A.

Idiopathic Disease(s) caused from unknown and/or spontaneous origins.

Illusion Misinterpretation of perception of something existing.

Inducer The primary patient who first experienced DI.

Infestation Presence of internal or external parasites, pathogens, or inorganic material so as to be harmful or bothersome.

Itch Unpleasant sensation producing the desire to scratch.

Label-line theory A hypothesis where different nerves with the same physiology in transmitting impulses along their axons can generate different sensations.

Lichenified Skin that has become thick and leathery.

Lichen simplex chronicus (LSC) Skin condition caused by chronic rubbing, itching, and scratching. This can be caused by skin allergies, psoriasis, eczema, neurosis, anxiety, depression, or other emotional problems.

Magnan's sign Cocaine addicts' clinical paranesthesia; a constantly moving body or powder/sand under the skin surface (V. Magnan, French psychiatrist, 1835–1916).

Morgellons Type of delusional infestation.

Naturopathic medicine Natural holistic therapy-based medicine. These include physical, spiritual, mental, emotional, genetic, social, and environmental elements. Therapies include massage, acupuncture, exercise, nutritional counseling, and herbal/essential oil treatments. This form of medicine can be overvalued as a replacement for mainstream medicine, whereas both should be complementary.

NSAID Nonsteroidal anti-inflammatory drug.

Occlusive dressing Air- and watertight trauma medical bandages used in first aid. These dressings generally made with waxy coating for a total seal with no absorbent properties.

Onychotillomania A psychodermatosis where there is repeated self-inflicted injury and trauma caused by an irresistible neurotic urge to pick, pull, or bite the fingernails.

Overvalued idea The overvalued idea, first described by Wernicke, refers to a solitary, abnormal belief that is neither delusional nor obsessional in nature, but which is preoccupying to the extent of dominating the sufferer's life (McKenna 1984).

PANDAS Pediatric autoimmune neuropsychiatric disorders associated with streptococcal infections (PANDAS). Cross-reactive antibodies when fighting of strep bacteria can cause obsessive-compulsive disorder (OCD) or tic disorder.

Paresthesia Sensations of pricking, tingling, or crawling/creeping on the skin with no apparent cause.

Pattern theory Sensation of itch encoded across many sensory receptors and the spinal cord which causes a collective pattern of neuronal activity that ultimately determines the "sensation experience."

Pareidolia To interpret meaningful images from random ambiguous visual patterns.

Perseverative Repeated negative thinking on the past or the future. Response actions to a stimulus are repeated even though the causative stimulus has disappeared.

PPI Proton pump inhibitor.

Prurigo nodularis (PN) A chronic inflammatory skin disease as an extremely itchy symmetrical rash seen on arms, legs, upper back, and/or abdomen.

Pruritus Itching.

Quality of life Ability to enjoy normal life activities that includes personal perception of well-being, the environment, and personal physical and mental health.

Recidivism From Latin to mean "recurring" or to "fall back." A person repeats an undesirable behavior despite experiencing negative consequences of that particular behavior.

Rosacea Blushing or flushing of the face with visible blood vessels. Small puss-filled bumps may also be present.

Scarification To scratch, cause small cuts, or pick at irritations on the skin. To lacerate.

Seborrheic dermatitis Scaly patches, red skin, and/or stubborn dandruff on the scalp. Also, can be present in oil regions of the body such as face, sides of nose, chest, eyelids, eyebrows, and/or ears.

SGA Second-generation antipsychotic.

SNRIs Serotonin–norepinephrine reuptake inhibitors.

Somatic Related to the body not the mind.

Somatic preoccupation A fixed focus on somatic symptoms rooted in underlying psychological distress such as depression, PTSD, or guilt.

Somatopsychic Mental illness that responds to ongoing underlying somatic disease (medical conditions) where the two become interdependent.

SSRIs Selective serotonin reuptake inhibitors.

Sunk cost fallacy When a person is reluctant to abandon a deleterious course of action, because they have heavily invested in it using time and resources, even when it is clear that abandonment of the action would be beneficial.

Switch cost effect Rapid transitions from one task to another involving significant cognitive effort. This causes loss of concentration, reduced performance, and mental and physical fatigue.

Syndrome A complex of symptoms indicating a specific condition with a cause that is not necessarily understood.

Tangential Diverging from previous lines of thought. Erratic thinking.

Terminal delusional state A rigid and severe delusion experienced over extended time. Validation is primary motivation for seeking care with total lack of patient insight. Unless persuaded, these patients are impossible to treat (Brown et al. 2014).

Trichotillomania A mental disorder that drives an irresistible recurrent urge to pull out hair from the scalp, eyebrows, or other body regions.

Urticaria Severe skin reactions that include hives, burning, stinging, and intense itching.

Village care Collaborative care by physicians, experts, family, and social networks.

Xerosis cutis Unusually dry skin.

Index